Advances in Computer Science and Technology

More information about this series at http://www.springer.com/series/13197

Chuan-Kun Wu • Dengguo Feng

Boolean Functions and Their Applications in Cryptography

Chuan-Kun Wu
State Key Lab of Information Security
Institute of Information Engineering
Chinese Academy of Sciences
Beijing, China

Dengguo Feng
Institute of Software
Chinese Academy of Sciences
Beijing, China

ISSN 2198-2686 ISSN 2198-2694 (electronic)
Advances in Computer Science and Technology
ISBN 978-3-662-48863-8 ISBN 978-3-662-48865-2 (eBook)
DOI 10.1007/978-3-662-48865-2

Library of Congress Control Number: 2015957583

Springer Heidelberg New York Dordrecht London

Printed on acid-free paper

Springer-Verlag GmbH Berlin Heidelberg is part of Springer Science+Business Media (www.springer.com)

Preface

Nonlinear Boolean functions are necessary building blocks in cryptography; they often play a key role in the design of many stream ciphers and block ciphers. In the design of stream ciphers, nonlinear combining functions are an important type to study, where the cryptographic properties of the combining Boolean function directly reflect the security vulnerability of the cipher, hence the study of cryptographic properties of Boolean functions directly relating to the security of the nonlinear combiners as a special model of stream ciphers. In block cipher algorithms, nonlinear functions are also one of the core components. In the design of many modern cryptographic algorithms, including the Advanced Encryption Standard (AES), the nonlinear cryptographic functions have been used. Therefore, the study of cryptographic properties of nonlinear Boolean functions not only helps cryptanalysis but also plays an important guidance in the design of cryptographic algorithms that have resistance against many cryptographic attacks.

Primary cryptographic properties of Boolean functions include nonlinearity, correlation immunity, and algebraic property; they represent different security measures against different cryptanalyses. The nonlinearity is a measure about the resistance of Boolean functions against linear or affine approximation attack, and this property can be extended to include the algebraic degree, linear structures, and propagation criterion. These extended properties seem to have less importance than the original definition of the nonlinearity and have attracted less study. The correlation immunity is a measure against correlation attack, and the algebraic immunity is a measure against algebraic attack. Note that these cryptographic properties are not independent of each other, they have some relations and some restrictions to each other, and this means that one cannot find a Boolean function with all these properties to reach the best extent; therefore, a Boolean function needs to have multiple cryptographic properties for practical use, and these cryptographic properties have to compromise to reach a state of conditional optimum, i.e., under the condition that certain cryptographic properties are met, the chances to have other specific cryptographic properties to be the best.

Apart from numerous research papers on cryptographic properties of Boolean functions in public literatures, there have been quite a few books about cryptographic Boolean functions, or covering some content about cryptographic Boolean functions.

This book studies a few hot issues about nonlinear functions in contemporary cryptography. It is realized that the study of cryptographic Boolean functions has been undergoing a continuous repaid development, and this book tends to reflect some research results of the authors in certain aspects of cryptographic properties of Boolean functions. This book is not meant to have a comprehensive coverage of the topics in this field; it is designed to perhaps compliment some missing material about cryptographic Boolean functions. Nevertheless, the book covers the primary cryptographic properties of Boolean functions.

The contents of this monograph are as follows: Chapter 1 introduces the basic concept of some fundamental cryptographic properties of Boolean functions, and these properties are closely related to different security measures of the Boolean functions. Chapter 2 studies some independence properties of Boolean functions, mostly the independence of Boolean functions of their variables, including algebraic independence, statistical independence, and algebraic degeneration. These independence properties are used in the forthcoming chapters. Chapter 3 deals with the nonlinear properties of Boolean functions, including algebraic degree, nonlinearity, and linear structures of Boolean functions. These nonlinear properties are fundamental for a Boolean function to be used in an encryption algorithm. Chapter 4 is about the correlation immunity of Boolean function, an interesting property measuring the resistance against correlation attack, with different attempts in constructing correlation-immune Boolean functions being made, and the concept of correlation immunity is also extended. Chapter 5 is about the algebraic immunity of Boolean functions, a security measure against algebraic attack that was proposed in recent years. Chapter 6 studies the symmetric property of Boolean functions, although the symmetric property is not a property to be pursued for cryptographic applications, and there are many interesting related properties; hence the results in this chapter are mostly of theoretical significance. Chapter 7 views cryptographic S-boxes as vectorial Boolean functions, with a special class being Boolean permutations, and studies their cryptographic properties and constructions. Chapter 8 attempts to give some applications of Boolean functions, where the applications are beyond the original target as in stream and block cipher design.

Although each chapter has a focus, which is often a specific cryptographic property, however other cryptographic properties are also considered when needed. For example, when considering the correlation immunity, the nonlinearity is also considered, and this is because the cryptographic properties are related to each other; pursing one property may have to sacrifice the requirement on other properties and at least have to reduce the requirement on other properties. This has been reflected in different chapters of this monograph.

Finally, the authors would like to express sincere thanks to Prof. Guanghong Sun, Hohai University, Nanjing, China, for his contribution to some of the materials in Chap. 3 and Prof. Wenying Zhang, Shandong Normal University, Jinan, China, for her contribution to most of the materials in Chap. 5. The authors are grateful to the publisher of the monograph and the editing team for their assistance and valuable comments to make this monograph available, and the first author appreciates the support by Natural Science Foundation of China under project number 61173134.

Beijing, China

Chuan-Kun Wu
Dengguo Feng

Contents

Notations

$GF^n(2)$:	n-Dimensional vector space over the binary field $GF(2)$.
$\oplus$:	The Exclusive-Or operation of Boolean values.
$\mathcal{F}_n$:	The set of Boolean functions in n variables.
$\mathcal{L}_n$:	The set of linear Boolean functions in n variables.
$\mathcal{A}_n$:	The set of affine Boolean functions in n variables.
$supp(f)$:	The support of function $f(x)$.
$\overline{supp}(f)$:	The complementary set of $supp(f)$, i.e., $GF^n(2) - supp(f)$.
$wt(f)$:	The Hamming weight of $f(x)$.
$nl(f)$:	The nonlinearity of $f(x)$.
$CI(f)$:	The correlation immunity of $f(x)$.
$AN(f)$:	The set of annihilators of $f(x)$.
$AD(f)$:	The algebraic degeneracy of $f(x)$.
$AI(f)$:	The algebraic immunity of $f(x)$.
$\langle w, x \rangle$:	The inner product of vectors w and x, i.e., $\langle w, x \rangle = w_1x_1 \oplus w_2x_2 \oplus \cdots \oplus w_nx_n$
$A\|B$:	A is a factor of B.
$Prob(A\|B)$:	The probability of event A happened given the condition that event B happened.
$\langle S \rangle$:	The linear span of the set S.

Chapter 1
Boolean Functions and Their Walsh Transforms

Boolean functions are fundamental building blocks for many cryptographic algorithms. This chapter introduces basic concepts and operations of Boolean functions, including Walsh transforms and basic cryptographic properties of Boolean functions.

1.1 Logic Gates and Boolean Variables

Our world is so complex. It has countless number of variant things. However, science tells us that everything is composed of very basic elements, namely, atoms. Compared with the variant things, the number of different atoms is very small. This is similar to the situation of modern computers. Today the computers are very powerful and complex. Regardless of how complicated they are structured, and how different they are from one to another, they are all composed of very basic logic circuits using very basic gates (operators) that only apply to binary numbers, i.e., 0 and 1.

Integrated circuits such as microprocessors, RAMs, interface chips, and so on are manufactured by putting tens of thousands of simple logic gates into a silicon chip. These circuits are then built into more powerful computers.

The minimum units of normal electronically devices are called *cells*. A cell can have two status,[1] high voltage or low voltage. When a cell has high voltage, it stands for a TRUE Boolean value and is represented by binary value 1, while when a cell has low voltage, it stands for a FALSE Boolean value and is represented by binary value 0. Connections of those cells in terms of their assigned values are performed by three fundamental logic gates, namely, OR, AND, and NOT. These

[1]For multiple logic circuits, a cell can have more than two status. Here we only consider the two status logic.

C.-K. Wu, D. Feng, *Boolean Functions and Their Applications in Cryptography*,
Advances in Computer Science and Technology, DOI 10.1007/978-3-662-48865-2_1

Table 1.1 Truth table of logic gates

A	B	AND ($A \wedge B$)	OR ($A \vee B$)	NOT ($\bar{A}$)	XOR ($A \oplus B$)
0	0	0	0	1	0
0	1	0	1	1	1
1	0	0	1	0	1
1	1	1	1	0	0

logic gates correspond to operations on binary numbers. By connecting these logic gates together in various ways, we have different kinds of logic circuits. The basic logic gates are the simplest logic circuits.

The most intuitive way to describe how the logic gates work is to use truth tables. Let A and B be Boolean variables representing the value of logic cells. The output of the logic gates can be illustrated in Table 1.1. Note that for simplicity, we use binary values 1 and 0 to represent the Boolean values TRUE and FALSE, respectively.

There are some other kinds of logic gates which are also very useful; however, they all can be represented as the composition of the above mentioned ones. The logic gate exclusive-or (XOR for short), denoted by $\oplus$, can be expressed using the above three basic logic gates as $A \oplus B = (A \vee B) \wedge (\bar{A} \vee \bar{B})$. The outcome of $A \oplus B$ is exactly the same as modulo 2 addition of the binary values of A and B. This property coincides with the addition operation over the finite field $GF(2) = \{0, 1\}$ with two elements and two operations: modulo 2 addition and modulo 2 multiplication. It is interesting to notice that by employing the logic gate AND, all the three basic logic gates can be represented using the logic gates XOR and AND. More precisely we have, apart from logic gate AND,

$$A \vee B = A \oplus B \oplus (A \wedge B), \quad \bar{A} = A \oplus 1.$$

Note that the logic gate OR also coincides with the multiplication operation over the finite field $GF(2)$. This enables us to study logic circuits using the functions of XOR and AND, i.e., Boolean functions, and the variables representing the values of logic cells are called Boolean variables.

One of the advantages of using Boolean variables and the Boolean operations XOR and AND is that these two Boolean operations can be treated as operations over the binary field $GF(2)$, and many of the results from finite fields can be used.

1.2 Boolean Functions and Their Representations

Let $GF^n(2)$ be the n-dimensional vector space over the binary field $GF(2)$. A function $f : GF^n(2) \longrightarrow GF(2)$ is called a *Boolean function* in n variables. We write it as $f(x) = f(x_1, x_2, \ldots, x_n)$, where x is the shorthand writing of vector $(x_1, x_2, \ldots, x_n)$. The vector of all the outputs of $f(x)$ is called the *truth table* of $f(x)$, which has dimension 2^n. Of course x has to follow a particular order when going through all the possible values in $GF^n(2)$. If we treat a binary vector as the binary representation of an integer, i.e.,

$$x = (x_1, x_2, \ldots, x_n) = \sum_{i=1}^{n} x_i 2^{n-i},$$

when the integer takes all the values from 0 to $2^n - 1$, then the corresponding vector goes through all the elements in $GF^n(2)$. Traditionally, we let the value of the binary representation of the integer to go from 0 incrementally to $2^n - 1$. If we collect all the vectors where $f(x)$ takes value 1, then the set of collections is called the *support of* $f(x)$, denoted by $supp(f) = \{x : f(x) = 1\}$. The complement of the support, denoted by $\overline{supp}(f) = \{0, 1\}^n - supp(f) = \{x : f(x) = 0\}$, is called the *annihilation set of* $f(x)$. It is trivial to verify that $\overline{supp}(f) = supp(1 \oplus f(x))$. The number of 1's in the truth table of $f(x)$ is called the Hamming weight of $f(x)$ and is denoted by $wt(f)$. It is obvious that $wt(f)$ is the number of elements in $supp(f)$. Function $f(x)$ is called *balanced* if $wt(f) = 2^{n-1}$, and it is called an *affine* function if there exist $a_0, a_1, \ldots, a_n \in GF(2)$ such that $f(x) = a_0 \oplus a_1x_1 \oplus \cdots \oplus a_nx_n$, where $\oplus$ means modulo 2 addition (equivalent to XOR operation). In particular, if $a_0 = 0$, it is also called a *linear* function. We will denote by $\mathcal{F}_n$, the set of all Boolean functions in n variables, by $\mathcal{L}_n$, the set of linear ones, and by $\mathcal{A}_n$, the set of affine ones. By these notations, we have

$$\mathcal{L}_n \subset \mathcal{A}_n \subset \mathcal{F}_n.$$

Let $f(x)$, $g(x)$ be two Boolean functions in n variables. The *distance* between $f(x)$ and $g(x)$, denoted by $d(f, g)$, is the number of coordinates with different values in their truth tables, or equivalently it can be written as $d(f, g) = wt(f \oplus g)$. This distance is also known as *Hamming distance*. By the properties of XOR operation, we get a relationship between the Hamming distance of two Boolean functions and the Hamming weight of them:

$$wt(f \oplus g) = wt(f) + wt(g) - 2wt(fg).$$

The *nonlinearity* of $f(x)$, denoted by $nl(f)$, is the minimum distance between $f(x)$ and all affine functions, i.e.,

$$nl(f) = \min\{d(f(x), l(x)) : l(x) \in \mathcal{A}_n\}. \quad (1.1)$$

Boolean variables $x = (x_1, x_2, \ldots, x_n)$ can be treated as probabilistic variables which take random values from $GF^n(2)$ equally likely with uniform probability distribution. From this treatment, each x_i is an independent variable over $\{0, 1\}$, and a Boolean function $f(x)$ is also treated as a function in random variables, which is hence a random variable over $\{0, 1\}$. We will use this treatment when discussing statistical properties.

There can be different arithmetical operations on Boolean functions. Let $f(x)$ and $g(x)$ be two Boolean functions in n variables. Then their addition $f(x) \oplus g(x)$ is the XOR operation of their corresponding outputs, and their multiplication $f(x)g(x)$ is the multiplication of their corresponding outputs. With these two operations, it is easy to verify that $\mathcal{F}_n$, $\mathcal{L}_n$, and $\mathcal{A}_n$ are all communicative rings.

There are a number of cryptographic properties that a practically applicable Boolean function is supposed to oppose. Some of the properties have certain levels of conflict, so they cannot meet the optimum status at the same time. Boolean functions satisfying multiple cryptographic properties have to make a reasonable compromise [6, 20, 23, 33].

1.2.1 Algebraic Normal Form

One way to represent a Boolean function is to write the function in terms of Boolean variables. When a Boolean function is written as

$$f(x) = c_0 \bigoplus_{1 \le i \le n} c_i x_i \bigoplus_{1 \le i < j \le n} c_{ij} x_i x_j \bigoplus \cdots \bigoplus c_{1,\ldots,n} x_1 x_2 \ldots x_n, \tag{1.2}$$

it is called the *algebraic normal form* representation of $f(x)$, or the *ANF* in brief, where $c_0, c_i, c_{ij}, \cdots, c_{1,\ldots,n}$ are coefficients having a value in $\{0, 1\}$. It can be proven that every Boolean function in n variables can be represented uniquely in the form of Eq. 1.2, i.e., a unique set of coefficients. So the representation of Eq. 1.2 is universal. For example, $f(x_1, x_2, x_3) = x_1 \oplus x_2 x_3$ is the algebraic normal form of a Boolean function in three variables.

When a Boolean function is represented in its algebraic normal form, it is the XOR of a number of terms, and each *term* is a multiplication of either zero (the constant) or more Boolean variables. The number of variables in one multiplicative term is called the *algebraic degree* (or simply the degree) of the term, and the algebraic degree of a Boolean function is the highest algebraic degree of its terms with a nonzero coefficient in its ANF representation. The algebraic degree of $f(x)$ is denoted as $deg(f)$. For instance, the degree of the above example $f(x) = x_1 \oplus x_2 x_3$ is 2, as in the algebraic normal form of $f(x)$, and the term of the highest possible degree with nonzero coefficient is $x_2 x_3$. The highest degree of a Boolean function in n variables is n, and for the case of $n = 3$, only when $x_1 x_2 x_3$ appears in the algebraic normal form of a function, the function is of the highest degree. Note that in the algebraic normal form of a Boolean function, every individual variable has a degree of at most one; this is because for a Boolean variable x, $x^2 = x$ is always true, regardless whether the value of x is 1 or 0.

When the algebraic normal form of a Boolean function has degree 0, the Boolean function is called a *constant*. It can be seen that in this case, for all the possible values of x, we always have that $f(x) = 1$ or $f(x) = 0$, but not both. When the algebraic normal form is of degree 1, the function is an *affine* Boolean function.

Particularly, when the constant term is 0, i.e., $f(0) = 0, f(x)$ is a *linear* Boolean function. For a linear Boolean function $f(x)$, it has the property that $f(x \oplus y) = f(x) \oplus f(y)$. Note that an affine Boolean function is either a linear one or a linear Boolean function XOR the constant 1. So traditionally affine Boolean functions are treated as having the same level of nonlinearity as linear functions.

1.2.2 Truth Table Representation

There are only finite numbers of inputs to a Boolean function; hence there are finite numbers of outputs. If each output corresponds to an input, then with a complete list of input-output pairs, the function can be uniquely determined. This is even simplified if we sort the inputs in such a fixed order: map each input $(x_1, x_2, \ldots, x_n)$ into an integer $x' = \sum_{i=1}^{n} x_i 2^{n-i}$ of which the binary representation of the integer is the same as the input, and let x' go from 0 to $2^n - 1$ incrementally, and then according to this order, the outputs of a Boolean function can represent the function. This is called the *truth table* representation. As for the above example, $f(x) = x_1 \oplus x_2x_3$, when the variables are ordered as $000, 001, 010, 011, 100, 101, 110, 111$, and according to this ordered inputs, the outputs of the function are listed as a vector 00011110, which is the truth table of the function.

1.2.3 Support Representation

Sometimes it is easier to record a Boolean function by the values of the inputs where the function takes value 1, particularly for those having very low Hamming weight.

Let $f(x) \in \mathcal{F}_n$. The *support* of $f(x)$ is defined as

$$supp(f) = \{x : f(x) = 1\} \tag{1.3}$$

It is seen that the support of a Boolean function is a set of vectors of dimension n, and the cardinality of $supp(f)$ satisfies

$$0 \leq |supp(f)| \leq 2^n. \tag{1.4}$$

For example, the support for function $f(x_1, x_2, x_3) = x_3 \oplus x_1x_3 \oplus x_2x_3 \oplus x_1x_2x_3$ is just $\{(001)\}$. For this function, the support representation is much simpler than the algebraic normal form as well as truth table representation.

If we list all the vectors in $supp(f)$ as a matrix X_f, then X_f is a $wt(f) \times n$ binary matrix whose rows represent the nonzero points of $f(x)$ and whose column represents the number of variables of $f(x)$. Then the matrix X_f is called the *characteristic matrix* of $f(x)$ which is unique when the rows are in a specific order, e.g., the order defined above.

If $f(x)$ is balanced, then its characteristic matrix has 2^{n-1} different rows and n columns. Further properties of Boolean functions can be observed from the properties of their characteristic matrices, as will be seen later in this book.

1.2.4 Minterm Representation

From the above, we know that the support of a Boolean function uniquely defines the Boolean function, i.e., there is a one-to-one mapping from the set of Boolean functions and the set of their supports. Given a Boolean function $f(x) \in \mathcal{F}_n$, and assume that its support is *supp(f)*, then *supp(f)* is a set of binary vectors. For any of these vectors, there exists a Boolean function whose support is a set having that vector only. More precisely, let $\alpha = (a_1, a_2, \ldots, a_n) \in GF^n(2)$, and define $x_i^{(a_i)} = 1$ if and only if $x_i = a_i$, where $a_i \in \{0, 1\}$ is a constant and x_i is a binary variable. Denote

$$x^\alpha = x_1^{(a_1)} x_2^{(a_2)} \cdots x_n^{(a_n)}. \tag{1.5}$$

Then it is easy to verify that the support of x^α is indeed α. Note that $x_i^{(a_i)} = (x_i \oplus a_i \oplus 1)$; hence, we can write x^α as

$$x^\alpha = (x_1 \oplus a_1 \oplus 1)(x_2 \oplus a_2 \oplus 1) \cdots (x_n \oplus a_n \oplus 1) = \prod_{i=1}^{n} (x_i \oplus a_i \oplus 1).$$

It is easy to note that for any $f(x) \in \mathcal{F}_n$, we have that

$$supp(f) = \bigcup_{\alpha \in supp(f)} supp(x^\alpha); \tag{1.6}$$

hence, we can write $f(x)$ as

$$\begin{aligned} f(x) = \bigoplus_{\alpha \in supp(f)} x^\alpha &= \bigoplus_{\alpha \in supp(f)} x_1^{(a_1)} x_2^{(a_2)} \cdots x_n^{(a_n)} \\ &= \bigoplus_{\alpha \in supp(f)} \prod_{i=1}^{n} (x_i \oplus a_i \oplus 1) \end{aligned} \tag{1.7}$$

Eq. 1.7 is called the *minterm representation* of $f(x)$, where each x^α is called a minterm of $f(x)$.

1.2.5 Representation Conversions

Given a Boolean function, it can be represented by either algebraic normal form, or truth table, or the support. As the function is not changed regardless whatever a representation is used, different representations should be all equivalent. Therefore, there should be a method of converting from one representation to another.

1.2.5.1 Algebraic Normal Form to Truth Table Conversion

Given a boolean function $f(x)$ in n variables, it is not hard to find a way of converting algebraic normal form into the truth table representation. This can be done by simply recording the outputs of the function by feeding input x, where x, when treated as a binary representation of integers, goes from 0 to $2^n - 1$ incrementally.

1.2.5.2 Truth Table to Support Conversion

The truth table of a Boolean function in n variables is a binary vector of dimension 2^n, and each coordinate of the vector corresponds to an input x which can be treated as an n-dimensional binary vector. Choose the vectors x corresponding to value 1 in the truth table of $f(x)$, and then they form the support of $f(x)$.

1.2.5.3 Support to Minterm Conversion

The definition of the minterm representation of a Boolean function actually gives a conversion: for each vector $\alpha \in supp(f)$, define the minterm x^α, then

$$f(x) = \bigoplus_{\alpha \in supp(f)} x^\alpha \tag{1.8}$$

is the minterm representation of $f(x)$.

1.2.5.4 Minterm to Algebraic Normal Form Conversion

When we remove the parentheses in the minterm representation of $f(x)$ as in Eq. 1.7, removing the same terms that appear for even number of times (since $\oplus$ is modulo 2 addition) and keeping one of each of the terms that appear for odd number of times, then the result is the algebraic normal form representation of $f(x)$.

Now we know that different representations of Boolean functions can be converted from one form to another. The above shows the possibility of conversion, but not necessarily the most efficient way of conversion. This book will not cover the efficiency issue of the conversion from one form of representations to another, and hence the possibility of conversion only shows the equivalence of these different representations.

1.2.5.5 Truth Table to Algebraic Normal Form Conversion

Apart from algebraic normal form, another popular form of representation is the truth table representation, and it often needs to convert a truth table representation into algebraic normal form; hence, we give the following steps with an example demonstrating how it works.

- Formulate the support of the function from the truth table.
  For example, if the truth table for the given function is (11000100), then the support is a collection of the index of the truth table with nonzero coordinates, i.e., $x = 0, 1, 5$, or in binary representation form, we have $supp(f) = \{000, 001, 101\}$.
- Convert the support to minterm representation.
  For the above example, we have

$$f(x) = \bigoplus_{c \in supp(f)} x_1^{(c_1)} x_2^{(c_2)} x_3^{(c_3)} = x_1^{(0)} x_2^{(0)} x_3^{(0)} \oplus x_1^{(0)} x_2^{(0)} x_3^{(1)} \oplus x_1^{(1)} x_2^{(0)} x_3^{(1)}.$$

- Expand the minterm form into a polynomial to get the algebraic normal form. Note that $x_i^{(c_i)} = x_i$ if $c_i = 1$ and $x_i^{(c_i)} = 1 \oplus x_i$ if $c_i = 0$.
  For the above example, we have

$$f(x) = 1 \oplus x_1 \oplus x_2 \oplus x_1 x_2 \oplus x_1 x_3 \oplus x_1 x_2 x_3.$$

Note: The above method of truth table to algebraic normal form conversion is by no means optimum. It can be seen from the example that the first two minterms can be combined as $(1 \oplus x_1)(1 \oplus x_2)$, and hence expansion is easier than to expand every minterm individually.

Example 1.1. Let $n = 3$, $f(x) = x_1x_2 \oplus x_1x_3 \oplus x_2x_3$, then when $x = (x_1, x_2, x_3)$ takes values (000) = 0, (001) = 1, (010) = 2, (011) = 3, (100) = 4, (101) = 5, (110) = 6, (111) = 7, the outputs of $f(x)$ form the vector (00010111) which is the truth table of $f(x)$. From the truth table, it is seen that when $x \in \{011, 101, 110, 111\}$, $f(x) = 1$, $supp(f) = \{011, 101, 110, 111\}$, and the minterm representation of $f(x)$ is $f(x) = \bar{x}_1x_2x_3 \oplus x_1\bar{x}_2x_3 \oplus x_1x_2\bar{x}_3 \oplus x_1x_2x_3$. Here we can use $\bar{x} = x \oplus 1$ to represent $x^{(0)}$.

1.2.6 Enumeration of Boolean Functions

It is sometimes very useful to know how many Boolean functions are there in a certain class. For Boolean functions in n variables, as there is a one-to-one relationship between a Boolean function in n variables and a binary vector of dimension 2^n (the truth table of the function), the number of all Boolean functions in n variables is 2^{2^n}, i.e.,

$$|\mathcal{F}_n| = 2^{2^n}.$$

It is seen that the number of Boolean functions increases dramatically with the increasing of the number of variables n. However, the number of affine Boolean functions is much smaller. Note that the XOR operation of any two affine Boolean functions will result in an affine Boolean function, so $\mathcal{A}_n$ forms a vector space. To determine the size of $\mathcal{A}_n$, it is sufficient if we can determine the dimension of $\mathcal{A}_n$. It is trivial to find a basis of $\mathcal{A}_n$ which is $1, x_1, x_2, \ldots, x_n$. These functions form a basis of $\mathcal{A}_n$, because they are elements in $\mathcal{A}_n$, and no one can be represented by others via linear combination (over binary field), and every function in $\mathcal{A}_n$ can be represented as a linear combination of these functions. So we have

$$|\mathcal{A}_n| = 2^{n+1},$$

and $|\mathcal{L}_n| = |\mathcal{A}_n|/2 = 2^n$.

Apart from linear Boolean functions and affine Boolean functions as two subclasses of Boolean functions, there are many other subclasses of Boolean functions with some other specific properties. For example, symmetric Boolean functions and Boolean functions reaching the highest possible nonlinearity (known as Bent functions) are another two subclasses of Boolean functions. Later on we will see more subclasses of Boolean functions grouped by some particular cryptographic properties. Studying the enumeration of those function with cryptographic properties has important cryptographic significance. For instance, it may be the case that a subclass of Boolean functions have very good cryptographic properties, but if the number of such functions is very small, then the application of these functions would be very limited. We will look at the enumeration problem later for some classes of cryptographic Boolean functions.

1.3 Walsh Transforms and Walsh Spectrum of Boolean Functions

Walsh transform [3] is an important tool to represent many properties of Boolean functions [30]. In particular, it has been shown to be a very useful tool in representing cryptographic properties of Boolean functions [10]. This section introduces the Walsh transform and its properties.

1.3.1 Walsh Functions and Walsh Transforms

Definition 1.1. The Walsh orthogonal family is a collection of Walsh functions defined over $GF^n(2)$

$$W(w, x) = (-1)^{\langle w, x\rangle} \tag{1.9}$$

where $w = (w_1, w_2, \ldots, w_n)$ and $x = (x_1, x_2, \ldots, x_n)$ are n-dimensional binary vectors, and $\langle w,\ x\rangle = w_1x_1 + w_2x_2 + \cdots + w_nx_n$ is the *inner product* of vectors w and x. Note that since both w and x are elements in $GF^n(2)$, the result of the inner product is an element in $GF(2)$, and hence the addition in the inner product is modulo 2 addition, which has the same effect as XOR operation. However, when the inner product is an exponent of -1, then there is no difference whether the addition is real or in the sense of modulo 2.

In the above, it does not make difference when treating each value of x_i and that of w_i as binary values or real values from $\{0, 1\}$, so we will not differentiate when a value in $\{0, 1\}$ is binary or real. Note that the Walsh functions are real-valued functions. It is easy to verify that the following properties hold, where a vector is treated equivalently as an integer in binary representation.

(1) Symmetric property: $W(w, x) = W(x, w)$ holds for any w and x.

(2) Orthogonal property: $\sum_{w=0}^{2^n-1} W(w, x)W(w, t) = \begin{cases} 2^n \text{ if } x = t \\ 0 \ \text{ else} \end{cases}$

From the orthogonal property of the Walsh orthogonal family, we get

$$\sum_{w=0}^{2^n-1} W(w, x) = \sum_{w=0}^{2^n-1} (-1)\langle w, x\rangle = \begin{cases} 2^n \text{ if } x = 0 \\ 0 \ \text{ if } x \neq 0 \end{cases} \tag{1.10}$$

More generally, we have

Lemma 1.1. *Let V be a vector subspace of $GF^n(2)$, and*

$$V^\perp = \{y :\ \forall x \in V, \langle x,\ y\rangle = 0\}$$

be the orthogonal vector space of V. Then we have

$$\sum_{w\in V} W(w, x) = \begin{cases} |V| \ \textit{if}\ w \in V^\perp \\ 0 \quad \textit{if}\ w \notin V^\perp \end{cases} \tag{1.11}$$

It can be seen that Eq. 1.10 is a special case of Eq. 1.11 when $V = GF^n(2)$.

Definition 1.2. Let $f(x) :\ GF^n(2) \longrightarrow R$ be a function defined over $GF^n(2)$ and whose values are from the set of real numbers R. Then the *Walsh transform* of $f(x)$ is defined as

$$S_f(w) = \sum_{x=0}^{2^n-1} f(x)W(w,x) = \sum_{x=0}^{2^n-1} f(x)(-1)^{\langle w,x\rangle} \tag{1.12}$$

and the Walsh inverse transform corresponding to Eq. 1.12 is

$$f(x) = 2^{-n}\sum_{w=0}^{2^n-1} S_f(w)W(w,x) = 2^{-n}\sum_{w=0}^{2^n-1} S_f(w)(-1)^{\langle w,x\rangle}. \tag{1.13}$$

Then the truth table of $S_f(w)$ is called the *Walsh spectrum* of $f(x)$.

Although Boolean functions are defined over $GF(2)$, they can well be treated as real-valued functions that take values 0 and 1 only. So the Walsh transform and the inverse transform also apply to Boolean functions.

A Boolean function can also be treated as a binary logical function that takes values TRUE (represented by value 1) and FALSE (represented by value 0). In the implementation of electronic circuits, it is often more convenient to use $\{-1,\ 1\}$ to represent the domain of binary functions than to use $\{0,\ 1\}$, and hence the following transform is used to map $\{0,\ 1\}$ to $\{-1,\ 1\}$:

$$\delta(f(x)) = (-1)^{f(x)}.$$

By this transform, the Boolean function $f(x)$ is then mapped to function $\delta(f(x))$ on the domain $\{-1,\ 1\}$. The Walsh transform can also apply to $\delta(f(x))$ as

$$S_{\delta(f(x))}(w) = \sum_{x=0}^{2^n-1} \delta(f(x))(-1)^{\langle w,x\rangle} = \sum_{x=0}^{2^n-1} (-1)^{f(x)+\langle w,x\rangle} \tag{1.14}$$

Since the Walsh transform of $\delta(f(x))$ is often represented using $f(x)$ as in Eq. 1.14, it is often represented as

$$S_{(f)}(w) = \sum_{x=0}^{2^n-1} (-1)^{f(x)+\langle w,x\rangle}.$$

In order to differentiate these two types of Walsh transforms, we call $S_{(f)}(w)$ as *type II Walsh transform* of $f(x)$, and call $S_f(w)$ as *type I Walsh transform* of $f(x)$.

Similar to the type I Walsh transform, the corresponding inverse of type II Walsh transform can be represented as

$$\delta(f(x)) = 2^{-n}\sum_{w=0}^{2^n-1} S_{(f)}(w)(-1)^{\langle w,x\rangle}. \tag{1.15}$$

Note that when the Boolean function $f(x)$ is treated as a real-valued function, the Walsh transform remains the same. By this treatment, the two functions can be converted to each other:

$$\delta(f(x)) = (-1)^{f(x)} = 1 - 2f(x)$$

Then, the type II Walsh transform of $f(x)$ can be converted from the type I Walsh transform of $f(x)$, i.e.,

$$\begin{aligned} S_{(f)}(w) &= \sum_{x=0}^{2^n-1}(-1)^{f(x)+\langle w,x\rangle} \\ &= \sum_{x=0}^{2^n-1}(1-2f(x))(-1)^{\langle w,x\rangle} \\ &= \sum_{x=0}^{2^n-1}(-1)^{\langle w,x\rangle} - 2\sum_{x=0}^{2^n-1}f(x)(-1)^{\langle w,x\rangle} \\ &= \begin{cases} 2^n - 2S_f(w) \text{ if } w = 0 \\ \quad -2S_f(w) \text{ else} \end{cases} \end{aligned} \tag{1.16}$$

On the other hand, the type I Walsh transform of $f(x)$ can be converted from its type II Walsh transform as

$$S_f(w) = \begin{cases} 2^{n-1} - \frac{1}{2}S_{(f)}(w) \text{ if } w = 0 \\ \quad -\frac{1}{2}S_{(f)}(w) \text{ else} \end{cases} \tag{1.17}$$

In the following discussion, we may not specifically name the types of Walsh transforms or spectrum, as the types can be identified from the notations.

1.3.2 Properties of Walsh Transforms

Now we give some properties of Walsh transforms on Boolean functions. Due to the easy conversion from one type to another, it is sufficient to use only one type, and the same properties can be established to a different type of the Walsh transform.

Theorem 1.1. *Let $f_1(x), f_2(x) \in \mathcal{F}_n$. Then*

$$S_{f_1 \oplus f_2}(w) = S_{f_1}(w) + S_{f_2}(w) - 2S_{f_1f_2}(w) \tag{1.18}$$

Proof. By the conversion between XOR operation and the addition over the real numbers, i.e., $a \oplus b = a + b - 2ab,\ a, b \in \{0, 1\}$, we have

$$
\begin{aligned}
S_{f_1 \oplus f_2}(w) &= \sum_{x=0}^{2^n-1} (f_1(x) \oplus f_2(x))(-1)^{\langle w,x\rangle} \\
&= \sum_{x=0}^{2^n-1} (f_1(x) + f_2(x) - 2f_1(x)f_2(x))(-1)^{\langle w,x\rangle} \\
&= \sum_{x=0}^{2^n-1} f_1(x)(-1)^{\langle w,x\rangle} + \sum_{x=0}^{2^n-1} f_2(x)(-1)^{\langle w,x\rangle} \\
&\quad -2\sum_{x=0}^{2^n-1} f_1(x)f_2(x)(-1)^{\langle w,x\rangle} \\
&= S_{f_1}(w) + S_{f_2}(w) - 2S_{f_1f_2}(w)
\end{aligned}
$$

□

Theorem 1.2. *Let* $f_1(x), f_2(x) \in \mathcal{F}_n$. *Then*

$$
S_{f_1f_2}(w) = 2^{-n} \sum_{\tau=0}^{2^n-1} S_{f_1}(\tau) S_{f_2}(w \oplus \tau) \tag{1.19}
$$

Proof. By the inverse Walsh transform, we have

$$
\begin{aligned}
S_{f_1f_2}(w) &= \sum_{x=0}^{2^n-1} f_1(x)f_2(x)(-1)^{\langle w,x\rangle} \\
&= \sum_{x=0}^{2^n-1} [2^{-n} \sum_{\tau=0}^{2^n-1} S_{f_1}(\tau)(-1)^{\langle \tau,x\rangle}][2^{-n} \sum_{\alpha=0}^{2^n-1} S_{f_2}(\alpha)(-1)^{\langle \alpha,x\rangle}](-1)^{\langle w,x\rangle} \\
&= 2^{-2n} \sum_{\tau=0}^{2^n-1} \sum_{\alpha=0}^{2^n-1} S_{f_1}(\tau) S_{f_2}(\alpha) \sum_{x=0}^{2^n-1} (-1)^{\langle (\tau \oplus \alpha \oplus w),x\rangle}
\end{aligned}
$$

By Eq. 1.10 it is known that, when $\alpha = w \oplus \tau$, we have

$$
\sum_{x=0}^{2^n-1} (-1)^{\langle (\tau \oplus \alpha \oplus w),x\rangle} = 2^n,
$$

else

$$
\sum_{x=0}^{2^n-1} (-1)^{\langle (\tau \oplus \alpha \oplus w),x\rangle} = 0.
$$

So we have

$$S_{f_1 f_2}(w) = 2^{-n} \sum_{\tau=0}^{2^n-1} S_{f_1}(\tau) S_{f_2}(w \oplus \tau)$$

Hence, the conclusion of Theorem 1.2 holds. □

From Theorem 1.2, we can easily get the following conclusion.

Theorem 1.3. *Let $f(x) \in \mathcal{F}_n$. Then the Walsh spectrum of $f(x)$ satisfies*

$$S_f(w) = 2^{-n} \sum_{\tau=0}^{2^n-1} S_f(\tau) S_f(w \oplus \tau) \tag{1.20}$$

On the other hand, if Eq. 1.20 holds, then $f(x)$ must be a Boolean function.

Proof. In Theorem 1.2, let $f_1(x) = f_2(x) = f(x)$, and notice that a Boolean function $f(x)$ must satisfy that $f^2 = f$, then Eq. 1.20 holds. On the other hand, by Theorem 1.2, the right-hand side of Eq. 1.20 equals $S_{f \cdot f}(w)$, which holds for every $w \in GF^n(2)$ if and only if $f^2 = f$, which holds if and only if $f(x)$ is a Boolean function. □

Theorem 1.3 can be used to judge whether a real-valued function is a Boolean function when only the Walsh spectrum is given. In this case, the Walsh spectrum uniquely determines a Boolean function. We do not usually treat the Walsh spectrum as a representation of Boolean functions, because to check whether the Walsh spectrum is from a Boolean function requires much computation.

Theorem 1.4. *Let $f(x) \in \mathcal{F}_n$. Then the Walsh transform of $f(x)$ satisfies*

$$2^{-n} \sum_{w=0}^{2^n-1} S_f(w) = f(0) \in \{0, 1\} \tag{1.21}$$

Proof. Take the summation of Eq. 1.12 for all $w \in GF^n(2)$, we have

$$\begin{aligned}
2^{-n} \sum_{w=0}^{2^n-1} S_f(w) &= \sum_{w=0}^{2^n-1} \sum_{x=0}^{2^n-1} f(x)(-1)^{\langle w,x \rangle} \\
&= 2^{-n} \sum_{x=0}^{2^n-1} f(x) \sum_{w=0}^{2^n-1} (-1)^{\langle w,x \rangle} \\
&= f(0) \qquad \text{(by Eq. 1.10)}
\end{aligned}$$

□

Theorem 1.5. *Let $f(x) \in \mathcal{F}_n$, D be an $n \times n$ invertible (nonsingular) matrix over $GF(2)$. Then*

$$S_{f(xD)}(w) = S_{f(x)}(w(D^{-1})^T) \tag{1.22}$$

where $(D^{-1})^T$ is the transpose of matrix D^{-1} which is the inverse of matrix D.

Proof.

$$\begin{aligned} S_{f(xD)}(w) &= \sum_{x=0}^{2^n-1} f(xD)(-1)^{\langle w,x\rangle} \\ &= \sum_{x=0}^{2^n-1} f(xD)(-1)^{\langle w,(xDD^{-1})\rangle} \\ &= \sum_{y=0}^{2^n-1} f(y)(-1)^{\langle w,(yD^{-1})\rangle} \qquad (y = xD) \\ &= \sum_{y=0}^{2^n-1} f(y)(-1)^{\langle w(D^{-1})^T,y\rangle} \\ &= S_{f(x)}(w(D^{-1})^T) \end{aligned}$$

□

Theorem 1.5 shows the relationship between the Walsh spectrums of two Boolean functions where one is induced from the other by variable invertible linear transform. Below are some properties with respect to the overall Walsh spectrum of Boolean functions.

Theorem 1.6 (Plancheral). *Let $f(x) \in \mathcal{F}_n$ and $S_f(w)$ be the Walsh spectrum of $f(x)$. Then we have*

$$\sum_{w=0}^{2^n-1} S_f^2(w) = 2^n \sum_{x=0}^{2^n-1} f^2(x) = 2^n wt(f) \tag{1.23}$$

Proof.

$$\begin{aligned} \sum_{w=0}^{2^n-1} S_f^2(w) &= \sum_{w=0}^{2^n-1}\Big[\sum_{x=0}^{2^n-1} f(x)(-1)^{\langle w,x\rangle} \sum_{y=0}^{2^n-1} f(y)(-1)^{\langle w,y\rangle}\Big] \\ &= \sum_{x=0}^{2^n-1}\sum_{y=0}^{2^n-1} f(x)f(y) \sum_{w=0}^{2^n-1}(-1)^{\langle w,(x\oplus y)\rangle} \\ &= 2^n \sum_{x=y} f(x)f(y) = 2^n wt(f) \end{aligned}$$

□

Theorem 1.6 tells that the sum of the squares of the Walsh spectrum of a Boolean function is 2^n times the Hamming weight of the function. If the Hamming weight of $f(x)$ is small, then the chances for $S_f(w)$ to have a big absolute value is also small.

Now we give a spectrum description of self-correlation function of Boolean functions. First we introduce the concept.

Definition 1.3. Let $f(x) \in \mathcal{F}_n$. Then

$$R_f(\tau) = \sum_{x=0}^{2^n-1} f(x)f(x \oplus \tau)$$

is called the *self-correlation function* of $f(x)$, where $\tau \in GF^n(2)$.

The self-correlation function of a Boolean function measures the common coordinates with value 1 between $f(x)$ and $f(x \oplus \tau)$. Note that for any fixed $\tau \in GF^n(2)$, when x goes through all the elements in $GF^n(2)$, $x \oplus \tau$ also goes through all the elements in $GF^n(2)$. So the truth table of $f(x \oplus \tau)$ is a permutation of the truth table of $f(x)$. Now we give an example to show how this works.

Example 1.2. Let $f(x) = x_1x_2 \oplus x_1x_3 \oplus x_2x_3 \in \mathcal{F}_3$. Then the truth table of $f(x)$ is (00010111). When x takes values (000, 001, 010, 011, 100, 101, 110, 111), the shifts $x \oplus \tau$ and the corresponding truth tables of $f(x \oplus \tau)$ are as follows:

τ	$x \oplus \tau$	$f(x \oplus \tau)$	$R_f(\tau)$
0	000, 001, 010, 011, 100, 101, 110, 111	00010111	4
1	001, 000, 011, 010, 101, 100, 111, 110	00101011	2
2	010, 011, 000, 001, 110, 111, 100, 101	10001101	2
3	011, 010, 001, 000, 111, 110, 101, 100	10001110	2
4	100, 101, 110, 111, 000, 001, 010, 011	01110001	2
5	101, 100, 111, 110, 001, 000, 011, 010	10110010	2
6	110, 111, 100, 101, 010, 011, 000, 001	11010100	2
7	111, 110, 101, 100, 011, 010, 001, 000	11101000	0

There is a good Walsh spectrum property of the self-correlation functions of Boolean functions.

Theorem 1.7 (Wiener-Khinchin). *Let $f(x) \in \mathcal{F}_n$, $R_f(\tau)$ be the self-correlation function of $f(x)$. Then we have*

$$R_f(\tau) = 2^{-n} \sum_{w=0}^{2^n-1} S_f^2(w)(-1)^{\langle w, \tau \rangle} \tag{1.24}$$

Proof.

$$
\begin{aligned}
R_f(\tau) &= \sum_{x=0}^{2^n-1} f(x)f(x\oplus\tau) \\
&= \sum_{x=0}^{2^n-1} \left(2^{-n}\sum_{w=0}^{2^n-1} S_{f(x)}(w)(-1)^{\langle w,x\rangle}\cdot 2^{-n}\sum_{w'=0}^{2^n-1} S_{f(x\oplus\tau)}(w')(-1)^{\langle w',x\rangle}\right) \\
&= 2^{-2n}\sum_{w=0}^{2^n-1}\sum_{w'=0}^{2^n-1} S_{f(x)}(w)S_{f(x\oplus\tau)}(w')\sum_{x=0}^{2^n-1}(-1)^{\langle (w\oplus w'),x\rangle} \\
&= 2^{-n}\sum_{w=w'} S_{f(x)}(w)S_{f(x\oplus\tau)}(w') \\
&= 2^{-n}\sum_{w=0}^{2^n-1} S_{f(x)}(w)S_{f(x\oplus\tau)}(w)
\end{aligned}
$$

where

$$
\begin{aligned}
S_{f(x\oplus\tau)}(w) &= \sum_{x=0}^{2^n-1} f(x\oplus\tau)(-1)^{\langle w,x\rangle} \\
&= \sum_{x=0}^{2^n-1} f(x)(-1)^{\langle w,x\rangle+\langle w,\tau\rangle} \quad (x\leftarrow x\oplus\tau) \\
&= (-1)^{\langle w,\tau\rangle}\sum_{x=0}^{2^n-1} f(x)(-1)^{\langle w,x\rangle} \\
&= (-1)^{\langle w,\tau\rangle} S_{f(x)}(w)
\end{aligned}
$$

Taking this into the above, we have Eq. 1.24. □

It is noted that all the above properties are given using the type I Walsh transform. As stated earlier, since there is convenient conversion between type I and type II Walsh transforms, any property given in one type can be converted to the other type of Walsh transform. However, sometimes the expression of certain properties in one type of Walsh transform is more compact than in the other type. The following is one such case where it uses the type II Walsh transform.

Theorem 1.8 (Parseval). *Let $f(x) \in \mathcal{F}_n$, $S_{(f)}(w)$ be the type II Walsh spectrum of $f(x)$. Then we have*

$$\sum_{w=0}^{2^n-1} S_{(f)}^2(w) = 2^{2n}. \tag{1.25}$$

Proof.

$$\begin{aligned}
\sum_{w=0}^{2^n-1} S_{(f)}^2(w) &= \sum_{w=0}^{2^n-1}[\sum_{x=0}^{2^n-1}(-1)^{f(x)+\langle w,x\rangle}\sum_{y=0}^{2^n-1}(-1)^{f(y)+\langle w,y\rangle}] \\
&= \sum_{x=0}^{2^n-1}\sum_{y=0}^{2^n-1}(-1)^{f(x)\oplus f(y)}\sum_{w=0}^{2^n-1}(-1)^{\langle w,(x\oplus y)\rangle} \\
&= 2^n\sum_{x=y}(-1)^{f(x)\oplus f(y)} = 2^{2n}.
\end{aligned}$$

□

1.3.3 Hadamard Matrices

Another representation of Walsh functions and Walsh transform is to use matrices. Denote

$$W_{(n)} = [W(w,x)] = [(-1)^{\langle w,x\rangle}],\ \ 0 \le w,\ x \le 2^n - 1 \tag{1.26}$$

as a $2^n \times 2^n$ binary matrix whose element on the i-th row and the j-th column is $W(i,j)$. Then when $n = 1$ and when $n = 2$, it is easy to formulate the matrices

$$W_{(1)} = \begin{bmatrix} 1 & 1 \\ 1 & -1 \end{bmatrix}$$

and

$$W_{(2)} = \begin{bmatrix} 1 & 1 & 1 & 1 \\ 1 & -1 & 1 & -1 \\ 1 & 1 & -1 & -1 \\ 1 & -1 & -1 & 1 \end{bmatrix}$$

The matrix $W_{(n)}$ is called a *Hadamard matrix* of order n, and the Hadamard matrices have the following property:

$$W_{(n)} = W_{(1)} \otimes W_{(n-1)} = W_{(n-1)} \otimes W_{(1)} = W_{(1)}^{[n]}. \tag{1.27}$$

where $\otimes$ means the *Kronecker product* defined as

$$\begin{bmatrix} a_{11} & a_{12} \\ a_{21} & a_{22} \end{bmatrix} \otimes \begin{bmatrix} b_{11} & b_{12} \\ b_{21} & b_{22} \end{bmatrix} = \begin{bmatrix} a_{11}\begin{bmatrix} b_{11} & b_{12} \\ b_{21} & b_{22} \end{bmatrix} & a_{12}\begin{bmatrix} b_{11} & b_{12} \\ b_{21} & b_{22} \end{bmatrix} \\ a_{21}\begin{bmatrix} b_{11} & b_{12} \\ b_{21} & b_{22} \end{bmatrix} & a_{22}\begin{bmatrix} b_{11} & b_{12} \\ b_{21} & b_{22} \end{bmatrix} \end{bmatrix}$$

$$= \begin{bmatrix} a_{11}b_{11} & a_{11}b_{12} & a_{12}b_{11} & a_{12}b_{12} \\ a_{11}b_{21} & a_{11}b_{22} & a_{12}b_{21} & a_{12}b_{22} \\ a_{21}b_{11} & a_{21}b_{12} & a_{22}b_{11} & a_{22}b_{12} \\ a_{21}b_{21} & a_{21}b_{22} & a_{22}b_{21} & a_{22}b_{22} \end{bmatrix},$$

and $W_{(1)}^{[n]}$ means the consecutive Kronecker product of $A_{(1)}$ for n times. Particularly, it is trivial to verify the equality of $W_{(2)} = W_{(1)} \otimes W_{(1)}$.

By the orthogonal property of the Walsh functions, it is easy to induce the corresponding properties of Hadamard matrices:

(1) $W_{(n)}$ is a symmetric matrix of order $2^n \times 2^n$.

(2) $W_{(n)} \cdot W_{(n)} = 2^n I(2^n)$, where $I(2^n)$ is an identity matrix of order 2^n, and the operator for normal matrix multiplication is denoted by a dot "$\cdot$" which can be omitted for convenience of writing.

(3) $W_{(n)}^{-1} = 2^{-n} W_{(n)}$.

Let $F = (f(0), f(1), f(2), \ldots, f(2^n - 1))$ be the truth table representation of Boolean function $f(x)$, and similarly we define S_f and $S_{(f)}$ to be the truth table of $S_f(w)$ and $S_{(f)}(w)$, respectively. Define $\delta(f) = (-1)^f$, and denote $\delta(F) = [\delta(f(0)), \delta(f(1)), \delta(f(2)), \ldots, \delta(f(2^n - 1))]$. Then the Walsh transform and the inverse Walsh transform can be represented by means of matrices as follows:

$$S_f = F \cdot W_{(n)} \tag{1.28}$$

$$F = 2^{-n} S_f \cdot W_n \tag{1.29}$$

$$S_{(f)} = \delta(F) \cdot W_{(n)} \tag{1.30}$$

$$\delta(F) = 2^{-n} S_{(f)} \cdot W_{(n)} \tag{1.31}$$

When Eqs. 1.28, 1.29, 1.30 and 1.31 can be used to compute the Walsh spectrum or the inverse Walsh spectrum, it is faster than to use the normal Walsh transform as represented in Eqs. 1.12, 1.13, 1.14 and 1.15.

With the matrix representation of Walsh transform, many results from Walsh transform can be migrated to this new representation. For example, the Plancheral formula can be expressed as

$$Trace(S_f^T S_f) = Trace(F^T F) \tag{1.32}$$

or

$$S_f S_f^T = FF^T. \tag{1.33}$$

Since the Hadamard matrix representation has a parallel theory as the Walsh transform, the transform hence is also called *Walsh-Hadamard* transform. The Hadamard matrix representation can be used to formulate fast computation of Walsh-Hadamard transform. In the following, for the simplicity of writing, we will use the term Walsh transform.

1.4 Basic Models of Stream Ciphers That Use Boolean Functions

Boolean functions are widely used in cryptography and cryptanalysis, as well as many other areas [5]. In the design of cryptographic algorithms, both stream ciphers [4, 26] and block ciphers use nonlinear functions, where nonlinear Boolean functions are an important class of cryptographic primitives [2, 25, 26]. However, the cryptographic properties of Boolean functions mostly come from the consideration of designing stream ciphers, as nonlinear Boolean functions are a core component in many stream ciphers [9], and any potential threat or attack to one of such models (which often lead to the attacks to other models) will lead to a security requirement, i.e., a new cryptographic property. In order to introduce the cryptographic properties or cryptographic requirements of Boolean functions, we first introduce the basic models of stream ciphers.

Stream ciphers are one of the most popular ciphers in traditional encryption as well as modern encryption processes using high-speed electronic devices. Because the speed of stream ciphers is normally faster than block ciphers, and much faster than public key ciphers, stream ciphers are a type of encryption mechanisms that are not likely to be replaced by any other ciphers. Irrespective of the high speed advantage in hardware and software, it is unfortunate that all the basic stream cipher models are under attacks to different degree, which has shown that their security strength is lower than the designed security strength (corresponding to brute force attack by testing all the possible keys). One of such example is the $A5$ algorithm which is a standard algorithm used in Global System for Mobile Communications (GSM).

The basic mechanism of stream cipher is very simple: treat original message as a stream of bits[2] and the key as a stream of bits, then the arithmetic performing

[2] A more general case is that both the message and the key are streams of units, where the unit can be a bit, byte, or even a larger data unit such as 32-bit block. However, the underneath arithmetic performing encryption of a message unit and a key unit should be bit-wise XOR. If the arithmetic between a message unit and a key unit is more complicated than bit-wise XOR, then it would be

the encryption is the XOR operation. The output of the XOR of message stream and the key stream is the ciphertext which is apparently a stream of bits. According to Shannon's theory [32], if the key stream is composed of pure random bits, then the encryption is perfect, i.e., from the ciphertext one gets no information about the original message. This system is known as *one-time pad*. Although the one-time-pad mechanism provides the best security in theory, practically however it is not easy to implement. The difficulty comes from the key management: when encryption is done by one party and the decryption is done by another, as the same key has to be used for encryption and decryption, both of the parties must have the same key in order to complete secure message transmission. Shannon has proved that, in order to achieve one-time-pad security, the size of the key should be at least the same as that of the message to be encrypted. When the message is large, to allow both of the communication parties to share a key of the required size can be difficult. In order to make it more practical, one solution is to use pseudorandom sequences instead of pure random sequences as encryption keys. A pseudorandom sequence is defined to have the following properties: (1) randomness and (2) reproduction. By randomness, it means that a pseudorandom sequence should look like a random sequence where it is hard to find rules about how the sequence is generated by looking at the sequence itself, or in other words, the bits 0 and 1 seem to appear at random. The reproduction property requires that given certain small amount of data (often means the initial key), the same pseudorandom sequence can be reproduced. The reproduction property has significant difference from random sequences, where the latter one cannot be reproduced for certain.

Given that, the basic model of stream cipher can be designed as follows: a pseudorandom sequence generator that generates a pseudorandom sequence given a seed key, and when a message stream is given, by using the XOR operation between the message stream and the generated pseudorandom sequence which is known as the *key stream*, the ciphertext stream is produced. In the decryption, the same pseudorandom key stream is used to XOR with the ciphertext stream to recover the original message. This basic model is depicted in Fig. 1.1.

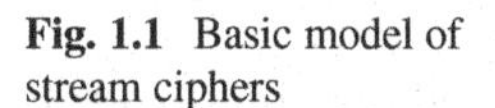

Fig. 1.1 Basic model of stream ciphers

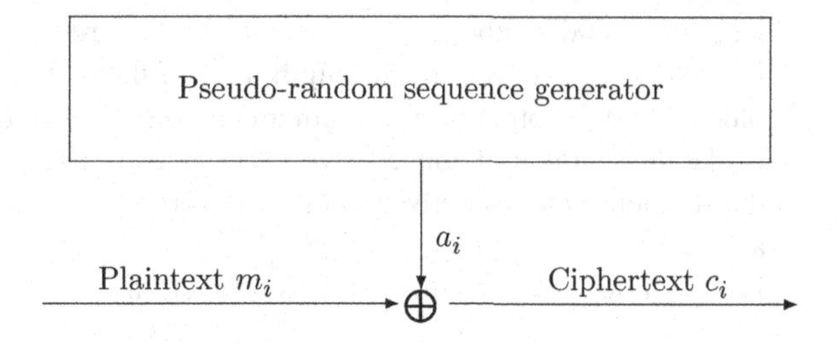

treated as a block cipher. When the arithmetic is a little bit more complicated than XOR but much simpler than traditional block ciphers, the system can be treated as a hybrid. For simplicity, our discussion will be based on the simple case where both the message and the key are streams of bits.

Denote $m = (m_i)$ as the sequence of plaintext, $k = (k_i)$ as the sequence of key stream, and $c = (c_i)$ as the sequence of ciphertext, where i indicates the index of bits. Then the encryption process can simply be written as

$$c_i = m_i \oplus k_i, \tag{1.34}$$

and the decryption process can be written as

$$m_i = c_i \oplus k_i. \tag{1.35}$$

With this model of cipher, the security relies on the property of the key stream. So the study of stream ciphers becomes mostly the study of pseudorandom sequences. Now we introduce a few traditional pseudorandom sequence generators.

1.4.1 Linear Feedback Shift Registers

We have seen that the design of pseudorandom sequences is to let the sequence have pseudorandomness. What is pseudorandomness? How to measure it? In 1967, Golomb [13] gives a description of randomness of binary sequences (composed of 0's and 1's). Note that a pseudorandom sequence must be a periodic sequence, or eventually a periodic sequence, as the generator of such a sequence is a finite-state machine which will eventually exhaust the states and move to one of the states appeared earlier, eventually resulting in a periodic sequence. If it is not period, then by cutting off some parts from the beginning, it must be a period sequence. Hence, Golomb's description of pseudorandomness is based on binary period sequences. It defines that a periodic binary sequence should satisfy the following properties in order to meet the pseudorandomness:

1. The number of 0's and 1's in one period should be equal or as close as possible.
2. In one period, the number of runs of length 1 should be about half of the total runs, and that of length 2 should be about quarter of the total runs, ... and that of length i should be about $\frac{1}{2^{i+1}}$ of that of the total runs.
3. For all the runs of any fixed length, half of them should be all-one runs (called blocks) and the other half of them are all-zero runs (called gaps). An all-one run is like the segment of consecutive 1's in ...011110..., and an all-zero run is like the segment of consecutive 0's in ...10000001..... Both of the cases are called a *run*.
4. For any integer k, the autocorrelation function

$$AC(k) = \frac{1}{p}\sum_{i=1}^{p}(-1)^{a_i \oplus a_{(i+k) \bmod p}}$$

of the sequence has two values: its values are the same for all $k \neq 0$, where p is the period of sequence (a_i).

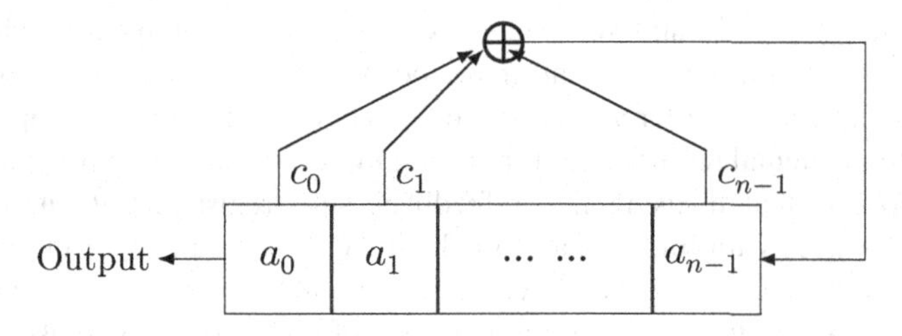

Fig. 1.2 Linear feedback shift register (LFSR)

In searching sequences that satisfy the above defined properties, one class seems to be so good and they are easy to generate, and a subclass of them meet the properties so well. This class of sequences are linear feedback shift register (LFSR) sequences, and the subclass of them are m-sequences. To put it in a simple way, a linear feedback shift register is a collection of memory cells, and when an electric pulse comes, the content of the cells shifts to the one on its left (or equivalently to the right). The cell on the most right then is filled with a linear combination of the original content of all the cells. This is illustrated in Fig. 1.2.

The sequence generated by the LFSR as shown in Fig. 1.2 can be written as

$$a_m = c_0 a_{m-n} \oplus c_1 a_{m-n-1} \oplus \cdots \oplus c_{n-1} a_{m-1}, \quad m \geq n, \tag{1.36}$$

where n is the length of the LFSR (also called the order of the LFSR). Corresponding to this linear feedback, there is a polynomial $f(x) = c_0 \oplus c_1 x \oplus \ldots \oplus c_{n-1} x^{n-1}$ which is called the generating polynomial of the LFSR [14]. When the generating polynomial of an LFSR is primitive,[3] from any nonzero initial state, the LFSR will produce a sequence with very good pseudorandom properties, and such a sequence is called an m-sequence. The m-sequences meet the pseudorandom properties very well and hence have wide applications not only in stream ciphers but more in spectrum communications.

Although the LFSR may produce sequences such as m-sequences that have very good randomness, when they are used as key streams, it is not secure. In 1969, Berlekamp and Massey developed an algorithm that can efficiently reconstruct an LFSR that can generate the whole sequence given a segment of the sequence. This algorithm is known as Berlekamp-Massey algorithm (or B-M algorithm for short). When the length of the segment is $2n$ or larger, where n is the order of the LFSR, the B-M algorithm can reconstruct an LFSR of order n that can generate the same period of sequence, although the actual period of the sequence can be up to $2^n - 1$. Given B-M algorithm, it is expected that the value of n should be very large so that to get $2n$ consecutive bits is practically not possible. This leads to a cryptographic

[3] A primitive polynomial $f(x)$ over $GF(2)$ of degree n is such that its minimum order is $2^n - 1$, i.e., $\min\{t : f(x)|(x^t - 1)\} = 2^n - 1$.

measurement of pseudorandom sequences – the linear complexity. The *linear complexity* of a sequence (segment or period) is the minimum value n such that there exists an LFSR of order n to generate the sequence. The linear complexity has become a fundamental requirement for cryptographic pseudorandom sequences.

It is apparent that in using linear feedback shift registers [29], it is hard to produce sequences with high linear complexity, unless the order of the register is very high which is not practical. However, it is possible to produce sequences with high linear complexity using other models. One of such models is to use nonlinear feedback shift registers, and from the name it is known that the feedback function is nonlinear. Among the nonlinear feedback shift register sequences, a special subset of them have maximum period 2^n, which means that from any initial state, the feedback function will change the state to all possible elements in $GF^n(2)$. This class of nonlinear feedback shift register sequences are called M-sequences. Nonlinear feedback shift register sequences are a large class of sequences, and most of them have high linear complexity; however, other properties such as randomness are not clear in most of the cases, and even for those with known randomness such as M-sequences, how to efficiently generate them is still not practically useful.

1.4.2 *Nonlinear Filtering Generators and Nonlinear Combiners*

In order to generate pseudorandom sequences with high linear complexity, while nonlinear feedback shift registers are possible choices, people found other more efficient ways of generating pseudorandom sequences using one or more LFSRs and a nonlinear Boolean function which act as filtering function or combining function. A nonlinear filtering generator is composed of an LFSR and a nonlinear function $f(x)$, where the content of the cells in the LFSR is taken as the input of $f(x)$, and the output of $f(x)$ is the final output of the generator. This is depicted in Fig. 1.3.

Another LFSR-based generator is nonlinear combiner which is composed of a few LFSRs, and the output of all the LFSRs forms the input of the nonlinear combining function $f(x)$, and the output of $f(x)$ is the final output of the generator. This is depicted in Fig. 1.4.

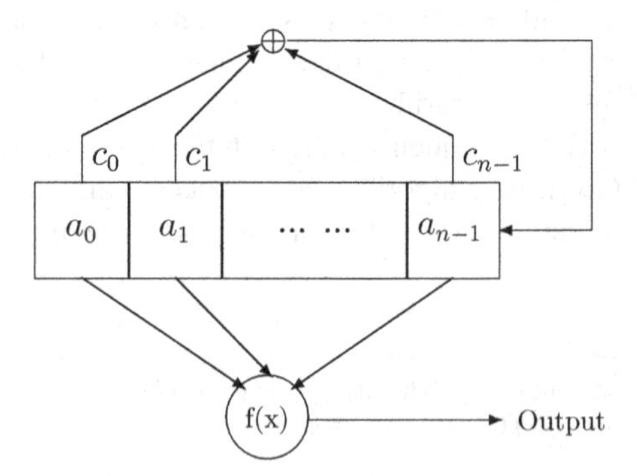

Fig. 1.3 Nonlinear filtering generator

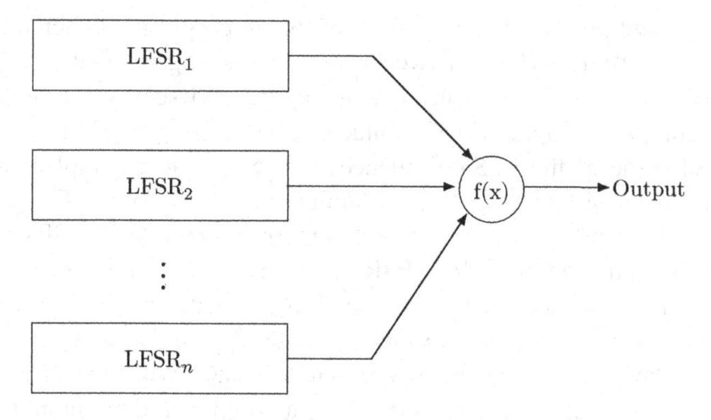

Fig. 1.4 Nonlinear combiner generator

There are other LFSR-based generators for pseudorandom sequences such as clock-controlled generators. Since they are not necessary for our introduction to cryptographic properties of Boolean functions, we are not going to introduce them.

The nonlinear filtering generators and the nonlinear combiners are somehow equivalent: A nonlinear filtering generator can be treated as a nonlinear combiner with all the LFSRs being the same but different initial states, and a nonlinear combiner can be treated as a nonlinear filtering generator based on a larger LFSR (the length of this hypothetic LFSR should be at least the minimum common divisor of the lengths of the LFSRs in the combiner). However, traditionally, the nonlinear Boolean function in a nonlinear filtering generator is called the filtering function, and the function in a nonlinear combiner is called the combining function.

1.5 Cryptographic Properties of Boolean Functions

There are many different kinds of attacks to the stream ciphers, and hence the Boolean functions used in the stream ciphers must have some required properties. These properties designed to make the ciphers secure against some known and potential attacks are known as cryptographic properties. Some of the very common cryptographic properties of Boolean functions are briefly described as follows, and we will look into them in more detail in the chapters later.

1.5.1 Algebraic Degree

Let us take the nonlinear combiner for consideration. With respect to linear complexity of the output segments, assuming each LFSR$_i$ has order n_i which means that the periodic sequences generated by this LFSR$_i$ will have linear complexity

n_i. We also assume that the orders of the LFSRs are co-prime of each other. Then the summation (bit-wise XOR) of two of the sequences generated by LFSR$_i$ and LFSR$_j$ will be $n_i + n_j$. The multiplication of the two sequences however will have linear complexity $n_i n_j$, which is much larger than $n_i + n_j$. In general, the summation of some of the LFSR sequences will have linear complexity the sum of the orders of those LFSRs, while the multiplication of those LFSR sequences will have linear complexity the product of the orders of those LFSRs. It is seen that the multiplication of t of the LFSR sequences will result in a sequence of much higher linear complexity than summation can achieve. This corresponds to a multiplicative term of degree t in the algebraic normal form of the combining function $f(x)$. Therefore, if we expect the output sequence of the nonlinear combiner to have high linear complexity, the corresponding nonlinear combining function is expected to be of high algebraic degree. This is why algebraic degree becomes one of the cryptographic measurements.

If the linear complexity of the nonlinear combiner generator sequences is the only cryptographic requirement to pursue, then we can let the nonlinear combining function $f(x)$ to be of the highest algebraic degree n, where n is the number of variables of $f(x)$, which is also the number of the LFSRs as in the nonlinear combiner model. However, practically there are other cryptographic requirements to meet, and some of the requirements may conflict. So to achieve a good compromise of all the required cryptographic properties, it has to sacrifice the level of some of the cryptographic requirements. For example, practically the Boolean functions used in cryptosystems do not reach the highest algebraic degree. However, the algebraic degree of the employed Boolean functions cannot be too low either. In general, the algebraic degree should be larger than $\frac{n}{2}$.

1.5.2 Balance

When a Boolean function is used in cryptography, it is expected that the output of the function is unbiased, or at least near unbiased, i.e., the chances for the output to be 0 are about the same as the chances for the output to be 1. When a Boolean function has equal chances to output 0 or 1 when the input variables go through all the possibilities, the function is called *balanced*. Obviously a balanced Boolean function $f(x)$ satisfies that $wt(f) = 2^{n-1}$. In the design of stream ciphers, Boolean functions are often required to be balanced, as this would give outputs with balanced number of 0's and 1's and looks more random.

When a Boolean function is balanced, the requirement of algebraic degree has to have some sacrifice. From the minterm representation, it can be seen that the number of minterms of $f(x)$ is $wt(f)$, and when the minterms are expanded to convert to the algebraic normal form, every such minterm contributes a product of all the variables $x_1x_2\cdots x_n$. Since every two same terms will vanish due to the XOR operation, we have that

Lemma 1.2. *Let $f(x) \in \mathcal{F}_n$. Then the term $x_1x_2\cdots x_n$ does not appear in the algebraic normal form of $f(x)$ if and only if $wt(f)$ is even, in which case $deg(f) \le n-1$. In particular, if $f(x)$ is balanced, then $deg(f) \le n-1$.*

1.5.3 Nonlinearity

For security reasons, pseudorandom sequences generated by linear shift registers are not supposed to be used in encryptions directly. This leads to the study of nonlinear shift register sequences and linear shift register sequences with a nonlinear filtering function or nonlinear combining function. In order to make sure that a function is nonlinear, in its algebraic normal form representation, there should be at least one nonlinear term. As defined earlier, the algebraic degree is a measurement about the nonlinear feature of a function; however, this is not ideal in many applications. For example, function $f(x) = x_1 + x_1x_2\cdots x_n$ in n variables has the highest possible algebraic degree n; however, when we use $f'(x) = x_1$ to approximate $f(x)$, then the difference is only one out of $2^n - 1$, which means that the linear approximation is very close. Then we define another nonlinear measurement, the nonlinearity of a Boolean function, to be the minimum distance between a given function to all linear functions. Since the set of linear functions and affine functions differ only by a constant, the concept of nonlinearity is extended to be the minimum distance of the given function to all the affine functions, denoted by $nl(f)$, i.e.,

$$nl(f) = \min_{l(x)\in\mathcal{A}_n} d(f(x), l(x)) = \min_{l(x)\in\mathcal{A}_n} wt(f(x) \oplus l(x)) \tag{1.37}$$

1.5.4 Linear Structure

A Boolean function $f(x) \in \mathcal{F}_n$ is said to have a *linear structure* $\alpha \in GF^n(2)$ if $f(x) \oplus f(x \oplus \alpha) \equiv c$, where c is a constant of $\{0, 1\}$. In particular α is called an *invariant linear structure* if $c = 0$ and a *complementary linear structure* if $c = 1$.

It is apparent that the linear structure is another extreme case as opposed to the propagation criterion. More study on linear structures can be found in Sect. 3.8.

1.5.5 Propagation Criterion

A Boolean function $f(x) \in \mathcal{F}_n$ may have no linear structures at all, i.e., for any $\alpha \in GF^n(2), f(x) \oplus f(x \oplus \alpha)$ is not a constant; instead, it may even be a balanced Boolean function.

A Boolean function $f(x) \in \mathcal{F}_n$ is said to satisfy the *propagation criterion* with respect to a nonzero vector α if $f(x) \oplus f(x \oplus \alpha)$ is balanced.

Let $f(x) \in \mathcal{F}_n$. If for any $\alpha \in GF^n(2)$ with $wt(\alpha) = 1, f(x) \oplus f(x \oplus \alpha)$ is always a balanced Boolean function, then $f(x)$ is said to satisfy the *strict avalanche criterion (SAC)*. Furthermore, if for any nonzero vector $\alpha \in GF^n(2), f(x) \oplus f(x \oplus \alpha)$ is always a balanced Boolean function, then $f(x)$ is said to be *perfect nonlinear*.

The strict avalanche criterion has attracted much study (see, e.g., [1, 7, 8, 11, 15, 18, 19, 21, 22, 31, 34, 35]). If a Boolean function $f(x)$ satisfies the strict avalanche criterion, then when any 1 bit of its input changes, exactly half of the output bits will change. If a Boolean function $f(x)$ is perfect nonlinear, then any change of its input will result in exactly half of the output bits changed.

Higher-order SAC is also an interesting cryptographic property, with construction of such functions [16, 17] being a challenging topic.

As a generalization of the concepts of avalanche criterion and perfect nonlinear, the following concept was proposed by Preneel et al. in [27].

A Boolean function $f(x)$ is said to satisfy the propagation criterion [28] of order k if it satisfies the propagation criterion with respect to all α with $1 \leq wt(\alpha) \leq k$ and is denoted by $PC(f) = k$. Apparently this concept is a generalization of the concept of strict avalanche criterion.

Note: *Strict avalanche criterion* (SAC) is equivalent to the propagation criterion of order 1 (i.e., $PC(f) = 1$), and *perfect nonlinear* defined in [24] is equivalent to the propagation criterion of order n (i.e., $PC(f) = n$).

1.5.6 Correlation Immunity

A boolean function $f(x) \in \mathcal{F}_n$ is said to be *correlation immune* of order k if $f(x)$ is statistically independent of any k of its variables. The correlation immunity is a security measure about how resistant a Boolean function is against correlation attack [12]. Both the statistical independence and the correlation immunity will be further studied later. Correlation immunity of Boolean functions will be further studied in Chap. 4.

1.5.7 Algebraic Immunity

Let $f(x) \in \mathcal{F}_n$; if there exists a nonzero Boolean function $g(x) \in \mathcal{F}_n$ such that $g(x)f(x) = 0$, then $g(x)$ is called an *annihilator* of $f(x)$. The minimum algebraic degree of the annihilators of $f(x)$ and of $f(x) \oplus 1$ is called the *algebraic immunity* of $f(x)$. The algebraic immunity is a measurement about how resistant a Boolean function is against algebraic attack. This topic is further studied in Chap. 5 in this book.

1.5.8 Remarks

There can be more properties of Boolean functions than those listed above, and some of them can be classified as cryptographic properties to be met by cryptographic functions and components in cryptographic algorithms, while others are just interesting properties. For example, the number of nonlinear terms of Boolean functions in their algebraic normal form is deemed to be a cryptographic requirement, although not much formal study on this has been done. On the other hand, the symmetric property of Boolean functions is not deemed to be a cryptographic property; instead it is treated as something that cryptographic function should somehow avoid, as in general case the symmetric property does not benefit the security of a cipher. However, due to the symmetric property being interesting, there are still much research on this class of Boolean functions. This book also has a chapter devoted to the cryptographic properties of symmetric functions, with a particular focus on the correlation immunity of symmetric Boolean functions.

References

1. Babbage S.: On the relevance of the strict avalanche criterion. Electron. Lett. **26**(7), 461–462 (1990)
2. Beal, M., Monaghan M.F.: Encryption using random Boolean functions. In: Cryptography and Coding, pp. 219–230. Clarendon, Oxford (1989)
3. Beauchamp, K.G.: Applications of Walsh and Related Functions. Academic, London (1984)
4. Beker, H., Piper, F.: Cipher Systems. Northwood Books, London (1982)
5. Carlet, C.: Boolean functions for cryptography and error correcting codes. In: Crama, Y., Hammer, P. (eds.) Boolean Models and Methods in Mathematics, Computer Science and Engineering, pp. 257–397. Cambridge University Press, Cambridge/New York (2010)
6. Charpin, P., Pasalic, E.: On propagation characteristics of resilient functions. In: Selected Areas in Cryptography. LNCS 2595, pp. 175–195. Springer, Berlin (2003)
7. Cusick, T.W.: Boolean functions satisfying a higher order strict avalanche criterion. In: Advances in Cryptology, Proceedings of Eurocrypt'93. LNCS 765, pp. 102–117. Springer, Berlin/New York (1994)
8. Cusick, T.W., Stanica, P.: Bounds on the number of functions satisfying the strict avalanche criterion. Inf. Process. Lett. **60**, 215–219 (1996)
9. Daemen, J., Govaerts, R., Vandewalle, J.: On the design of high speed self-synchronizing stream ciphers. In: Proceedings of ISCC/ISITA'92, Singapore, pp. 279–283 (1992)
10. Feng, D.G.: Spectral Theory and Its Applications in Cryptography. Science Press, Beijing (2000) (in Chinese)
11. Forre, R.: The strict avalanche criterion: spectral properties of Boolean functions and an extended definition. In: Advances in Cryptology, Proceedings of Crypto'88. LNCS 403, pp. 450–468. Springer, Berlin (1990)
12. Forre, R.: A fast correlation attack on nonlinearly feedforward filtered shift-register sequences. In: Advances in Cryptology, Proceedings of Eurocrypt'89. LNCS 434, pp. 586–595. Springer, Berlin (1990)
13. Golomb, S.W.: Shift Register Sequences. Holden-Day, San Francisco (1967)
14. Herlestam, T.: On functions of linear shift register sequences. In: Advances in Cryptology, Proceedings of Eurocrypt'85. LNCS 219, pp. 119–129. Springer, Berlin (1986).

15. Kim, K., Matsumoto, T., Imai, H.: A recursive construction method of S-boxes satisfying strict avalanche criteria. In: Advances in Cryptology, Proceedings of Crypto'90. LNCS 537, pp. 564–574. Springer, Berlin (1991)
16. Kurosawa, K., Satoh, T.: Generalization of higher order SAC to vector output Boolean functions. In: Advances in Cryptology, Proceedings of Asiacrypt'96. LNCS 1163, pp. 218–231. Springer, Berlin (1996)
17. Kurosawa, K., Satoh, T.: Design of SAC/PC(l) of order k Boolean functions and three other cryptographic criteria. In: Advances in Cryptology, Proceedings of Eurocrypt'7. LNCS 1233, pp. 434–449. Springer, Berlin (1997)
18. Lloyd, S.: Counting functions satisfying a higher order strict avalanche criterion. In: Advances in Cryptology, Proceedings of Eurocrypt'89. LNCS 434, pp. 63–74. Springer, Berlin (1990)
19. Lloyd, S.: Characterizing and counting functions satisfying the strict avalanche criterion of order (n-3). In: Cryptography and Coding II, pp. 165–172. Clarendon Press, Oxford (1992)
20. Lloyd, S.: Counting binary functions with certain cryptographic properties. J. Cryptol. **5**(2), 107–131 (1992)
21. Lloyd, S.: Balanced uncorrelatedness and the strict avalanche craterion. Dist. Appl. Math. **41**, 223–233 (1993)
22. Maitra, S.: Highly nonlinear balanced Boolean functions with good local and global avalanche characteristics. Inf. Process. Lett. **83**, 281–286 (2002)
23. Maitra, S., Sarkar, P.: Cryptographically significant Boolean functions with five valued Walsh spectra. Theor. Comput. Sci. **276**, 133–146 (2002)
24. Meier, W., Staffelbach, O.: Nonlinearity criteria for cryptographic functions. In: Advances in Cryptology, Proceedings of Eurocrypt'89. LNCS 434, pp. 549–562. Springer, Berlin (1990)
25. O'Connor, L.J.: An analysis of product ciphers based on the properties of Boolean functions. Ph.D. Thesis, University of Waterloo (1992)
26. Piper, F.C.: Stream ciphers. In: Cryptography. LNCS 149, pp. 181–188. Springer, Berlin/New York (1983)
27. Preneel, B., et al.: Propagation characteristics of Boolean bent functions. In: Advances in Cryptology, Proceedings of Eurocrypt'90. LNCS 473, pp. 161–173. Springer, Berlin (1991)
28. Preneel, B., et al.: Boolean functions satisfying propagation criteria. In: Advances in Cryptology, Proceedings of Eurocrypt'91. LNCS 547, pp. 141–152. Springer, Berlin (1991).
29. Ronse, C.: Feedback Shift Registers. Springer, Berlin/New York (1984)
30. Sarkar, P.: A note on the spectral characterization of Boolean functions. Inf. Process. Lett. **74**, 191–195 (2000)
31. Seberry, J., Zhang, X.M., Zheng, Y.: Improving the strict avalanche characteristics of cryptographic functions. Inf. Process. Lett. **50**, 37–41 (1994)
32. Shannon, C.E.: Communication theory of secrecy systems. Bell Syst. Tech. J. **28**, 59–88 (1949)
33. Yang, Y.X., Guo, B.: Further enumerating Boolean functions of cryptographic significance. J. Cryptol. **8**, 115–122 (1995)
34. Youssef, A.M., Tavares, S.E.: Comment on bounds on the number of functions satisfying the strict avalanche criterion. Inf. Process. Lett. **60**, 271–275 (1996)
35. Youssef, A.M., Cusick, T.W., Stănică, P., Tavars, S.E.: New bound on the number of functions satisfying the strict avalanche criterion. In: Proceedings of the Third Annual Workshop on Selected Areas in Cryptography, Ottawa, pp. 49–56 (1996)

Chapter 2
Independence of Boolean Functions of Their Variables

This chapter studies a few different independences of Boolean functions of their variables, including algebraic independence, statistical independence, and algebraic degeneracy.

2.1 Introduction

Take a Boolean function as a network with n inputs and one output, then there might be some relationship between the inputs and the output. The relationship between the inputs and the output of a Boolean function may yield useful information in breaking an encryption algorithm when such a Boolean function is a core component of the algorithm. This relationship can be strongly dependent, or lightly dependent, or even independent in some sense. For example, differential cryptanalysis [1, 2] uses the relationship between the differentials of inputs and that of the corresponding outputs of round functions of a block cipher. When such a round function is treated as a multi-input multi-output Boolean function, the independence of the variables (inputs) of the function with the outputs (coordinate functions) may largely affect the differential cryptanalysis.

This chapter studies different kinds of independence of Boolean functions of their variables. Some previous studies can be found in [5].

2.2 The Algebraic Independence of Boolean Functions of Their Variables

For a Boolean function $f(x)$ in n variables, it may not depend on all of its input variables, i.e., some of the variables may not contribute to the output of $f(x)$. This is often the case for the nonlinear feedforward generators, where the nonlinear filtering

C.-K. Wu, D. Feng, *Boolean Functions and Their Applications in Cryptography*,
Advances in Computer Science and Technology, DOI 10.1007/978-3-662-48865-2_2

function may not use all of the values from each of the cells of the LFSR. In this case, the function is said to be algebraically independent of those variables. More formally, we give the following definition:

Definition 2.1. Let $f(x) \in \mathcal{F}_n$. If the value of $f(x)$ is not affected by the value of x_i, i.e.,

$$f(x_1, \cdots, x_{i-1}, 0, x_{i+1}, \cdots, x_n) = f(x_1, \cdots, x_{i-1}, 1, x_{i+1}, \cdots, x_n)$$

holds for any $(x_1, \cdots, x_{i-1}, x_{i+1}, \cdots, x_n) \in GF^{n-1}(2)$, then $f(x)$ is said to be *algebraically independent* of x_i or simply *independent* of x_i.

For the general case, we have

Definition 2.2. Let $f(x) \in \mathcal{F}_n$. Denote by $\Delta(i_1, i_2, \ldots, i_k) = \{x \in GF^n(2) : x_j = 0 \text{ if } j \notin \{i_1, i_2, \ldots, i_k\}\}$. If

$$f(x \oplus \alpha) = f(x)$$

holds for all $\alpha \in \Delta(i_1, i_2, \ldots, i_k)$, then $f(x)$ is said to be *independent* of variables $x_{i_1}, x_{i_2}, \ldots, x_{i_k}$. For simplicity, we simply call the function to be independent of $x_{i_1}, x_{i_2}, \ldots, x_{i_k}$.

Definition 2.2 is a generalization of Definition 2.1. As an example, $f(x) = x_1$ is independent of variables $x_2, x_3, \ldots, x_n$, because any assignment of these variables will not affect the value of $f(x)$ which only depends on the value of x_1.

Mitchell [3] called the functions defined above as *degenerate* functions and recommended that in cryptographic applications, degenerate functions should be avoided. We will generalize this concept to a more general case in the next section.

By this definitions above we naturally have

Theorem 2.1. *Boolean function $f(x)$ is independent of a subset of its variables $x_{i_1}, x_{i_2}, \ldots, x_{i_k}$, if and only if $x_{i_1}, x_{i_2}, \ldots, x_{i_k}$ do not appear in the algebraic normal form of $f(x)$.*

Proof: The sufficiency is obvious: if $x_{i_1}, x_{i_2}, \ldots, x_{i_n}$ do not appear in the algebraic normal form of $f(x)$, then any change of their values will not affect the value of $f(x)$. So we only need to prove the necessity.

Assume the contrary; for simplicity we assume that $f(x)$ is independent of x_i, and in the mean time, x_i appears in the algebraic normal form (ANF) of $f(x)$, then the ANF of $f(x)$ can be transformed into the form $f(x) = g(x) \oplus x_i h(x)$, where both $g(x)$ and $h(x)$ are independent of x_i, and $h(x) \neq 0$. Then there must exist α such that $h(\alpha) = 1$. Define $e_i \in GF^n(2)$ be the vector whose i-th coordinate is 1 and 0 elsewhere. Since $h(x)$ is independent of x_i, we have that $h(\alpha \oplus e_i) = 1$; hence, we have

$$\begin{aligned} f(\alpha \oplus e_i) &= g(\alpha \oplus w_i) \oplus (a_i \oplus 1)h(\alpha \oplus e_i) \\ &= g(\alpha) \oplus a_i \oplus 1 \end{aligned}$$

On the other hand, we have $f(\alpha) = g(\alpha) \oplus a_i h(\alpha) = g(\alpha) \oplus a_i$, which yields that $f(\alpha) = f(\alpha) \oplus 1$ which is a contradiction. This contradiction shows the correctness of the necessity, and hence the conclusion of the theorem follows. □

Apart from Theorem 2.1, we have the following judgment about the independence of a Boolean function of its variables in terms of Walsh spectrum.

Theorem 2.2. *Let* $f(x) \in \mathcal{F}_n$. *Then* $f(x)$ *is independent of* $x_{i_1}, x_{i_2}, \ldots, x_{i_k}$, *if and only if*

$$S_f(w) = 0$$

holds for every $w \in GF^n(2)$ *with* $\sum_{j=1}^{k} w_{i_j} \neq 0$.

Proof: For the simplicity of writing and without loss of generality, we prove that $f(x)$ is independent of $x_1, x_2, \ldots, x_k$ if and only if $S_f(w) = 0$ holds for every w with $\sum_{i=1}^{k} w_i \neq 0$. For convenience of writing, denote $w \in GF^n(2)$ as $w = (w_{(1)}, w_{(2)})$, where $w_{(1)} = (w_1, w_2, \ldots, w_k)$ and $w_{(2)} = (w_{k+1}, w_{k+2}, \ldots, w_n)$. Similarly we denote $x_{(1)} = (x_1, x_2, \ldots, x_k)$, $x_{(2)} = (x_{k+1}, x_{k+2}, \ldots, x_n)$, and $a_{(1)} = (a_1, a_2, \ldots, a_k)$.

Necessity: By Definition 2.2, assume that $f(x)$ is independent of $x_1, x_2, \ldots, x_k$ then $f(x) = f(0, x_{(2)})$. So we have

$$\begin{aligned} S_f(w) &= \sum_{x=0}^{2^n-1} f(x)(-1)^{\langle w, x\rangle} \\ &= \sum_{a_{(1)}=0}^{2^k-1} \left(\sum_{x:\, x_{(1)}=a_{(1)}} f(x)(-1)^{\langle (w_{(1)}, w_{(2)}),\, (x_{(1)}, x_{(2)})\rangle} \right) \\ &= \sum_{x_{(2)} \in GF^{n-k}(2)} f(0, x_{(2)})(-1)^{\langle w_{(2)}, x_{(2)}\rangle} \sum_{x_{(1)} \in GF^k(2)} (-1)^{\langle w_{(1)}, x_{(1)}\rangle} \end{aligned}$$

By Eq. 1.10, for any $w_{(1)} \neq 0$, we have that

$$\sum_{x_{(1)} \in GF^k(2)} (-1)^{\langle w_{(1)}, x_{(1)}\rangle} = 0,$$

and hence $S_f(w) = 0$.

Sufficiency: By the inverse Walsh transform, we have

$$\begin{aligned} f(x \oplus \alpha) &= 2^{-n} \sum_{w=0}^{2^n-1} S_f(w)(-1)^{\langle w,\, (x \oplus \alpha)\rangle} \\ &= 2^{-n} \sum_{w=0}^{2^n-1} S_f(w)(-1)^{\langle w, x\rangle + \langle w, \alpha\rangle} \end{aligned}$$

Let $W(i_1, i_2, \ldots, i_k) = \{w \in GF^n(2) : \sum_{j=1}^{k} w_{i_j} \neq 0\}$. Then for any $\alpha \in \Delta(i_1, i_2, \ldots, i_k)$, if $w \notin W(i_1, i_2, \ldots, i_k)$, then we have $\langle w, \alpha\rangle = 0$, hence $S_w(-1)^{\langle w,x\rangle+\langle w,\alpha\rangle} = S_w(-1)^{\langle w,x\rangle}$. If $w \in W(i_1, i_2, \ldots, i_k)$, by the assumption of the theorem, i.e., assume that $S_f(w) = 0$ holds for all $w \in W(i_1, i_2, \ldots, i_k)$, we have the equality $S_w(-1)^{\langle w,x\rangle+\langle w,\alpha\rangle} = S_w(-1)^{\langle w,x\rangle} = 0$. Therefore the above can be written as

$$
\begin{aligned}
f(x \oplus \alpha) &= 2^{-n} \sum_{w=0}^{2^n-1} S_f(w)(-1)^{\langle w,x\rangle+\langle w,\alpha\rangle} \\
&= 2^{-n} \sum_{w=0}^{2^n-1} S_f(w)(-1)^{\langle w,x\rangle} \\
&= f(x)
\end{aligned}
$$

By Definition 2.2, this means that $f(x)$ is independent of $x_{i_1}, x_{i_2}, \ldots, x_{i_k}$. □

There is a close connection between the self-correlation function $R_f(\tau)$ of a Boolean function $f(x)$ (see Definition 1.3) and the independence of the Boolean function of its variables. First we give

Lemma 2.1. *The self-correlation function satisfies that $R_f(\tau) = wt(f)$ if and only if $f(x \oplus \tau) = f(x)$ holds for all $x \in GF^n(2)$, i.e., $f(x \oplus \tau)$ and $f(x)$ are the same Boolean function in variable x.*

Proof: Since the truth tables of $f(x)$ and that of $f(x \oplus \tau)$ are all 2^n-dimensional binary vectors, and it is easy to see that $R_f(\tau) = wt(f)$ if and only if wherever $f(x) = 1, f(x \oplus \tau) = 1$ also holds. This means that the truth table of $f(x \oplus \tau)$ is the same as the truth table of $f(x)$, so we have $f(x \oplus \tau) = f(x)$. □

Theorem 2.3. *A necessary and sufficient condition for $f(x) \in \mathcal{F}_n$ to be independent of its variables $x_{i_1}, x_{i_2}, \ldots, x_{i_k}$ is that*

$$R_f(\tau) = wt(f) \tag{2.1}$$

holds for all $\tau \in \Delta(i_1, i_2, \ldots, i_k)$.

Proof: Now assume that $f(x)$ is independent of $x_{i_1}, x_{i_2}, \ldots, x_{i_k}$. Then by Definition 2.2, for any $\tau \in \Delta(i_1, i_2, \ldots, i_k)$, $f(x \oplus \tau) = f(x)$ must hold, and hence we have $R_f(\tau) = wt(f)$. On the other hand, if for all $\tau \in \Delta(i_1, i_2, \ldots, i_k)$ we have $R_f(\tau) = wt(f)$, then by Lemma 2.1, we have $f(x \oplus \tau) = f(x)$, which means that $f(x)$ is independent of $x_{i_1}, x_{i_2}, \ldots, x_{i_k}$. This proves the theorem. □

Definition 2.3. Let $f(x) \in \mathcal{F}_n$. Then

$$R_f(\tau_1, \tau_2, \ldots, \tau_k) = \sum_{x=0}^{2^n-1} f(x)f(x \oplus \tau_1)f(x \oplus \tau_2)\cdots f(x \oplus \tau_k) \tag{2.2}$$

is called the *k-fold self-correlation function* of $f(x)$, where $\tau_i \in GF^n(2)$, $i = 1, 2, \ldots, k$.

The concept of k-fold self-correlation function is a generalization of the self-correlation function as defined in Definition 1.3. When $k = 1$, the onefold self-correlation function is simply called the self-correlation function. Similar to the case of self-correlation function, we have

Lemma 2.2. *$R_f(\tau_1, \tau_2, \ldots, \tau_k) = wt(f)$ if and only if*

$$f(x) = f(x \oplus \tau_1) = f(x \oplus \tau_2) = \cdots = f(x \oplus \tau_k)$$

holds for all $\tau_i \in GF^n(2)$, $i = 1, 2, \ldots, k$.

Denote by e_i be the vector over $GF^n(2)$ whose i-th coordinate is 1 and 0 elsewhere. By Lemma 2.2 we have

Theorem 2.4. *A necessary and sufficient condition for $f(x) \in \mathcal{F}_n$ to be independent of its variables $x_{i_1}, x_{i_2}, \ldots, x_{i_k}$ is that*

$$R_f(e_1, e_2, \ldots, e_k) = wt(f). \tag{2.3}$$

Proof: The proof of the necessity is similar to the proof of Theorem 2.3. Here we give a slightly different proof of the sufficiency.

Assume Eq. 2.3 holds, and then by Lemma 2.2, $f(x) = f(x \oplus e_1) = f(x \oplus e_2) = \cdots = f(x \oplus e_k)$ holds for all $x \in GF^n(2)$. So for any $\tau \in \Delta(i_1, i_2, \ldots, i_k)$, there must exist $a_i \in \{0, 1\}$ such that $\tau = a_1 e_{i_1} \oplus a_2 e_{i_2} \oplus \cdots \oplus a_k e_{i_k}$. Therefore,

$$\begin{aligned} f(x \oplus \tau) &= f(x \oplus a_1 e_{i_1} \oplus a_2 e_{i_2} \oplus \cdots \oplus a_k e_{i_k}) \\ &= f((x \oplus a_2 e_{i_2} \oplus \cdots \oplus a_k e_{i_k}) \oplus a_1 e_{i_1}) \\ &= f(x \oplus a_2 e_{i_2} \oplus \cdots \oplus a_k e_{i_k}) \\ &= \cdots \\ &= f(x) \end{aligned}$$

This means that $f(x)$ is independent of $x_{i_1}, x_{i_2}, \ldots, x_{i_k}$. □

Note that for any fixed $\tau \in GF^n(2)$, $x \oplus \tau$ is a permutation of $GF^n(2)$, this is because when x goes through all the elements in $GF^n(2)$ in a fixed order, $x \oplus \tau$ will also go through all the elements in $GF^n(2)$ in a different order. For $f(x) \in \mathcal{F}_n$, if $f(x \oplus \tau_1) = f(x \oplus \tau_2)$, then $f(x)$ is said to be indistinguishable with permutations $x \oplus \tau_1$ and $x \oplus \tau_2$. This can be rewritten as $f(x \oplus \tau_1 \oplus \tau_2) = f(x)$, and in this case, $\tau_1 \oplus \tau_2$ is called an *invariant* of $f(x)$. There is a Walsh spectrum description of the invariants of Boolean functions as described below.

Theorem 2.5. *Let $\tau = e_{i_1} \oplus e_{i_2} \oplus \cdots \oplus e_{i_k}$. Then $f(x \oplus \tau) = f(x)$ if and only if*

$$S_f(w) = 0$$

holds for all $w \in GF^n(2)$ with $\langle w, \tau \rangle = 1$.

Proof: *Necessity*: Assume $f(x \oplus \tau) = f(x)$ holds, where $\tau = e_{i_1} \oplus e_{i_2} \oplus \cdots \oplus e_{i_k}$. Then

$$\begin{aligned} S_f(w) &= \sum_{x=0}^{2^n-1} f(x)(-1)^{\langle w,x\rangle} \\ &= \sum_{x=0}^{2^n-1} f(x \oplus \tau)(-1)^{\langle w,x\rangle} \\ &= \sum_{x=0}^{2^n-1} f(x)(-1)^{\langle w,(x\oplus\tau)\rangle} \\ &= (-1)^{\langle w,\tau\rangle} \sum_{x=0}^{2^n-1} f(x)(-1)^{\langle w,x\rangle} \\ &= (-1)^{\langle w,\tau\rangle} S_f(w) \end{aligned}$$

Therefore, for any $w \in GF^n(2)$ with $\langle w, \tau\rangle = 1$, the above yields $S_f(w) = -S_f(w)$ and hence $S_f(w) = 0$ must hold.

Sufficiency: If $S_f(w) = 0$ always hold for any $w \in GF^n(2)$ with $\langle w, \tau\rangle = 1$, then by Plancherel formula (Theorem 1.6)

$$\sum_{w=0}^{2^n-1} S_f^2(w) = 2^n wt(f)$$

we have $wt(f) = 2^{-n} \sum_{w:\ \langle w,\tau\rangle=0} S_f^2(w)$. Then by the Wiener-Khinchin formula (Theorem 1.7), we have

$$\begin{aligned} R_f(\tau) &= 2^{-n} \sum_{x=0}^{2^n-1} S_f^2(w)(-1)^{\langle w,\tau\rangle} \\ &= 2^{-n} \sum_{w:\ \langle w,\tau\rangle=0} S_f^2(w) \\ &= wt(f) \end{aligned}$$

By Lemma 2.1 we have $f(x \oplus \tau) = f(x)$. □

When $f(x)$ is independent of variables $x_{i_1}, x_{i_2}, \ldots, x_{i_k}$, the function can be treated as a function in only $n-k$ variables. Denote by

$$\begin{aligned} g(y) &= g(y_1, y_2, \ldots, y_{n-k}) \\ &= f(x_1, \cdots, x_{i_1-1}, 0, x_{i_1+1}, \cdots, x_{i_k-1}, 0, x_{i_k+1}, \cdots, x_n) \end{aligned}$$

Then the Walsh spectrum of $f(x)$ can be expressed as

$$\begin{aligned} S_f(w) &= \sum_{x=0}^{2^n-1} f(x)(-1)^{\langle w,x\rangle} \\ &= 2^k \sum_{x:\ \sum_j x_{i_j}=0} f(x)(-1)^{\langle w,x\rangle} \\ &= 2^k \sum_{y\in GF^{n-k}(2)} g(y)(-1)^{\langle w^{(1)},y\rangle} \\ &= 2^k S_g(w^{(1)}) \end{aligned}$$

where $w^{(1)} \in GF^{n-k}(2)$ is the vector generated from w after removing all the i_j-th coordinates of $w, j = 1, 2, \ldots, k$. This means that

$$dim(\langle\{w : S_f(w) \neq 0\}\rangle) = dim(\langle\{w^{(1)} : S_g(w^{(1)}) \neq 0\}\rangle)$$

where $\langle S\rangle$ is the smallest vector space containing set S, which is also called the linear span of S, and $dim(V)$ means the dimension of vector space V. Obviously the right-hand side of the above equation is at most $n-k$; hence, we have

Theorem 2.6. *Let $f(x) \in \mathcal{F}_n$. If $f(x)$ is independent of k of its variables, then the linear span of the set of inputs (vectors) with nonzero Walsh values is at most $n-k$.*

2.3 The Degeneracy of Boolean Functions

One of the applications of Boolean functions in stream ciphers is to act as a combining function or the like. For security consideration (and due to Berlekamp-Massey algorithm), the combining functions should be nonlinear and ideally should have high nonlinearity. However, from a different viewpoint, a nonlinear function may be treated as the composition of a collection of linear functions and a nonlinear function. By function composition, we mean that the output of one or more functions become the input of another. By this function deposition (the inverse process of composition), it may be possible to find a simpler nonlinear component of the functions. If the number of linear functions can be smaller than the number of the original inputs, then the original function is said to be degenerate. More formally we have

Definition 2.4. Let $f(x) \in \mathcal{F}_n$. If there exists an $n \times k$ matrix D over $GF(2)$ and $g(y) \in \mathcal{F}_k$ such that

$$f(x) = g(xD) = g(y), \tag{2.4}$$

where $k < n$, then $f(x)$ is said to be *algebraically degenerate* or *degenerate* in brief. If k is the minimum value such that Eq. 2.4 holds, i.e., there does not exist an $n \times (k-1)$ matrix D' and a Boolean function $h(y) \in \mathcal{F}_{k-1}$ such that $f(x) = h(xD')$ holds, then $g(y)$ is called an *algebraically degenerated function* of $f(x)$ or a *degenerated function* of $f(x)$ in brief. The value $n-k$ is called the *degree of degeneracy* of $f(x)$ and is denoted as

$$AD(f) = n - k.$$

If there does not exist such an $n \times k$ matrix D with $k < n$ that $f(x) = g(xD)$ holds, i.e., the minimum value of the above k is $k = n$, then $f(x)$ is said to be *algebraically nondegenerate*.

It should be noted that a Boolean function in n variables cannot be equal to a Boolean function in k $(k < n)$ variables, so the equality of Definition 2.4 only means the algebraic representation. A simple such example is $f(x) = x_1$, in which it can be treated as a Boolean function in any number of variables, depending where it is defined, but it is always equivalent to a Boolean function in one variable, in terms of algebraic representation in the sense of linear transformation on its variables. Another such example is $f(x) = x_1 \oplus x_2 \oplus \cdots \oplus x_n$, in which it is also equivalent to a Boolean function in one variable, and again this equivalence is in the sense of algebraic representation by a linear transformation on their variables. In the forthcoming discussion, the equivalence of Boolean functions in different numbers of variables is always in this sense, unless specified otherwise.

The degeneracy property of Boolean functions was also studied in [7]. It is easy to see that if a Boolean function is independent of some of its variables, then the Boolean function is degenerate.

Theorem 2.7. *Let $f(x) \in \mathcal{F}_n$. If $f(x)$ is independent of some of its variables, then $f(x)$ is degenerate.*

Proof: Without loss of generality, suppose that $f(x)$ is independent of x_1. Let D be a $n \times (n-1)$ matrix generated by deleting the first column of the identity matrix I_n, and let $y = xD$. Then it is trivial to verify that the ANF of $f(x)$ can be replaced by the ANF of a Boolean function with y as its $n-1$ variables. By Definition 2.4, $f(x)$ is degenerate. □

By the definition above it is easy to induce the following results.

Lemma 2.3. *Let $f_1(x) = f(xA)$ be two Boolean functions in n variables, where A is an $n \times n$ nonsingular binary matrix. Then we have $deg(f) = deg(f_1)$.*

Proof: By $f_1(x) = f(xA)$ it is known that each variable of $f(x)$ is a linear combination of the variables $x_1, x_2, \ldots, x_n$. Take the linear combinations into the algebraic normal form representation of $f(x)$; no minterm with degree higher than $deg(f)$ can be produced. This means that $deg(f_1) \le deg(f)$. Note that $f(x) = f_1(xA^{-1})$ holds, so we have $deg(f) \le deg(f_1)$; hence, we must have $deg(f) = deg(f_1)$ and the result of Lemma 2.3 holds. □

By Lemma 2.3 we have

Theorem 2.8. *Let $f(x) \in \mathcal{F}_n$ and $g(y) \in \mathcal{F}_k$ be an algebraically degenerated function of $f(x)$. Then we have $deg(f) = deg(g)$.*

Proof: By Definition 2.4, there must exist an $n \times k$ binary matrix D such that $f(x) = g(xD) = g(y)$ holds. Let $deg(g) = t$; and assume a term of $g(y)$ with the highest degree is $y_{i_1} y_{i_2} \cdot y_{i_t}$, and then by replacing each y_{i_j}, $j = 1, 2, \ldots, t$, with a linear combination of $x_1, x_2, \ldots, x_n$, it will result in a polynomial of degree no more than t. This means that by replacing the variables of each y_i, $i = 1, 2, \ldots, k$ with linear combinations of $x_1, x_2, \ldots, x_n$, the algebraic degree of the resulted function will be no more than t, i.e., we have $deg(f) \leq deg(g)$. Note that by Definition 2.4, the rank of D in Eq. 2.4 must be k, so there must exist a $k \times k$ binary matrix P such that $PD = [I_k, 0_{k\times(n-k)}]^T = D'$, where I_k is the $k \times k$ identity matrix. This means that $f_1(x) = f(xP^{-1}) = g(xD') = g(x_1, x_2, \ldots, x_t)$. It is obvious that the degree of $f_1(x)$ is the same as that of $g(xD')$, because they have the same representation in variables $x_1, x_2, \ldots, x_t$. By Lemma 2.3, $deg(f) = deg(f_1)$, so the conclusion of the theorem holds. □

By Theorem 2.8 we immediately get

Corollary 2.1. *Let $f(x) \in \mathcal{F}_n$. If $deg(f) = n$, then $f(x)$ is algebraically nondegenerate.*

This section will develop a method to find the algebraically degenerated function $g(y)$ of a given Boolean function $f(x)$, if $f(x)$ is algebraically degenerate. First we introduce a concept about coset decomposition.

Definition 2.5. Let V be a vector subspace of $GF^n(2)$. For any $\alpha \in GF^n(2) \setminus V$, $V_1 = \alpha \oplus V = \{\alpha \oplus x : x \in V\}$ is called a *coset* of V. The decomposition

$$GF^n(2) = \bigcup_{\alpha \in GF^n(2) \setminus V} \alpha \oplus V \tag{2.5}$$

is called the *coset decomposition* of $GF^n(2)$ with respect to V, and α is called a *coset leader* of V for the coset V_1. Apparently any element (vector) in V_1 can be a coset leader, but they are equivalent in terms of coset decomposition.

Definition 2.5 can be generalized to a general vector space with a general vector operation which is not necessarily the bit-wise exclusive-or. However, we rarely use the general case in this book.

From the coset decomposition of the whole vector space $GF^n(2)$ as represented in Eq. 2.5, it is easy to verify that

(1) $\alpha \oplus V = \beta \oplus V$ if and only if $\alpha \oplus \beta \in V$;
(2) if $\alpha \oplus \beta \notin V$, then $(\alpha \oplus V) \cap (\beta \oplus V) = \phi$, where ϕ is the empty set;
(3) V is a factor of the total number of vectors in $GF^n(2)$ which is 2^n.

There is a sophisticated theory of coset-related issues, and they are beyond the coverage of this book. By using the some known results from the theory of coset and coset decomposition of vector spaces, we can get the following result:

Theorem 2.9. *Let $f(x) \in \mathcal{F}_n$. Denote $V = \langle\{w : S_f(w) \neq 0\}\rangle$ be the linear span of the nonzero spectrum points of $f(x)$. Assume that dim$(V) = k$, and let $h_1, h_2, \ldots, h_k$ be a basis of V. Denote $H = [h_1^T, h_2^T, \ldots, h_k^T]$ which is an $n \times k$ matrix. Then there exists a Boolean function in k variables $g(y)$ such that*

$$g(y) = g(xH) = f(x) \tag{2.6}$$

Proof: By the expression of the inverse Walsh transform as in Eq. 1.13, we can write $f(x)$ as

$$f(x) = 2^{-n} \sum_{w=0}^{2^n-1} S_f(w)(-1)^{\langle w,x\rangle} = 2^{-n} \sum_{w\in V} S_f(w)(-1)^{\langle w,x\rangle}.$$

In the above, only $w \in V$ needs to be considered; hence, for any $x \in V^{\perp}$, $\langle w, x\rangle = 0$ always hold. Hence for any $\alpha \in GF^n(2)$ and $x \in V^{\perp}$, we have

$$f(x \oplus \alpha) = 2^{-n} \sum_{w\in V} S_f(w)(-1)^{\langle w, (x\oplus\alpha)\rangle} = 2^{-n}(-1)^{\langle w,\alpha\rangle} \sum_{w\in V} S_f(w).$$

This means that $f(x)$ is a constant on every coset of $V^{\perp}$. Let S be a set of all the coset leaders of $V^{\perp}$. Then we can establish a mapping φ from S to $GF^k(2)$ as:

$$\varphi(\alpha) = \alpha H$$

where H is the matrix as described in the theorem. It is easy to see that, for any $\alpha_1, \alpha_2 \in S$, if $\varphi(\alpha_1) = \varphi(\alpha_2)$, then $(\alpha_1 - \alpha_2)H = 0$. This means that $(\alpha_1 - \alpha_2) \in V^{\perp}$; therefore, $\alpha_1 = \alpha_2$; hence, φ is a one-to-one mapping. On the other hand, since $|S| = 2^k = |GF^k(2)$, hence φ must be a bijection. Therefore, the function $g(y)$ can be defined as

$$g(y) = f(\varphi^{-1}(y)), \quad y \in GF^k(2)$$

Then for any $\alpha \in S$, we have

$$g(\alpha H) = f(\varphi^{-1}(\alpha H)) = f(\alpha).$$

Therefore, for any $x \in GF^n(2)$, there must exist $\alpha \in S$ and $\beta \in V^{\perp}$ such that $x = \alpha \oplus \beta$; hence, $xH = (\alpha \oplus \beta)H = \alpha H$, and consequently we have

$$f(x) = f(\alpha) = g(\alpha H) = g(xH).$$

□

It can also be shown [4] that the dimension of the vector space V is the least number k so that $f(x)$ has an algebraically degenerated function in $\mathcal{F}_k$.

Corollary 2.2. *Let $f(x) \in \mathcal{F}_n$, A be an $n \times n$ nonsingular matrix, and let $g(x) = f(xA)$. Then $AD(g) = AD(f)$.*

Theorem 2.10. *Let $f(x) \in \mathcal{F}_n$. Then $dim(\langle\{w : S_f(w) \neq 0\}\rangle) = k$ if and only if $f(x)$ can be algebraically degenerated into a function in k variables.*

Proof: *Sufficiency*: If $f(x)$ can be algebraically degenerated into a function $g(y)$ in k variables, i.e., there exists a binary $n \times k$ matrix D such that

$$g(y_1, y_2, \ldots, y_k) = g(xD) = f(x_1, x_2, \ldots, x_n)$$

holds. Do the coset decomposition of $GF^n(2)$ with respect to $Ker(D)$ as

$$GF^n(2) = \bigcup_{i=1}^{k} \alpha_i \oplus Ker(D),$$

where

$$Ker(D) = \{x \in GF^n(2) : xD = 0\}$$

is the *kernel* of D (here D can be treated as a linear mapping defined by $GF^n(2) \to GF^k(2)$ as: $x \to xD$), and then we have

$$\begin{aligned}
S_f(w) &= \sum_{x=0}^{2^m-1} f(x)(-1)^{\langle w,x\rangle} \\
&= \sum_{i=1}^{r} \sum_{x \in Ker(D)} f(\alpha_i \oplus x)(-1)^{\langle w,(\alpha_i \oplus x)\rangle} \\
&= \sum_{i=1}^{r} \sum_{x \in Ker(D)} g(\alpha_i D)(-1)^{\langle w,x\rangle}(-1)^{\langle w,\alpha_i\rangle} \\
&= \sum_{i=1}^{r} g(\alpha_i D)(-1)^{\langle w,\alpha_i\rangle} \sum_{x \in Ker(D)} (-1)^{\langle w,x\rangle} \\
&= \begin{cases} 2^{n-r} \sum_{i=1}^{r} g(\alpha_i D)(-1)^{\langle w,\alpha_i\rangle} & \text{if } w \in (Ker(D))^{\perp} \\ 0 & else \end{cases}
\end{aligned}$$

where $r = rank(D)$. Obviously $r \leq k$; hence, we have

$$dim(\langle w : S_f(w) \neq 0\}\rangle) \leq dim((Ker(D))^{\perp}) \leq k.$$

This means that if $f(x)$ can be algebraically degenerated into a function in k variables, then the dimension of the linear span of its nonzero Walsh spectrum points is at most k. On the other hand, from Theorem 2.9, it is known that if $dim(\langle\{w : S_f(w) \neq 0\}\rangle) = k$, then $f(x)$ can be algebraically degenerated into a function in k variables. Combining the above, we have the conclusion of the theorem. □

Theorem 2.11. *If for any $x \in supp(f)$, we have $wt(x) \geq m$, i.e.,*

$$\min_{x \in supp(f)} wt(x) \geq m,$$

then

$$dim(\langle\{w : S_f(w) \neq 0\}\rangle) \geq m.$$

Proof: Assume the contrary, then by Theorem 2.10 we know that $f(x)$ can be algebraically degenerated into a function in k ($k \leq m-1$) variables $g(y_1, y_2, \ldots, y_k)$, i.e., $g(y_1, y_2, \ldots, y_k) = g(xD) = f(x)$, where D is an $n \times k$ binary matrix over $GF(2)$ with rank equal k. For any $y \in GF^k(2)$, there must exist $x = yD^-$ such that $xD = y$, where D^- is a generalized inverse matrix of D satisfying that $DD^-D = D$ (for the existence of generalized inverses of binary matrices, see, e.g., [6]). From a variety of generalized matrices of D, we can select such a D^- that it has exactly k columns, and then for any $y \in GF^k(2)$, the Hamming weight of $x = yD^-$ is smaller than or equal to k, and by the assumption, we have $f(x) = f(xD^-) = g(y) = 0$. This yields a contradiction, which means that the conclusion of the theorem must hold. □

2.4 Images of Boolean Functions on a Hyperplane

In Sect. 2.2, independence of a Boolean function with some of its variables has been studied. If we treat the space of variables $x = (x_1, x_2, \ldots, x_n)$ to be an n-dimensional vector space $GF^n(2)$, then every coordinate x_i is a direction, and e_i is the unit vector in that direction. When one direction remains a constant, say $x_i = c$, where $c \in \{0, 1\}$, then the vector space becomes $\{(x_1, x_2, \ldots, x_{i-1}, c, x_{i+1}, \ldots, x_n) \in GF^n(2)\}$, which forms an $(n-1)$-dimensional subspace of $GF^n(2)$. This subspace is called a *hyperplane* of $GF^n(2)$. Denote

$$H_i = \{(x_1, x_2, \ldots, x_{i-1}, c, x_{i+1}, \ldots, x_n) \in GF^n(2)\}$$

be the hyperplane in $GF^n(2)$ defined by $x_i = c$; then to be specific, we call H_i *the hyperplane of $GF^n(2)$ in the direction e_i*. More specifically, when $c = 0$, H_i is called a *linear hyperplane*, and when $c = 1$, H_i is called an *affine hyperplane*. In order to differentiate these two cases, we will denote $H_i^{(0)}$ as the linear hyperplane and denote

$H_i^{(1)}$ as the affine hyperplane. Otherwise, H_i denotes either a linear hyperplane or an affine hyperplane where these two cases do not make difference.

For any Boolean function $f(x) \in \mathcal{F}_n$, the restriction of $f(x)$ on the hyperplane $H_i^{(c)}$ is defined as

$$f_{e_i}^{(c)}(x) = f(x_1, x_2, \ldots, x_{i-1}, c, x_{i+1}, \ldots, x_n),$$

i.e., the restriction of $f(x)$ under the condition of $x_i = c$. It is trivial to verify that both $f_{e_i}^{(0)}(x)$ and $f_{e_i}^{(1)}(x)$ are independent of x_i, and $f_{e_i}^{(0)}(x) = f_{e_i}^{(1)}(x)$ if and only if $f(x)$ is independent of x_i, in which case we also have $f(x) = f_{e_i}^{(0)}(x)$.

More generally, let $\alpha = (a_1, a_2, \ldots, a_n) \in GF^n(2)$, then we can define a hyperplane of $GF^n(2)$ in the direction α as

$$H_\alpha^{(c)} = \{x \in GF^n(2) : \langle x, \alpha \rangle = c\},$$

where $\langle x, \alpha \rangle = a_1x_1 \oplus a_2x_2 \oplus \cdots \oplus a_nx_n$ and $c \in GF(2)$. Now the restriction of $f(x) \in \mathcal{F}_n$ on the hyperplane $H_\alpha^{(c)}$ is defined as

$$f_\alpha^{(c)}(x) = f(x)|_{\langle x, \alpha \rangle = c}.$$

Similarly, it can be proven that the restriction of $f(x)$ on the hyperplane $H_\alpha^{(c)}$ is independent of $\langle \alpha, x \rangle$, and $f_\alpha^{(0)}(x) = f_\alpha^{(1)}(x) = f(x)$ if and only if $f(x)$ is independent of $\langle \alpha, x \rangle$.

The restriction of a Boolean function on a hyperplane is also called the *image* of the function on the hyperplane, which is equivalent to a Boolean function in $n - 1$ variables, i.e., the image of a Boolean function in n variables on a hyperplane is degenerate to a Boolean function in $n - 1$ variables. When $\alpha = e_i$, i.e., when the hyperplane is defined by $x_i = c$, the image of $f(x)$ on the hyperplane

$$H_i^{(c)} = \{(x_1, x_2, \ldots, x_{i-1}, c, x_{i+1}, \ldots, x_n) \in GF^n(2)\}$$

becomes $f(x_1, x_2, \ldots, x_{i-1}, c, x_{i+1}, \ldots, x_n)$ which is degenerate to

$$f_1(x_1, x_2, \ldots, x_{i-1}, x_{i+1}, \ldots, x_n)$$

which is a Boolean function in $\mathcal{F}_{n-1}$. However, in a general case, it is not so intuitive to treat the image of a Boolean function on the hyperplane H_α as a Boolean function in $n - 1$ variables. In fact, for any $\alpha \neq 0$, we can always find an $n \times n$ invertible binary matrix A with α^T as its first column vector, where α^T is the transpose of α which is a column vector. Write $(y_1, y_2, \ldots, y_n) = (x_1, x_2, \ldots, x_n)A$, then we have $y_1 = \langle x, \alpha \rangle$, and we can write $(x_1, x_2, \ldots, x_n) = (y_1, y_2, \ldots, y_n)A^{-1}$. By taking this into the representation of $f(x)$, we get $f(x) = f(yA^{-1}) = g(y)$. The image of $f(x)$ on the hyperplane $H_\alpha^{(c)}$ is equivalent to the image of $g(y)$ on the hyperplane $H_1^{(c)}$;

since the image of $g(y)$ on the hyperplane $H_1^{(c)}$ is degenerate to a Boolean function in $n-1$ variables, so is the image of $f(x)$ on the hyperplane $H_\alpha^{(c)}$.

It is trivial to verify that the image of a Boolean function on a hyperplane H_i yields a function independent of variable x_i. The image of a Boolean function on a general hyperplane may not be independent of any variable; however, it must be an algebraically degenerate function.

Theorem 2.12. *Let $f(x) \in \mathcal{F}_n$. Then $f(x)$ is algebraically degenerate if and only if there exists a hyperplane H_α such that $f(x)$ is equivalent to its image on the hyperplane.*

Proof: If $f(x)$ is algebraically degenerate, by Definition 2.4, there must exist an $n \times k$ matrix D over $GF(2)$ and $g(y) \in \mathcal{F}_k$ such that $f(x) = g(xD) = g(y)$ holds, where $k < n$. Let D_1 be an $n \times (n-k)$ binary matrix such that $[D : D_1]$ makes a nonsingular matrix. Let α be a column vector of D_1 (say the first column, which exists since $k < n$). Then $f(x) = g(xD)$ is independent of $\langle \alpha, x \rangle$, which means that $f(x) = f_\alpha^{(0)}(x) = f_\alpha^{(1)}(x)$.

On the other hand, assume that $f(x)$ is equivalent to its image on a hyperplane $H_\alpha^{(c)}$, i.e., $f(x)$ is independent of $\langle \alpha, x \rangle$. Let D_1 be an $n \times (n-1)$ matrix such that $D = [\alpha : D_1]$ makes a nonsingular matrix, and let $y = xD$, then $x = yD^{-1}$, replacing each x_i in the ANF of $f(x)$ by the linear combination of $y_1, y_2, \ldots, y_n$; we get a Boolean function $g(y) \in \mathcal{F}_n$, such that $f(x) = g(y)$. Note that $y = xD$ implies that $y_1 = \langle \alpha, x \rangle$, by the assumption that $f(x)$ is independent of $\langle \alpha, x \rangle$, and we know that $g(y)$ is independent of y_1 and hence is equivalent to its image on hyperplane H_1. □

2.5 Derivatives of Boolean Functions

In the following discussion, let $f(x) \in \mathcal{F}_n$ be a Boolean function in n variables, let $\alpha \in GF^n(2)$ be an arbitrary vector, and let $e_i \in GF^n(2)$ be such a vector that only its i-th coordinate is 1 and 0 elsewhere.

Definition 2.6. Let

$$\Delta_\alpha(f) = f(x) \oplus f(x \oplus \alpha).$$

Then $\Delta_\alpha(f)$ is called *the derivative function of $f(x)$ with respect to α* or *the derivative of $f(x)$* in brief.

When $\alpha = e_i$ for some $1 \leq i \leq n$, we denote

$$\Delta_i(f) = f(x) \oplus f(x \oplus e_i)$$

and call it a *normal derivative of $f(x)$* with respect to e_i.

It is easy to verify that, when $\alpha = 0$ is an all-zero vector, $\Delta_\alpha(f) = \Delta_0(f) = 0$ is always a zero function for any Boolean function $f(x)$. This is a very trivial case, and in the discussion below, we will only consider the cases when $\alpha \neq 0$, unless specified otherwise.

Theorem 2.13. *Let $\alpha, \beta \in GF^n(2)$ be nonzero vectors. Then we have*

$$\Delta_\beta(\Delta_\alpha(f)) = \Delta_\alpha(\Delta_\beta(f)) = f(x) \oplus f(x \oplus \alpha) \oplus f(x \oplus \beta) \oplus f(x \oplus \alpha \oplus \beta).$$

Proof: By Definition 2.6, the conclusion can trivially be verified. □

Theorem 2.13 means that the derivative operation is commutative.

Definition 2.7. Let $\alpha, \beta \in GF^n(2)$. Then

$$\Delta_{\alpha,\beta}(f) = \Delta_\beta(\Delta_\alpha(f))$$

is called the *second-order derivative function of $f(x)$ with respect to α and β.*

To generalize Definition 2.7, for a set of vectors $A = \{\alpha_1, \alpha_2, \ldots, \alpha_t\} \subseteq GF^n(2)$, denote $f_1(x) = \Delta_{\alpha_1}(f), f_2(x) = \Delta_{\alpha_2}(f_1), f_3(x) = \Delta_{\alpha_3}(f_2), \cdots, f_t(x) = \Delta_{\alpha_t}(f_{t-1})$, and then it is easy to prove that the generated function $f_t(x)$ does not depend on the order of the vectors in A but only depends on A as a set counting repeat vectors. So it is reasonable to denote $f_t(x)$ to be $\Delta_A(f)$.

Definition 2.8. The above described function $\Delta_A(f)$ is called the *high-order derivative function of $f(x)$ with respect to A.*

Theorem 2.14. *Let $\alpha_i, \alpha_j \in A$ and $\alpha_i = \alpha_j$ for some $i \neq j$, and then for any Boolean function $f(x) \in \mathcal{F}_n$, we have $\Delta_A(f) = 0$.*

Proof: We only need to prove that $\Delta_{\alpha_1}(\Delta_{\alpha_2}(f)) = 0$ holds. By Theorem 2.13 and the assumption that $\alpha_i = \alpha_j$, the conclusion is true. □

Theorem 2.14 shows that taking the derivative operation twice on a same direction yields a zero function. However, having duplicate vectors in A is not a necessity for $\Delta_A(f) = 0$.

Theorem 2.15. *If the vectors in A are linearly dependent, then for any $f(x) \in \mathcal{F}_n$, we have $\Delta_A(f) = 0$.*

Proof: Without loss of generality, let $A = \{\alpha_1, \alpha_2, \cdots, \alpha_t\}$ and $\alpha_i = \alpha_1 \oplus \alpha_2 \oplus \cdots \oplus \alpha_{i-1}$, where $1 < i \leq t$. Then $\Delta_{\alpha_1,\alpha_2,\ldots,\alpha_i}(f) = \Delta_{\alpha_i}(\Delta_{\alpha_1,\alpha_2,\ldots,\alpha_{i-1}}(f))$. By Theorem 2.13, we get

$$\Delta_{\alpha_1,\alpha_2,\ldots,\alpha_{i-1}}(f) = f(x) \oplus \bigoplus_{j=1}^{i-1} f(x \oplus \alpha_j) \oplus \bigoplus_{1 \leq j_1 < j_2 \leq i-1} [f(x \oplus \alpha_{j_1} \oplus f(x \oplus \alpha_{j_2})) \oplus \cdots \oplus f(x \oplus \alpha_1 \oplus \alpha_2 \oplus \cdots \oplus \alpha_{i-1})] \quad (2.7)$$

By the assumption that $\alpha_i = \alpha_1 \oplus \alpha_2 \oplus \cdots \oplus \alpha_{i-1}$, we have

$$\Delta_{\alpha_1,\alpha_2,\ldots,\alpha_{i-1}}(f)(x \oplus \alpha_i) = \Delta_{\alpha_1,\alpha_2,\ldots,\alpha_{i-1}}(f)(x \oplus \alpha_1 \oplus \alpha_2 \oplus \cdots \oplus \alpha_{i-1}). \qquad (2.8)$$

It is easy to verify that every term in Eq. 2.8 is the same as a term in Eq. 2.7, and they appear almost in the reverse order. This means that

$$\Delta_{\alpha_1,\alpha_2,\ldots,\alpha_{i-1}}(f)(x \oplus \alpha_i) = \Delta_{\alpha_1,\alpha_2,\ldots,\alpha_{i-1}}(f).$$

So we have

$$\Delta_{\alpha_i}(\Delta_{\alpha_1,\alpha_2,\ldots,\alpha_{i-1}}(f)) = \Delta_{\alpha_1,\alpha_2,\ldots,\alpha_{i-1}}(f)(x \oplus \alpha_i) \oplus \Delta_{\alpha_1,\alpha_2,\ldots,\alpha_{i-1}}(f)(x) = 0$$

and the conclusion of Theorem 2.15 holds. □

Theorem 2.15 shows that even when A has no duplicate vectors, the high-order derivative $\Delta_A(f)$ may lead any Boolean function into a zero derivative. However, when the vectors in A are linearly independent, the case is different.

Theorem 2.16. *If $A \subseteq GF^n(2)$ is a collection of linearly independent vectors, then there exists a Boolean function $f(x) \in \mathcal{F}_n$ such that $\Delta_A(f)$ is not a constant function, i.e., $\Delta_A(f) \not\equiv c$, $c \in GF(2)$, if and only if the linear span of A is not $GF^n(2)$.*

Proof: Necessity: Assume that A has n linearly independent vectors $\alpha_1, \alpha_2, \ldots, \alpha_n$. Let D be a matrix composed by all the vectors in A as its column vectors, and then D is a nonsingular matrix. Write $f(x) = g(yD^{-1})$, or equivalently $g(y) = f(xD)$, and then we have

$$\Delta_{\alpha_i}(f) = f(x) \oplus f(x \oplus \alpha_i) = g(yD^{-1}) \oplus g((y \oplus \alpha_i)D^{-1}).$$

Note that $e_iD = \alpha_i$, so $\alpha_iD^{-1} = e_i$, so the above becomes

$$\Delta_{\alpha_i}(f) = g(yD^{-1}) \oplus g(yD^{-1} \oplus e_i).$$

Note that the function $g(yD^{-1})$ can be written as $g(yD^{-1}) = y_ig_1(\bar{y}_i) \oplus g_2(\bar{y}_i)$, where $\bar{y}_i = (y_1, y_2, \ldots, y_{i-1}, y_{i+1}, \ldots, y_n) \in GF^{n-1}(2)$; it is easy to verify that $\Delta_{\alpha_i}(f) = g_1(\bar{y}_i)$ is a Boolean function depending only on $n-1$ variables and is independent of y_i. Since the above equations hold for $i = 1, 2, \ldots, n$, so the high-order derivative $\Delta_A(f)$ must be a Boolean function independent of any variable, and such a Boolean function can only be a constant.

Sufficiency: By the assumption, if the linear span of A is not $GF^n(2)$, then there must exist a nonzero vector $\alpha \in GF^n(2)$ that is linearly independent of all the vectors in A, i.e., α is not a linear combination of the vectors in A. Let D be a matrix composed by the vectors in A as column vectors, let $f(x) = \langle\alpha, x\rangle$, and then by the transformation $y = x \cdot [\alpha : D]$, we can write $f(x) = g(y) = y_1$, which is independent of $y_i = \langle \alpha_i, x\rangle$ for all $\alpha_i \in A$; hence, we have $\Delta_{\alpha_i}(f) = f(x)$, and eventually we have $\Delta_A(f) = f(x)$ which is not a constant. □

The above theorem shows that the derivative operation will eventually lead a Boolean function into a constant, and if one more derivative applies, it will yield a zero, which is what Theorem 2.15 implies. In a general case, let $A \subseteq GF^n(2)$ be a basis of a k-dimensional vector subspace of $GF^n(2)$, and then the high-order derivative operation may lead some of the Boolean functions into zero, but not all, as has been shown in Theorem 2.16. More specifically we have

Theorem 2.17. *If $A \subseteq GF^n(2)$ is a basis of a k-dimensional vector subspace of $GF^n(2)$, where $k < n$, then for any $f(x) \in \mathcal{F}_n$, we have*

$$deg(\Delta_A(f)) \leq n - k$$

and there must exist a Boolean function $f_1(x) \in \mathcal{F}_n$ such that $deg(\Delta_A(f_1)) = n - k$.

Proof: First we prove the case when $k = 1$. Let α be the element in A. Without loss of generality, we assume the first coordinate of α is not zero. Then $D = [\alpha^T, e_2^T, \ldots, e_n^T]$ is a nonsingular matrix. Let $g(x) = f(xD)$, we know that

$$\begin{aligned}\Delta_A(f) &= \Delta_\alpha(f) = f(x \oplus \alpha) \oplus f(x)\\ &= g(xD^{-1} \oplus \alpha D^{-1}) \oplus g(xD^{-1})\\ &= g(xD^{-1} \oplus e_1) \oplus g(xD^{-1})\end{aligned}$$

Denote $g(xD^{-1}) = g_1(x)$, and then by the proof of Theorem 2.16, we have known that the algebraic degree of $\Delta_1(g_1) = g_1(x \oplus e_1) \oplus g_1(x)$ is less than or equal to $deg(g_1) - 1$, where the equality holds if and only if in the ANF of x_1, the degree of the term containing x_1 is the degree of x_1. Note that $deg(x_1) = deg(y) = deg(f)$, the conclusion holds for this case.

Now assume that the conclusion holds for the cases up to $k-1$. Denote a $(k-1)$-th-order derivative of $f(x)$ as $f'(x)$ (not unique and is subject to how the derivative is done). Then a k-th-order derivative of $f(x)$ is the same as applying a derivative on $f'(x)$. By the same proof as the case when $k = 1$, it can be proven that the conclusion of the theorem also holds for the case k. By the principle of mathematical deduction, the conclusion of the theorem holds. □

From the above discussion, we can see that the derivative operation on Boolean functions will decrease the algebraic degree and also make the derivative function to be algebraically degenerate.

Theorem 2.18. *If $A \subseteq GF^n(2)$ is a basis of a k-dimensional vector subspace of $GF^n(2)$, where $k < n$, then for any $f(x) \in \mathcal{F}_n$, the algebraic degeneracy of the derivative function must satisfy*

$$AD(\Delta_A(f)) \geq k$$

and there must exist a Boolean function $f_1(x) \in \mathcal{F}_n$ such that $AD(\Delta_A(f_1)) = k$.

Proof: First we prove the case when $k = 1$. Let α be the element in A. If $\alpha = e_1$, then $f(x)$ can always be written as $f(x) = x_1 f_1(x) \oplus f_2(x)$, where $f_1(x)$ and $f_2(x)$ are both independent of x_1. So we have

$$\begin{aligned}\Delta_1(f) &= f(x \oplus e_1) \oplus f(x)\\ &= ((x_1 \oplus 1) f_1(x) \oplus f_2(x)) \oplus x_1 f_1(x) \oplus f_2(x)\\ &= f_1(x)\end{aligned}$$

Since $f_1(x)$ is independent of x_1, by Theorem 2.7 it is known that $\Delta_1(f)$ is degenerate.

In a general case, since α is a nonzero vector, there must exist a nonsingular matrix D such that $\alpha \cdot D = e_1$. Write $y = xD$, and denote $g(x) = f(xD)$ and then $f(x \oplus \alpha) = g((x \oplus \alpha)D) = g(xD \oplus \alpha \cdot D) = g(xD \oplus e_1)$. Hence, we have

$$\begin{aligned}\Delta_\alpha(f) &= f(x \oplus \alpha) \oplus f(x)\\ &= g(xD \oplus e_1) \oplus g(xD)\\ &= g(y \oplus e_1) \oplus g(y)\\ &= \Delta_1(g)\end{aligned}$$

which is a degenerate Boolean function as has been shown above.

Now assume that the conclusion holds for the cases up to $k-1$. Denote a $(k-1)$-th-order derivative of $f(x)$ as $f'(x)$ (not unique and is subject to how the derivative is done). Then a k-th-order derivative of $f(x)$ is the same as applying a derivative on $f'(x)$. By the same proof as the case when $k = 1$, it can be proven that the conclusion of the theorem also holds for the case k. By the principle of mathematical deduction, the conclusion of the theorem holds. □

2.6 The Statistical Independence of Boolean Functions of Their Variables

A binary variable $x \in GF^n(2)$ is supposed to take all the 2^n possible values. However, in many instances, only a certain number of its particular values are taken into consideration. It matters what value it takes. So in general, we assume that the variable may take any possible value in $GF^n(2)$ with equal probability. This means that we treat x as a probabilistic variable with uniform probability distribution. By this treatment, any Boolean function with this variable, $f(x)$, is also a probabilistic event which has certain probability to be true (when its value equals 1) or false (when its value equals 0). Similarly we may study conditional probabilities, for example, when the precondition is that the variable takes a particular set of values in $GF^n(2)$, the probability that $f(x)$ has value 0 or value 1 is a conditional probability.

There are such Boolean functions, although they are not independent of any of their variables; however, statistically they seem to be not affected by some of their variables, i.e., the probability of such a function to take a certain value (0 or 1) is not affected by any predefined value of these variables. In this case, the function is said to be *statistically independent* of these variables. For example, the Boolean function in three variables $f(x) = x_1x_2 \oplus x_3$ takes value 1 if and only if $x \in \{(110), (001), (101), (011)\}$. So $Prob(f(x) = 1) = Prob(f(x) = 0) = \frac{1}{2}$, i.e., $f(x)$, is balanced, where $Prob(A)$ represents the probability that event A occurs. When $x_1 = 1$ is fixed, then $f(x)$ takes value 1 if and only if $x = (110)$ or $x = (101)$, and $f(x) = 0$ if and only if $x = (100)$ or $x = (111)$. Obviously, under the condition that $x_1 = 1$, let x_2 and x_3 be free variables, and then $f(x)$ is still balanced, i.e.,

$$Prob(f(x) = 1|x_1 = 1) = Prob(f(x) = 0|x_1 = 1) = \frac{1}{2},$$

where $Prob(A|B)$ represents the probability that event A occurs given the condition that B has occurred. It is easy to verify that $f(x)$ is also balanced under the condition that $x_1 = 0$. This means that regardless whatever a fixed value is assigned to x_1, the probability that $f(x)$ takes a certain value (0 or 1) remains the same, i.e.,

$$Prob(f(x) = a|x_1 = b) = Prob(f(x) = a)$$

holds for any $a, b \in \{0, 1\}$. Therefore, function $f(x)$ is statistically independent of x_1. It is easy to verify that this $f(x)$ is also statistically independent of x_2 but not of x_3. Apparently $f(x)$ is not independent of any of its variables, as all the variables appear in the algebraic normal form of $f(x)$. Now we give a formal definition of statistical independence.

Definition 2.9. Let $f(x) \in \mathcal{F}_n$. Treat each x_i as an independent binary variable which takes values from $GF(2)$ at random. If the probability of $f(x)$ to take a particular value is not affected by the precondition that x_i is assigned certain value, i.e.,

$$Prob(f(x) = b|x_i = a) = Prob(f(x) = b),$$

where $a, b \in \{0, 1\}$, $Prob(Z)$ means the probability that event Z occurs, and $Prob(A|B)$ means the conditional probability for event A to occur given that the event B occurred, then $f(x)$ is called to be *statistically independent* of x_i. More generally, if for some $1 \leq i_1 < i_2 < \cdots < i_k \leq n$,

$$Prob\,(f(x) = b|(x_{i_1}, x_{i_2}, \ldots, x_{i_k}) = (a_1, a_2, \ldots, a_k)) = Prob(f(x) = b) \qquad (2.9)$$

holds for any $b \in GF(2)$ and $(a_1, a_2, \ldots, a_k) \in GF^k(2)$, then $f(x)$ is called to be *statistically independent* of variables $x_{i_1}, x_{i_2}, \ldots, x_{i_k}$.

As in Definition 2.9, if $f(x)$ is statistically independent of $x_{i_1}, x_{i_2}, \ldots, x_{i_k}$, then any assignment of values to $x_{i_1}, x_{i_2}, \ldots, x_{i_k}$ will not affect the probability for $f(x)$ to take a certain value. This means that for a nonconstant Boolean function $f(x) \in \mathcal{F}_n$, we have

$$\begin{aligned} & Prob(f(x) = b|(x_{i_1}, x_{i_2}, \ldots, x_{i_k}) = (a_1, a_2, \ldots, a_k)) \\ & = Prob(f(x) = b) \neq 0 \end{aligned}$$

where $b \in \{0, 1\}$ and $(a_1, a_2, \ldots, a_k) \in GF^k(2)$.

Note that a function may be statistically independent of its variables x_{i_1}, x_{i_2}, ..., x_{i_k} individually but still not be statistically independent of the set of variables $x_{i_1}, x_{i_2}, \ldots, x_{i_k}$. For example, $f(x_1, x_2, x_3) = x_1x_3 \oplus x_2x_3 \oplus x_3$ is statistically independent of x_1 and of x_2 individually but not statistically independent of x_1, x_2 as a group. However, we have a simplified view on the statistical independence.

Theorem 2.19. *$f(x)$ is statistically independent of $x_{i_1}, \ldots, x_{i_k}$ if and only if for every $a, b \in GF(2)$ and for every nonzero vector $(c_1, \ldots, c_k) \in GF^k(2)$, we have*

$$Prob\left(f(x) = b \,\middle|\, \bigoplus_{j=1}^{k} c_j x_{i_j} = a\right) = Prob(f(x) = b).$$

By the probability congruent

$$\begin{aligned} Prob(f(x) = b) \cdot Prob((x_{i_1}, x_{i_2}, \ldots, x_{i_k}) &= (a_1, a_2, \ldots, a_k)|f(x) = b) \\ = \; & Prob((x_{i_1}, x_{i_2}, \ldots, x_{i_k}) = (a_1, a_2, \ldots, a_k)) \\ & \cdot Prob(f(x) = b|(x_{i_1}, x_{i_2}, \ldots, x_{i_k}) = (a_1, a_2, \ldots, a_k)) \end{aligned}$$

we know that $f(x)$ is statistically independent of $x_{i_1}, x_{i_2}, \ldots, x_{i_k}$ if and only if

$$\begin{aligned} & Prob((x_{i_1}, x_{i_2}, \ldots, x_{i_k}) = (a_1, a_2, \ldots, a_k)|f(x) = b) \\ & = Prob((x_{i_1}, x_{i_2}, \ldots, x_{i_k}) = (a_1, a_2, \ldots, a_k)) \\ & = 2^{-k}. \end{aligned}$$

So we have

Theorem 2.20. *Let $f(x) \in \mathcal{F}_n$ be a nonconstant Boolean function. Then a necessary and sufficient condition for $f(x)$ to be statistically independent of $x_{i_1}, x_{i_2}, \ldots, x_{i_k}$ is*

$$Prob((x_{i_1}, x_{i_2}, \ldots, x_{i_k}) = (a_1, a_2, \ldots, a_k)|f(x) = b) = 2^{-k} \tag{2.10}$$

holds for any $b \in \{0, 1\}$ and $(a_1, a_2, \ldots, a_k) \in GF^k(2)$.

Theorem 2.20 restricts $f(x)$ to be a nonconstant. What about a trivial Boolean function, i.e., a constant function? It is not so intuitive, particularly when b is not the same constant as $f(x)$, because in this case the precondition $f(x) = b$ does not hold, and to consider a probability under such a condition does not seem to make sense. However, logically Eq. 2.10 still holds, as traditionally we consider an event under a false condition to be true regardless how meaningless it seems to be. So the restriction of the considered function to be nonconstant is not necessary but for easy understanding.

Theorem 2.20 means that if $f(x)$ is statistically independent of variables $x_{i_1}, x_{i_2}, \ldots, x_{i_k}$, then the i_1-th, i_2-th, $\cdots$, i_k-th coordinates of the vectors in *supp(f)* (as well as those in $\overline{supp}(f) = GF^k(2) - supp(f)$) form $GF^k(2)$ or multiple copies of $GF^k(2)$. However, the inverse statement requires both *supp(f)* and $\overline{supp}(f)$ to satisfy the property. We show that if *supp(f)* satisfies the following:

$$\begin{aligned}
&Prob((x_{i_1}, x_{i_2}, \ldots, x_{i_k}) = (a_1, a_2, \ldots, a_k) | x \in supp(f)) \\
&= Prob((x_{i_1}, x_{i_2}, \ldots, x_{i_k}) = (a_1, a_2, \ldots, a_k)) \\
&= 2^{-k},
\end{aligned}$$

then $\overline{supp}(f)$ also satisfies the property, i.e.,

$$\begin{aligned}
&Prob((x_{i_1}, x_{i_2}, \ldots, x_{i_k}) = (a_1, a_2, \ldots, a_k) | x \in \overline{supp}(f)) \\
&= Prob((x_{i_1}, x_{i_2}, \ldots, x_{i_k}) = (a_1, a_2, \ldots, a_k)) \\
&= 2^{-k}.
\end{aligned}$$

This leads to the condition for judging whether $f(x)$ is statistically independent of $x_{i_1}, x_{i_2}, \ldots, x_{i_k}$ to be simpler. In fact,

$$\begin{aligned}
&Prob((x_{i_1}, x_{i_2}, \ldots, x_{i_k}) = (a_1, a_2, \ldots, a_k) | x \in \overline{supp}(f)) \\
&= Prob((x_{i_1}, x_{i_2}, \ldots, x_{i_k}) = (a_1, a_2, \ldots, a_k) | f(x) = 0) \\
&= Prob(f(x) = 0 | (x_{i_1}, x_{i_2}, \ldots, x_{i_k}) = (a_1, a_2, \ldots, a_k)) \\
&\quad \cdot Prob((x_{i_1}, x_{i_2}, \ldots, x_{i_k}) = (a_1, a_2, \ldots, a_k)) / Prob(f(x) = 0) \\
&= \frac{Prob((x_{i_1}, x_{i_2}, \ldots, x_{i_k}) = (a_1, a_2, \ldots, a_k))}{Prob(f(x) = 0)} \\
&\quad [1 - Prob(f(x) = 1 | (x_{i_1}, x_{i_2}, \ldots, x_{i_k}) = (a_1, a_2, \ldots, a_k))] \\
&= \frac{Prob((x_{i_1}, x_{i_2}, \ldots, x_{i_k}) = (a_1, a_2, \ldots, a_k))}{Prob(f(x) = 0)} \\
&\quad \left[1 - \frac{Prob((x_{i_1}, x_{i_2}, \ldots, x_{i_k}) = (a_1, a_2, \ldots, a_k) | f(x) = 1) \cdot Prob(f(x) = 1)}{Prob((x_{i_1}, x_{i_2}, \ldots, x_{i_k}) = (a_1, a_2, \ldots, a_k))}\right] \\
&= \frac{Prob((x_{i_1}, x_{i_2}, \ldots, x_{i_k}) = (a_1, a_2, \ldots, a_k))}{Prob(f(x) = 0)} [1 - Prob(f(x) = 1)] \\
&= Prob((x_{i_1}, x_{i_2}, \ldots, x_{i_k}) = (a_1, a_2, \ldots, a_k)) \\
&= 2^{-k}.
\end{aligned}$$

Hence, we have

Theorem 2.21. *Let $f(x) \in \mathcal{F}_n$. Then $f(x)$ is statistically independent of $x_{i_1}, x_{i_2}, \ldots, x_{i_k}$ if and only if*

$$Prob((x_{i_1}, x_{i_2}, \ldots, x_{i_k}) = (a_1, a_2, \ldots, a_k)|x \in supp(f)) = 2^{-k}$$

holds for any $(a_1, a_2, \ldots, a_k) \in GF^n(2)$.

An equivalent statement of Theorem 2.21 is to use $\overline{supp}(f)$ instead of $supp(f)$; both of the statements give a simpler condition for judging if $f(x)$ is statistically independent of $x_{i_1}, x_{i_2}, \ldots, x_{i_k}$.

It is noted that $Prob((x_{i_1}, x_{i_2}, \ldots, x_{i_k}) = (a_1, a_2, \ldots, a_k)|x \in supp(f))$ means that the i_1-th, i_2-th, ..., i_k-th coordinates of those $x \in supp(f)$ form $GF^n(2)$ or multiple copies of $GF^n(2)$. Below is another measurement about when this condition is met.

Lemma 2.4. *Let A be a set of m-dimensional binary vectors. Then*

$$\{(y_{i_1}, \ldots, y_{i_j}) : y \in A\}$$

contains an equal number of even weight vectors and odd weight vectors for every $1 \leq j \leq m$ and every possible j coordinates $1 \leq i_1 < i_2 < \cdots < i_j \leq m$, if and only if A is $GF^m(2)$ or contains multiple copies of $GF^m(2)$.

Proof: When $m = 2$, it is easy to check the correctness of the conclusion. Assume the conclusion is true for $m - 1$. Then the validity of the conclusion for m is proved as follows.

Denote by

$$A_0 = \{\alpha = (a_1, \ldots, a_m) \in A : a_m = 0\},$$

$$A_1 = \{\alpha = (a_1, \ldots, a_m) \in A : a_m = 1\}.$$

Assume the contrary that the first $m - 1$ coordinates of vectors in A_0 do not form $GF^{m-1}(2)$ or multiple copies of $GF^{m-1}(2)$. Then by the assumption we know that there must exist $1 \leq i_1 < i_2 < \cdots < i_j \leq m - 1$ such that in the set

$$A_0(i_1, i_2, \ldots, i_j) = \{(a_{i_1}, a_{i_2}, \ldots, a_{i_j}) : \alpha = (a_1, \ldots, a_m) \in A\}$$

the number s_0 of odd weight vectors is different from the number t_0 of even weight vectors. Similarly we can define the set $A_1(i_1, i_2, \ldots, i_j)$ in which we assume that there are s_1 odd weight vectors and t_1 even weight vectors. By the assumption above we must have $s_0 + s_1 = t_0 + t_1$. From the definition of A_0 and A_1, it is known that $A_0(i_1, i_2, \ldots, i_j, m)$ contains s_0 odd weight vectors and t_0 even weight vectors, and $A_1(i_1, i_2, \ldots, i_j, m)$ contains t_1 odd weight vectors and s_1 even weight vectors. Therefore,

$$A(i_1, i_2, \ldots, i_j, m) = A_0(i_1, i_2, \ldots, i_j, m) \bigcup A_1(i_1, i_2, \ldots, i_j, m)$$

contains $s_0 + t_1$ odd weight vectors and $s_1 + t_0$ even weight vectors. By the assumption $s_0 \neq t_0$ and the fact that $s_0 + s_1 = t_0 + t_1$, we have $s_0 + t_1 \neq s_1 + t_0$.

This means that the number of odd weight vectors and that of even weight vectors in $A_0(i_1, i_2, \ldots, i_j, m)$ are different which contradicts with the previous assumption. This means that the first $m-1$ coordinates of A_0 form $GF^{m-1}(2)$ or multiple copies of $GF^{m-1}(2)$. Similarly it can be proven that the first $m-1$ coordinates of A_1 form $GF^{m-1}(2)$ or multiple copies of $GF^{m-1}(2)$. Therefore, $A = A_0 \cup A_1$ is $GF^m(2)$ or multiple copies of $GF^m(2)$, and hence, the conclusion of Lemma 2.4 is true. □

Denote by $\Delta(i_1, i_2, \ldots, i_k) = \{x \in GF^n(2) : x_j = 0 \text{ if } j \notin \{i_1, i_2, \ldots, i_k\}\}$. Then we have

Theorem 2.22. *Let $f(x) \in \mathcal{F}_n$. Then a necessary and sufficient condition for $f(x)$ to be statistically independent of $x_{i_1}, x_{i_2}, \ldots, x_{i_k}$ is that, for any nonzero $w \in \Delta(i_1, i_2, \ldots, i_k)$, we have $S_f(w) \neq 0$.*

Proof: First it is noted that

$$S_f(w) = \sum_{x=0}^{2^n-1} f(x)(-1)^{\langle w,x\rangle} = \sum_{x \in supp(f)} (-1)^{\langle w,x\rangle}.$$

If $f(x)$ is statistically independent of $x_{i_1}, x_{i_2}, \ldots, x_{i_k}$, then by Lemma 2.4 we know that the i_1-th, i_2-th, ..., i_k-th coordinates of all the vectors in $\Delta(i_1, i_2, \ldots, i_k)$ form $GF^k(2)$ or multiple copies of $GF^k(2)$. Hence, for any nonzero $w \in \Delta(i_1, i_2, \ldots, i_k)$, by Eq. 1.10 we have $\sum_{x \in supp(f)}(-1)^{\langle w,x\rangle} = 0$, i.e., $S_f(w) = 0$. On the other hand, if $S_f(w) = 0$ holds for any nonzero $w \in \Delta(i_1, i_2, \ldots, i_k)$, then the i_1-th, i_2-th, ..., i_k-th coordinates of all the vectors in $\Delta(i_1, i_2, \ldots, i_k)$ satisfy the conditions of Lemma 2.4; hence, by Lemma 2.4, they form $GF^k(2)$ or multiple copies of $GF^k(2)$; hence, for any $(a_1, a_2, \ldots, a_k) \in GF^k(2)$, we have

$$Prob((x_{i_1}, x_{i_2}, \ldots, x_{i_k}) = (a_1, a_2, \ldots, a_k) | x \in supp(f)) = 2^{-k}.$$

By Theorem 2.21 we know that $f(x)$ is statistically independent of $x_{i_1}, x_{i_2}, \ldots, x_{i_k}$. □

Comparing the definition and properties of the concept of independence and that of statistical independence, we note that if a Boolean function is independent of some variables, then it must be statistically independent of those variables. In this sense the statistical independence is a weaker relationship about independence.

2.7 The Statistical Independence of Two Individual Boolean Functions

In Sect. 2.6 we considered the independence of Boolean functions of their variables. As a generalization of this relationship, we may consider the independence of two distinct Boolean functions. Let $f(x)$ and $g(x)$ be two Boolean functions of the same number of variables. The conditional probability $Prob(f(x) = a | g(x) = b)$, where $a, b \in \{0, 1\}$, means that among the set of input x's that satisfy $g(x) = b$, the probability that x also satisfies $f(x) = a$. For example, when $g(x)$ has a fixed value, say $g(x) = 1$, then the probability of $f(x)$ to take value 1 is a conditional probability, denoted by $Prob(f(x) = 1|g(x) = 1)$. Note that when the free variable $x \in GF^n(2)$ has uniform probability distribution, then it is easy to see that $Prob(f(x) = 1) = \frac{wt(f)}{2^n}$, and $Prob(f(x) = 1|g(x) = 1) = \frac{wt(fg)}{wt(g)}$.

2.7.1 Properties of the Statistical Independence of Boolean Functions

First we introduce the concept of statistical independence of Boolean functions.

Definition 2.10. Let $f(x)$ and $g(x)$ be two Boolean functions in n variables. Treat x as a random variable over $GF^n(2)$. If for any $a, b \in \{0, 1\}$, we always have

$$Prob(f(x) = a|g(x) = b) = Prob(f(x) = a), \tag{2.11}$$

Then $f(x)$ is called to be *statistically independent* of $g(x)$.

Definition 2.10 considers when the value of $f(x)$ is affected by the value of $g(x)$ as a precondition, i.e., whether the conditional probability equals the absolute probability. If the condition is satisfied, is the inverse true as well? That is, if $f(x)$ is statistically independent of $g(x)$, is $g(x)$ statistically independent of $f(x)$ as well? We have

Theorem 2.23. *Let $f(x)$ and $g(x)$ be two Boolean functions in n variables, and $a, b \in \{0, 1\}$. If*

$$Prob(f(x) = a|g(x) = b) = Prob(f(x) = a)$$

holds, then

$$Prob(g(x) = b|f(x) = a) = Prob(g(x) = b)$$

also holds.

Proof: For the convenience of description, we denote the two events by $A = \{f(x) = a\}$ and $B = \{g(x) = b\}$. Then the condition of Theorem 2.23 indicates that the events satisfy that $Prob(A|B) = Prob(A)$. By the multiplication rule of probability:

$$Prob(AB) = Prob(A|B)Prob(B) = Prob(B|A)Prob(A),$$

we have

$$Prob(B|A) = \frac{Prob(A|B)Prob(B)}{Prob(A)}.$$

Since $Prob(A|B) = Prob(A)$, we have $Prob(B|A) = Prob(B)$, and this means that $Prob(g(x) = b|f(x) = a) = Prob(g(x) = b)$, and hence the conclusion is true. □

Theorem 2.23 means that if $f(x)$ is statistically independent of $g(x)$, then the inverse is also true, i.e., $g(x)$ is also statistically independent of $f(x)$. Therefore, the statistical independence is a mutual relationship between two Boolean functions.

Theorem 2.24. *Let $f_1(x)$, $f_2(x)$, and $g(x)$ be all Boolean functions in n variables. If $f_1(x)$ and $f_2(x)$ are all statistically independent of $g(x)$, then $f_1(x) \oplus f_2(x)$ is also statistically independent of $g(x)$.*

Proof: By Definition 2.10, we only need to prove that the following holds for all $a, b \in \{0, 1\}$:

$$Prob(f_1(x) \oplus f_2(x) = a|g(x) = b) = Prob(f_1(x) \oplus f_2(x) = a).$$

First we consider the case when $a = b = 0$. We have

$$\begin{aligned}
&Prob(f_1(x) \oplus f_2(x) = 0|g(x) = 0)\\
&= Prob(f_1(x) = 0|g(x) = 0) \cdot Prob(f_2(x) = 0|g(x) = 0)\\
&\quad + Prob(f_1(x) = 1|g(x) = 0) \cdot Prob(f_2(x) = 1|g(x) = 0)\\
&= Prob(f_1(x) = 0) \cdot Prob(f_2(x) = 0) + Prob(f_1(x) = 1) \cdot Prob(f_2(x) = 1)\\
&= Prob(f_1(x) \oplus f_2(x) = 0).
\end{aligned}$$

Similarly, when $a = 1,\ b = 0$, we have

$$\begin{aligned}
&Prob(f_1(x) \oplus f_2(x) = 1|g(x) = 0)\\
&= Prob(f_1(x) = 0|g(x) = 0) \cdot Prob(f_2(x) = 1|g(x) = 0)\\
&\quad + Prob(f_1(x) = 1|g(x) = 0) \cdot Prob(f_2(x) = 0|g(x) = 0)\\
&= Prob(f_1(x) = 0) \cdot Prob(f_2(x) = 1) + Prob(f_1(x) = 1) \cdot Prob(f_2(x) = 0)\\
&= Prob(f_1(x) \oplus f_2(x) = 1).
\end{aligned}$$

This proves that when $b = 0$, regardless whether $a = 0$ or $a = 1$, the following always holds:

$$Prob(f_1(x) \oplus f_2(x) = a|g(x) = b) = Prob(f(x) \oplus f_2(x) = a).$$

Similarly it can be proven that when $b = 1$, the following also holds:

$$Prob(f_1(x) \oplus f_2(x) = a|g(x) = 1) = Prob(f_1(x) \oplus f_2(x) = a).$$

By Definition 2.10, $f_1(x) \oplus f_2(x)$ is statistically independent of $g(x)$. □

Theorem 2.25. *Let $f(x)$ and $g(x)$ be statistically independent Boolean functions. If $f(x)$ is a balanced, then $f(x) \oplus g(x)$ is also balanced.*

Proof: By the multiplication rule of probability, we have that

$$Prob(f(x) = 1|g(x) = 1) \cdot Prob(g(x) = 1) = Prob(f(x) = 1, g(x) = 1).$$

Since $f(x)$ is statistically independent of $g(x)$, we have $Prob(f(x) = 1|g(x) = 1) = Prob(f(x) = 1)$, and hence the above equation becomes

$$Prob(f(x) = 1) \cdot Prob(g(x) = 1) = Prob(f(x) = 1, g(x) = 1),$$

i.e.,

$$\frac{wt(f)}{2^n} \cdot \frac{wt(g)}{2^n} = \frac{wt(fg)}{2^n}.$$

Since $f(x)$ is balanced, i.e., $wt(f) = 2^{n-1}$, then the above equation becomes $wt(fg) = \frac{wt(g)}{2}$; therefore,

$$\begin{aligned} wt(f \oplus g) &= wt(f) + wt(g) - 2wt(fg) \\ &= 2^{n-1} + wt(g) - 2 \cdot \tfrac{wt(g)}{2} \\ &= 2^{n-1}. \end{aligned}$$

This proves that $f(x) \oplus g(x)$ is balanced. □

By Definition 2.10 it is trivial to prove that

Theorem 2.26. *Let $f(x)$ and $g(x)$ be two Boolean functions that are statistically independent of each other, then for any $a, b \in \{0, 1\}$, $f(x) \oplus a$ and $g(x) \oplus b$ are statistically independent of each other.*

Denote by $SI(f)$ the set of Boolean functions that are statistically independent of $f(x)$. Theorem 2.26 indicates that $SI(f) = SI(f \oplus 1)$. This means that any process of finding the statistically independent functions need only consider the case when their Hamming weight is no more than 2^{n-1}.

2.7.2 How to Judge When Two Boolean Functions Are Statistically Independent

The above has given some properties of the statistical independence of Boolean functions. Apart from the original definition, there is no progress on how to judge more efficiently when two Boolean functions are statistically independent of each. This section will give some more efficient or more practical means to judge whether two given Boolean functions are statistically independent of each other.

From Definition 2.10 we know that to judge whether two Boolean functions are statistically independent of each other, we need to check the validity of Eq. 2.11 for every possible values of $a, b \in \{0, 1\}$. The following theorem tells that we can in fact reduce the process by checking only one of the cases.

Theorem 2.27. *Let $f(x)$ and $g(x)$ be two Boolean functions in n variables. Then a sufficient and necessary condition for $f(x)$ to be statistically independent of $g(x)$ is that, for any fixed values $a_0, b_0 \in \{0, 1\}$, we have*

$$Prob(f(x) = a_0|g(x) = b_0) = Prob(f(x) = a_0).$$

Proof: By Definition 2.10 it is known that the necessity is obvious. Now the sufficiency is given below. Assume the above equation is true for $a_0 = b_0 = 1$. Since $Prob(f(x) = 1|g(x) = 1) = Prob(g(x) = 1)$, and the fact that both $Prob(f(x) = 0|g(x) = 1) + Prob(f(x) = 1|g(x) = 1) = 1$ and $Prob(f(x) = 0) + Prob(f(x) = 1) = 1$ are true, we have

$$Prob(f(x) = 0|g(x) = 1) = Prob(f(x) = 0).$$

This indicates that if Eq. 2.11 is true for $(a, b) = (1, 1)$, then it is also true for $(a, b) = (0, 1)$. Given the above, by the complete probability formula, we have

$$\begin{aligned} Prob(f(x) = 0|g(x) = 1) \cdot Prob(g(x) = 1) \\ +Prob(f(x) = 0|g(x) = 0) \cdot Prob(g(x) = 0) \\ = Prob(f(x) = 0) \end{aligned}$$

and by the real meaning of probability of Boolean functions (assuming that variable x has uniform probability distribution):

$$\begin{aligned} Prob(g(x) = 1) &= \frac{wt(g)}{2^n}, \\ Prob(g(x) = 0) &= 1 - \frac{wt(g)}{2^n}, \end{aligned}$$

the above becomes

$$\begin{aligned} &Prob(f(x) = 0) \cdot wt(g) \\ &\quad + Prob(f(x) = 0|g(x) = 0) \cdot [2^n - wt(g)] \\ &= 2^n Prob(f(x) = 0) \end{aligned}$$

Therefore, we have $Prob(f(x) = 0|g(x) = 0) = Prob(f(x) = 0)$. This indicates that if Eq. 2.11 is true for $(a, b) = (0, 1)$, then it is true for $(a, b) = (0, 0)$. Similarly, we can prove that if Eq. 2.11 is true for $(a, b) = (0, 0)$, then it is true for $(a, b) = (1, 0)$; and if Eq. 2.11 is true for $(a, b) = (1, 0)$, then it is true for $(a, b) = (1, 1)$. This proves that Theorem 2.27 is true. □

Compared with Definition 2.10, Theorem 2.27 simplifies the condition of judging if two Boolean functions are statistically independent of each other. However, this probabilistic method is not very comfortable in real applications.

Now we consider a special case: what about the statistical independence relationship of a constant with a Boolean functions? Without loss of generality, let's assume that $f(x) \equiv 1$. Then by Theorem 2.27, if we can prove that $Prob(f(x) = a|g(x) = b)$ holds for some $a, b \in \{0, 1\}$, then we can conclude that $f(x)$ and $g(x)$ are statistically independent of each other. Obviously regardless what the value of b is, the equality equals 1 if $a = 1$ and 0 if $a = 0$. This proves that $f(x)$ and $g(x)$ are statistically independent. Similarly when $f(x) \equiv 0$, it can be proven similarly. Therefore, we have

Theorem 2.28. *A constant Boolean function is statistically independent of all the Boolean functions of the same number of variables.*

It is noted that in the proof of Theorem 2.28, we naturally assume that $g(x) = b$ is possible. However, in an extreme case when $g(x) \equiv b \oplus 1$ is a constant, the event $(f(x) = a|g(x) = b)$ is under the condition of an impossible event which does not make sense. In this case, Theorem 2.27 is very helpful which tells that we can simply choose another value for $g(x)$, and by choosing $b' = b \oplus 1$, the conclusion about the statistical independence of constant Boolean functions $f(x)$ and $g(x)$ becomes very clear.

In a general case, a Boolean function is not statistically independent of itself, as the conditional probability $Prob(f(x) = a|f(x) = a)$ is always 1. However, when $Prob(f(x) = a) = 1$ holds, which means that $f(x) \equiv a$ is a constant Boolean function, the above conditional probability is acceptable, and hence constant Boolean functions (0 and 1) are the only ones that are statistically independent of themselves. In the following discussion, without being stated explicitly, the Boolean functions stated are all nonconstant. However, it is easy to verify that many of the results are also true for constant functions as well.

Below is another sufficient and necessary condition for judging if two Boolean functions are independent of each other.

Theorem 2.29. *Let $f(x)$, $g(x)$ be two Boolean functions in n variables. Then $f(x)$ is statistically independent of $g(x)$ if and only if*

$$wt(fg) = \frac{wt(f)wt(g)}{2^n} \tag{2.12}$$

Proof: Necessity:

$$\begin{aligned} &Prob(f(x) = 1|g(x) = 1) \\ &= \frac{Prob(f(x)=1, g(x)=1)}{Prob(g(x)=1)} \\ &= \frac{wt(fg)/2^n}{wt(g)/2^n} = \frac{wt(fg)}{wt(g)}. \end{aligned}$$

Note that the probability of $f(x) = 1$ to hold is $Prob(f(x) = 1) = \frac{wt(f)}{2^n}$. Assume that $f(x)$ is statistically independent of $g(x)$, then the above two probabilities are equal, i.e., $\frac{wt(fg)}{wt(g)} = \frac{wt(f)}{2^n}$, which indicates that $wt(fg) = \frac{wt(f)wt(g)}{2^n}$.

Similarly if we consider the probability of event $f(x) = 0$ given that the event $g(x) = 0$ has occurred, since $Prob(f(x) = 0|g(x) = 0) = Prob(f(x) = 0)$, we have

$$wt(\bar{f})wt(\bar{g}) = 2^n wt(\bar{f}\bar{g}), \tag{2.13}$$

where $\bar{f}(x) = f(x) \oplus 1$. Note that $wt(f \oplus g) = wt(f) + wt(g) - 2wt(fg)$; hence,

$$\begin{aligned}
wt(\bar{f}\bar{g}) &= wt(1 \oplus f \oplus g \oplus fg) \\
&= wt(1 \oplus f) + wt(g \oplus fg) - 2wt((1 \oplus f)(g \oplus fg)) \\
&= wt(1 \oplus f) + wt(g \oplus fg) - 2wt(g \oplus fg) \\
&= [2^n - wt(f)] - [wt(g) + wt(fg) - 2wt(fg)] \\
&= 2^n - wt(f) - wt(g) + wt(fg)
\end{aligned}$$

Since $wt(\bar{f}) = 2^n - wt(f)$ and $wt(\bar{g}) = 2^n - wt(g)$, by Eq. 2.13 we have

$$[2^n - wt(f)][2^n - wt(g)] = 2^n[2^n - wt(f) - wt(g) + wt(fg)].$$

Simplifying the above we have $2^n wt(fg) = wt(f)wt(g)$ which is Eq. 2.12.

Sufficiency: It is easy to prove that condition $wt(fg) = \frac{wt(f)wt(g)}{2^n}$ is also sufficient for $Prob(f(x) = 1|g(x) = 1) = Prob(f(x) = 1)$ to hold, and by Theorem 2.27, $f(x)$ is statistically independent of $g(x)$. □

2.7.3 Construction of Statistically Independent Boolean Functions

The above has discussed the properties and alternative representation of judging when two Boolean functions are statistically independent. Now our concentration will be to find an efficient and effective way to construct statistically independent functions given a proper Boolean function (at the moment constant is out of consideration). First of all, it needs to know the existence of Boolean functions that are statistically independent of the given one and then to find a way to construct them. By Eq. 2.12 it is easy to see that the following is true.

Theorem 2.30. *If the Hamming weight of $f(x)$ is odd, then the only functions that are statistically independent of $f(x)$ are constants.*

Proof: Assume that $g(x)$ is statistically independent of $f(x)$. Since $wt(f)$ is an odd number, we must have $\gcd(wt(f), 2) = 1$, where gcd() means the greatest common divisor of its inputs. By Eq. 2.12 we have that $2^n|wt(f)wt(g)$; therefore, $2^n|wt(g)$ must hold; hence, we must have that $g(x) \equiv 1$ or $g(x) \equiv 0$. □

Theorems 2.28 and 2.30 give clear picture about the statistical independence of constant functions and nonconstant functions with odd Hamming weight. What remains unclear is the statistical independence of nonconstant Boolean functions of even Hamming weight, which will be discussed below.

Suppose that $f(x)$ is such a Boolean function, then how to construct a Boolean function $g(x)$ such that they are statistically independent of each other? By Theorem 2.29, it suffices if $g(x)$ satisfies (2.12). The following algorithm will be able to find such statistically independent Boolean functions of a given one:

Algorithm 2.1 Input: Boolean function $f(x)$ in n variables.

(1) Check if $f(x)$ is a constant. If so, then output "all," meaning that all the Boolean functions are the outputs and then exits.
(2) Check if $f(x)$ has odd Hamming weight. If so, then output 0 and 1 and exits.
(3) Partition $supp(f)$ and $supp(\bar{f})$ evenly into $t = \gcd(2^n, wt(f))$ groups.
(4) From each of the partitions, select k $(1 \leq k < t)$ partitioned groups to form the support of $g(x)$.
(5) Output $g(x)$.

With regard to the effectiveness and efficiency of the algorithm, it is trivial to see that as long as the partition of both $supp(f)$ and $supp(\bar{f})$ is finished, the algorithm is almost finished (in the case only one output is expected). We are more concerned about the correctness of the algorithm, i.e., whether the output of the algorithm really yields Boolean functions that are statistically independent of the input $f(x)$. This can be guaranteed by the following theorem:

Theorem 2.31. *An output function $g(x)$ generated by Algorithm 2.7.3 is indeed statistically independent of $f(x)$.*

Proof: By Theorem 2.29, it suffices to prove that Eq. 2.12 holds. From the steps of Algorithm 2.7.3, it can be seen that, in set $supp(g)$, there are k partitioned groups of size $\frac{wt(f)}{\gcd(2^n,wt(f))}$ (the number of elements in the group) chosen from $supp(f)$, and these are the instances of variable x where both $f(x)$ and $g(x)$ take value 1; hence,

$$wt(fg) = k.\frac{wt(f)}{\gcd(2^n, wt(f))}.$$

It is also noted from the algorithm that there are k groups of size $\frac{2^n - wt(f)}{\gcd(2^n,wt(f))}$ chosen from $supp(\bar{f})$ included in $supp(g)$, and they together form all the elements of $supp(g)$; hence,

$$wt(g) = wt(fg) + k.\frac{2^n - wt(f)}{\gcd(2^n, wt(f))} = \frac{2^n k}{\gcd(2^n, wt(f))}.$$

Therefore, we have

$$
\begin{aligned}
Prob(f(x) = 1|g(x) = 1) &= \frac{wt(fg)}{wt(g)} \\
&= \frac{k.wt(f)}{\gcd(2^n, wt(f))} \Big/ \frac{2^n k}{\gcd(2^n, wt(f))} \\
&= \frac{wt(f)}{2^n} \\
&= Prob(f(x) = 1).
\end{aligned}
$$

This shows that Eq. 2.11 holds, and hence the theorem is true. □

Note that in the proof of Theorem 2.31, the value of $t = \gcd(2^n, wt(f))$ seems to be of no importance, and one suspect that it is not necessary to partition both $supp(f)$ and $supp(\bar{f})$ into t groups. However, it is easy to verify that t is the maximum possible value that both $supp(f)$ and $supp(\bar{f})$ can be partitioned evenly into t groups, because $\gcd(|supp(f)|, |supp(\bar{f})|) = t$. This means that any r even partition of both $supp(f)$ and $supp(\bar{f})$ will yield that r is a factor of t.

Theorem 2.31 states the correctness of Algorithm 2.7.3. Another question is where there are functions that are statistically independent of $f(x)$ but beyond the coverage of Algorithm 2.7.3, i.e., they cannot be constructed by Algorithm 2.7.3? The following theorem gives a confirmative answer:

Theorem 2.32. *The output of Algorithm 2.7.3 will exhaust all the Boolean functions that are statistically independent of the input function $f(x)$.*

Proof: It needs to prove that any Boolean function $g(x)$ that is statistically independent of $f(x)$ can be constructed by the steps of Algorithm 2.7.3. For the convenience of writing, we denote $d = \gcd(2^n, wt(f))$. Let $g(x)$ be an output of Algorithm 2.7.3. By Theorem 2.29 we have

$$
\begin{aligned}
wt(fg) &= \frac{wt(f)wt(g)}{2^n} \\
&= \frac{wt(f)}{d} \cdot \frac{d.wt(g)}{2^n}.
\end{aligned}
$$

Let $k = \frac{d.wt(g)}{2^n}$, and then k must be an integer (otherwise it will yield that $wt(fg)$ is not an integer which is impossible). Then the above equation means that there are $k.\frac{wt(f)}{d}$ elements from $supp(f)$ that are in $supp(g)$, or this can equivalently be understood as when the elements of $supp(f)$ are partitioned evenly into groups of size $\frac{wt(f)}{d}$, there are k such groups in $supp(g)$. More precisely, the size of $supp(f) \cap supp(g)$ is $k.\frac{wt(f)}{d}$. If it can be proven that the rest $k.\frac{wt(\bar{f})}{d}$ elements in $supp(g)$ are also in $supp(\bar{f})$, then $g(x)$ is indeed an output of Algorithm 2.7.3. It is easy to verify that the following holds:

$$k.\frac{wt(f)}{d} + k.\frac{wt(\bar{f})}{d} = k.\frac{2^n}{d} = \frac{d.wt(g)}{2^n}.\frac{2^n}{d} = wt(g).$$

This shows that $k.\frac{wt(f)}{d}$ elements from $supp(f)$ and $k.\frac{wt(\bar{f})}{d}$ elements from $supp(\bar{f})$ form all the elements of $supp(g)$. Hence, the conclusion of the theorem holds. □

In order to demonstrate how Algorithm 2.7.3 works, here we give a small example.

Example 2.1. Let $f(x) = x_1 \oplus x_2x_3 \oplus x_1x_2x_3 \oplus x_4 \oplus x_1x_4 \oplus x_2x_4 \oplus x_1x_2x_4$ be a Boolean function in four variables. Then it is easy to verify that $supp(f) = \{0001, 0011, 0110, 0111, 1000, 1001, 1010, 1011, 1100, 1101, 1110, 1111\}$ and $supp(\bar{f}) = \{0000, 0010, 0100, 0101\}$. Partition both $supp(f)$ and $supp(\bar{f})$ into $\gcd(2^n, wt(f)) = 4$ groups. There will be many different ways of partitioning $supp(f)$, and here we choose one partition, the one that any consecutive three elements in $supp(f)$ form a group. Then we choose k groups from both $supp(f)$ and $supp(\bar{f})$, respectively, as the support of $g(x)$. There are many different ways of choosing k groups as well. For a very simple case, when $k = 1$, we get the following outputs as $g(x)$:

$$g_1(x) = 1 \oplus x_1 \oplus x_2 \oplus x_1x_2 \oplus x_3 \oplus x_1x_3 \oplus x_3x_4 \oplus x_1x_3x_4$$

$$g_2(x) = x_3 \oplus x_1x_3 \oplus x_4 \oplus x_1x_4 \oplus x_2x_4 \oplus x_1x_2x_4 \oplus x_3x_4 \oplus x_1x_3x_4$$

$$g_3(x) = x_2 \oplus x_1x_2 \oplus x_4 \oplus x_1x_4$$

$$g_4(x) = x_2x_3 \oplus x_1x_2x_3 \oplus x_4 \oplus x_1x_4 \oplus x_2x_4 \oplus x_2x_3x_4$$

$$g_5(x) = 1 \oplus x_2 \oplus x_3 \oplus x_2x_3 \oplus x_4 \oplus x_1x_4 \oplus x_2x_4 \oplus x_1x_2x_4 \oplus x_3x_4 \oplus x_1x_3x_4$$

$$g_6(x) = x_1 \oplus x_1x_2 \oplus x_3 \oplus x_2x_3 \oplus x_3x_4 \oplus x_1x_3x_4$$

$$g_7(x) = x_1 \oplus x_2 \oplus x_1x_3 \oplus x_2x_3 \oplus x_2x_4 \oplus x_1x_2x_4$$

$$g_8(x) = x_1 \oplus x_1x_2 \oplus x_1x_3 \oplus x_1x_2x_3 \oplus x_2x_4 \oplus x_1x_2x_4$$

$$g_9(x) = 1 \oplus x_1 \oplus x_2 \oplus x_3 \oplus x_2x_3 \oplus x_1x_2x_3 \oplus x_4 \oplus x_1x_4 \oplus x_2x_4 \oplus x_3x_4 \oplus x_1x_3x_4 \oplus x_2x_3x_4$$

$$g_{10}(x) = x_1x_2 \oplus x_3 \oplus x_2x_3 \oplus x_1x_2x_3 \oplus x_1x_2x_4 \oplus x_3x_4 \oplus x_1x_3x_4 \oplus x_2x_3x_4$$

$$g_{11}(x) = x_2 \oplus x_1x_3 \oplus x_2x_3 \oplus x_1x_2x_3 \oplus x_2x_4 \oplus x_2x_3x_4$$

$$g_{12}(x) = x_1x_2 \oplus x_1x_3 \oplus x_2x_4 \oplus x_2x_3x_4$$

$$g_{13}(x) = 1 \oplus x_1 \oplus x_2 \oplus x_1x_2 \oplus x_3 \oplus x_1x_3 \oplus x_2x_3 \oplus x_4 \oplus x_1x_4 \oplus x_2x_4 \oplus x_3x_4 \oplus x_1x_3x_4 \oplus x_2x_3x_4$$

$$g_{14}(x) = x_3 \oplus x_1x_3 \oplus x_2x_3 \oplus x_1x_2x_4 \oplus x_3x_4 \oplus x_1x_3x_4 \oplus x_2x_3x_4$$

$$g_{15}(x) = x_2 \oplus x_1x_2 \oplus x_2x_3 \oplus x_2x_4 \oplus x_2x_3x_4$$

$$g_{16}(x) = x_1x_2x_3 \oplus x_2x_4 \oplus x_2x_3x_4$$

Example 2.1 shows one possibility of partitioning *supp(f)*. For each of the other different partitions, there will be another 16 functions produced. Similar cases are there when $k = 2, 3$. Note that when $k = 2$, the output functions are balanced.

2.7.4 Enumeration of Statistically Independent Boolean Functions

For a given Boolean function $f(x)$, if it is statistically independent of a Boolean function, there may exist many Boolean functions that are all statistically independent of $f(x)$. How many Boolean functions in n variables can be statistically independent of a given Boolean function? Theorem 2.30 tells that if the Hamming weight of $f(x)$ is odd, then only constants are statistically independent of $f(x)$. What about Boolean functions with even Hamming weight? We have

Theorem 2.33. *Let $f(x)$ be a Boolean function in n variables, and its Hamming weight is $wt(f) = t \cdot 2^k$, where $0 < k < n$, $t < 2^{n-k}$ is an odd number. Then there exist*

$$\sum_{r=1}^{2^k-1} \binom{t \cdot 2^k}{t \cdot r} \binom{2^n - t \cdot 2^k}{r \cdot (2^{n-k} - t)} \tag{2.14}$$

Boolean functions that are statistically independent of $f(x)$.

Proof: For any $r < 2^k$, we check how many Boolean functions are $g(x)$ with Hamming weight $r \cdot 2^{n-k}$ and are statistically independent of $f(x)$. Denote by $a = t \cdot 2^k$, $b = r \cdot 2^{n-k}$, $c = t \cdot r$. Then the number of 1's in the truth table of $f(x)$ is a. For those corresponding values of x where $f(x) = 1$, there are $\binom{a}{c}$ ways of letting the value of $g(x)$ to be 1 for c times. Fix the Hamming weight of $g(x)$ to be b, and then $g(x)$ needs to have another $b-c$ 1's for the other $2^n - a$ possible values of x. In this way the number of possible candidates of $g(x)$'s truth table is $\binom{a}{c}\binom{2^n-a}{b-c}$. Now we prove that function $g(x)$ constructed as above is indeed statistically independent of $f(x)$. From the construction it is easy to see that

$$Prob(f(x) = 1 | g(x) = 1) = \frac{c}{b}$$

$$Prob(f(x) = 1) = \frac{a}{2^n}$$

and note that $\frac{c}{b} = \frac{a}{2^n}$, by Theorem 2.27 it is known that $g(x)$ and $f(x)$ are indeed statistically independent of each other. The above is the number of such functions ($g(x)$) when number r is fixed, and the sum of those numbers for all possible r is the number of Boolean functions that are statistically independent of $f(x)$, i.e., $\sum_{r=1}^{2^k-1} \binom{a}{c}\binom{2^n-a}{b-c}$. □

Treat a Boolean function in n variables as a 2^n-dimensional binary vector, any such vector with $t \cdot 2^k$ 1's in its truth table corresponds to a Boolean function of Hamming weight $t \cdot 2^k$. Using the enumeration given in Theorem 2.33, sum those numbers for all possible t and k will result in the number of Boolean function pairs that are statistically independent of each other. We call such a pair as a statistically independent Boolean function pair. From the constructional proof of Theorem 2.33, it is known that the above enumeration counts statistically independent Boolean function pairs counting their orders. We should not distinguish the case when $f(x)$ is statistically independent of $g(x)$ and when $g(x)$ is statistically independent of $f(x)$, and hence when their order is not counted, we have

Theorem 2.34. *The number of statistically independent Boolean function pairs in n variables is*

$$\sum_{k=1}^{n-1} \sum_{t=1}^{2^{n-k}-1} \sum_{r=1}^{2^k-1} \binom{2^n}{t \cdot 2^k} \binom{t \cdot 2^k}{r \cdot t} \binom{2^n - t \cdot 2^k}{r \cdot (2^{n-k} - t)} \Big/ 2 \tag{2.15}$$

Equations 2.14–2.15 are the cases for general Boolean functions, and the expressions are a bit complicated. When we consider balanced Boolean functions, the expressions can be simplified. Balanced Boolean functions are a class of very important functions which have wide applications in practice. By Theorem 2.29, a necessary and sufficient condition for two Boolean functions to be independent is Eq. 2.12. If they are all balanced, then the condition can be simplified as $wt(fg) = 2^{n-2}$. Similar to the proof of Theorem 2.33 we have

Theorem 2.35. *Given a balanced Boolean function $f(x)$ in n variables, then the number of balanced Boolean functions that are statistically independent of $f(x)$ is $\binom{2^{n-1}}{2^{n-2}}^2$ and the number of balanced statistically independent Boolean function pairs are $\binom{2^n}{2^{n-1}}\binom{2^{n-1}}{2^{n-2}}^2/2$.*

Proof: This is actually the special case of Theorems 2.33 and 2.34 when $k = n-1$, $t = 1$, and $r = 2^{n-1}$. □

With the enumerations above, it is trivial to induce the probability that two Boolean functions chosen at random are statistically independent of each other. We mainly mention the special case when all the Boolean functions considered are balanced. We have

Corollary 2.3. *The probability that two balanced Boolean functions in n variables chosen at random are independent of each other is*

$$\frac{\binom{2^{n-1}}{2^{n-2}}^2}{2\binom{2^n}{2^{n-1}}} \approx \frac{1}{4\sqrt{\pi}} \cdot e^{\frac{-13}{3 \cdot 2^{n+2}}}.$$

Proof: It can be seen directly from Theorem 2.35 that the probability that two balanced Boolean functions in n variables chosen at random are independent of each other is

$$\frac{\binom{2^{n-1}}{2^{n-2}}^2}{2\binom{2^n}{2^{n-1}}}.$$

By Stirling approximation that $n! \approx \sqrt{2\pi}n^{n+\frac{1}{2}}e^{-n+\frac{1}{12n}}$, the above becomes

$$\begin{aligned}\frac{\binom{2^{n-1}}{2^{n-2}}^2}{2\binom{2^n}{2^{n-1}}} &= \frac{(2^{n-1}!)^4}{2\cdot 2^n!\cdot(2^{n-2}!)^4}\\ &\approx \frac{[\sqrt{2\pi}(2^{n-1})^{2^{n-1}+\frac{1}{2}}\cdot e^{-2^{n-1}+\frac{1}{12\cdot 2^{n-1}}}]^4}{2\cdot\sqrt{2\pi}(2^n)^{2^n+\frac{1}{2}}\cdot e^{-2^n+\frac{1}{12\cdot 2^n}}\cdot[\sqrt{2\pi}(2^{n-2})^{2^{n-2}+\frac{1}{2}}\cdot e^{-2^{n-2}+\frac{1}{12\cdot 2^{n-2}}}]^4}\\ &= \frac{1}{4\sqrt{\pi}}\cdot e^{\frac{13}{3\cdot 2^{n+2}}}.\end{aligned}$$

□

It can be seen that with the increase of n, the above probability does not reduce to zero. Instead, it is convergent to $\frac{1}{4\sqrt{\pi}} \approx 0.141$.

2.7.5 *On the Statistical Independence of a Group of Boolean Functions*

The above studies the statistical independence between two Boolean functions. Although Theorems 2.24 and 2.25 involve more than two Boolean functions, the final relationship is still between two Boolean functions. Now we consider the pairwise relationship between a group of Boolean functions. For the simplest case, we first consider when there are three such functions.

Let $f_1(x)$, $f_2(x)$, and $f_3(x)$ be Boolean functions in n variables. If $f_1(x)$ is statistically independent of $f_2(x)$, and $f_2(x)$ is statistically independent of $f_3(x)$, is $f_1(x)$ statistically independent of $f_3(x)$? That is, does the statistical independence relationship has transferability? The answer is unfortunately no. An extreme case is when $f_3(x) = f_1(x) \oplus 1$. Even in a general case, a simple example can convenience this. For example, $f_1(x) = x_1$ is statistically independent of $f_2(x) = x_1 \oplus x_2$, and $f_2(x)$ is statistically independent of $f_3(x) = x_1 \oplus x_1x_3 \oplus x_2x_3$; however, $f_1(x)$ and $f_3(x)$ are not statistically independent of each other. Are there cases where a group of Boolean functions is pairwise statistically independent? If we let $f_4(x) = x_1 \oplus x_1x_2 \oplus x_3 \oplus x_2x_3$, then $f_1(x)$, $f_2(x)$ and $f_4(x)$ form such a group. What are the other constraints for the Boolean functions to be pairwise statistically independent? This section studies the statistical independence of a group of Boolean functions.

Definition 2.11. Let $f_1(x), f_2(x), \ldots, f_m(x) \in \mathcal{F}_n$. Treating x as a random variable over $GF^n(2)$ with uniform probability distribution. If

$$\begin{aligned} & Prob((f_1(x), f_2(x), \ldots, f_m(x)) = (a_1, a_2, \ldots, a_m)) \\ = \ & Prob(f_1(x) = a_1)Prob(f_2(x) = a_2) \cdots Prob(f_m(x) = a_m) \end{aligned} \tag{2.16}$$

holds for all $(a_1, a_2, \ldots, a_m) \in GF^m(2)$, then the function group

$$\{f_1(x), f_2(x), \ldots, f_m(x)\}$$

is called a *statistically independent Boolean function family*.

Note that when $m = 2$, the statistically independent Boolean function family is composed of two Boolean functions that are statistically independent of each other. Given this definition, it is ready to answer the above question about when a set of Boolean functions can be pairwise statistically independent.

Theorem 2.36. *Let $\{f_1(x), f_2(x), \ldots, f_m(x)\}$ be a statistically independent Boolean function family in $\mathcal{F}_n$. Then any of its subset is a statistically independent Boolean function family.*

Proof: Without loss of generality, we prove that $\{f_1(x), f_2(x), \ldots, f_k(x)\}$ forms a statistically independent Boolean function family, where $k < m$. For any $(a_1, a_2, \ldots, a_k) \in GF^k(2)$, since $\{f_1(x), f_2(x), \ldots, f_m(x)\}$ forms a statistically independent Boolean function family, we have

$$\begin{aligned} & Prob((f_1(x), f_2(x), \ldots, f_k(x)) = (a_1, a_2, \ldots, a_k)) \\ = & \sum_{(a_{k+1},\ldots,a_m) \in GF^{m-k}(2)} \begin{array}{l} Prob((f_1(x), \ldots, f_k(x)) = (a_1, \ldots, a_k), \\ \quad (f_{k+1}(x), \ldots, f_m(x)) = (a_{k+1}, \ldots, a_m)) \end{array} \\ = & \sum_{(a_{k+1},\ldots,a_m) \in GF^{m-k}(2)} \prod_{i=1}^{k} Prob(f_i(x) = a_i) \prod_{j=k+1}^{m} Prob(f_j(x) = a_j) \\ = & \prod_{i=1}^{k} Prob(f_i(x) = a_i) \sum_{(a_{k+1},\ldots,a_m) \in GF^{m-k}(2)} \prod_{j=k+1}^{m} Prob(f_j(x) = a_j) \\ = & \prod_{i=1}^{k} Prob(f_i(x) = a_i) \left[\begin{array}{l} \sum_{a_{k+1} \in \{0,1\}} Prob(f_{k+1}(x) = a_{k+1}) \cdots \\ \sum_{a_m \in \{0,1\}} Prob(f_m(x) = a_m) \end{array} \right] \\ = & \prod_{i=1}^{k} Prob(f_i(x) = a_i) \end{aligned}$$

By Definition 2.11, $\{f_1(x), f_2(x), \ldots, f_k(x)\}$ forms a statistically independent Boolean function family, and hence the conclusion of the theorem follows. □

As a direct corollary of Theorem 2.36, we have

Corollary 2.4. *Let $\{f_1(x), f_2(x), \ldots, f_m(x)\}$ be a statistically independent Boolean function family in $\mathcal{F}_n$, and then they are pairwise statistically independent, i.e., any two functions $f_i(x)$ and $f_j(x)$ are statistically independent of each other if $i \neq j$.*

Corollary 2.4 reflects a property of statistically independent Boolean function family. However, the inverse is not true, i.e., when a group of Boolean functions satisfy that they are pairwise statistically independent, these functions may not constitute a statistically independent Boolean function family. One such counterexample is that, given two statistically independent Boolean functions $f_1(x)$ and $f_2(x)$, we can construct $f_3(x) = f_1(x) \oplus f_2(x)$, and it is easy to verify that both $f_1(x)$ and $f_2(x)$ are also statistically independent of $f_3(x)$; however, the group $\{f_1(x), f_2(x), f_3(x)\}$ does not form a statistically independent Boolean function family, because $Prob((f_1(x), f_2(x), f_3(x)) = (0, 0, 1)) = 0$ may not equal $Prob(f_1(x) = a_1) \cdot Prob(f_2(x) = a_2) \cdot Prob(f_3(x) = a_3)$.

In fact, given a statistically independent Boolean function family, we will show that more pairwise statistically independent functions can be constructed. First we give a further study on the statistically independent Boolean function families.

Theorem 2.37. *Let $f_1(x), f_2(x), \ldots, f_m(x) \in \mathcal{F}_n$ be a statistically independent Boolean function family. Then for any $m \times m$ matrix A over $GF(2)$,*

$$(g_1(x), g_2(x), \ldots, g_m(x)) = (f_1(x), f_2(x), \ldots, f_m(x))A$$

forms a statistically independent Boolean function family, if and only if A is nonsingular.

Proof: *Sufficiency*: Since

$$(g_1(x), g_2(x), \ldots, g_m(x)) = (f_1(x), f_2(x), \ldots, f_m(x))A,$$

for any $(a_1, a_2, \ldots, a_m) \in GF^m(2)$, we have

$$\begin{aligned}
&Prob((g_1(x), g_2(x), \ldots, g_m(x)) = (a_1, a_2, \ldots, a_m)) \\
&= Prob((f_1(x), f_2(x), \ldots, f_m(x))A = (a_1, a_2, \ldots, a_m)) \\
&= Prob((f_1(x), f_2(x), \ldots, f_m(x)) = (a_1, a_2, \ldots, a_m)A^{-1}) \\
&= Prob((f_1(x), f_2(x), \ldots, f_m(x)) = (b_1, b_2, \ldots, b_m)) \\
&= \prod_{i=1}^{m} Prob(f_i(x) = b_i)
\end{aligned}$$

where $(b_1, b_2, \ldots, b_m) = (a_1, a_2, \ldots, a_m)A^{-1}$. Let $A = [\alpha_1^T, \alpha_2^T, \ldots, \alpha_m^T]$, and denote by $\Delta_i = \{(b_1, b_2, \ldots, b_m) : (b_1, b_2, \ldots, b_m)\alpha_i^T = a_i\}$. Then we have

$$\begin{aligned} Prob(g_i(x) = a_i) &= Prob((f_1, f_2, \ldots, f_m)\alpha_i^T = a_i) \\ &= \sum_{(b_1, b_2, \ldots, b_m) \in \Delta_i} Prob((f_1, f_2, \ldots, f_m) = (b_1, b_2, \ldots, b_m)) \end{aligned}$$

and hence

$$\begin{aligned} &\prod_{i=1}^{m} Prob(g_i = a_i) \\ &= \prod_{i=1}^{m} \Big(\sum_{(b_1, b_2, \ldots, b_m) \in \Delta_i} Prob((f_1, f_2, \ldots, f_m) = (b_1, b_2, \ldots, b_m))\Big) \\ &= \prod_{i=1}^{m} \Big(\sum_{(b_1, b_2, \ldots, b_m) \in \Delta_i} \prod_{j=1}^{m} Prob(f_j = b_j)\Big) \end{aligned}$$

The above is equivalent to the sum for $(b_1, b_2, \ldots, b_m) \in \Delta_i$ which is uniquely determined by

$$(b_1, b_2, \ldots, b_m)(\alpha_1^T, \alpha_2^T, \ldots, \alpha_m^T) = (a_1, a_2, \ldots, a_m),$$

i.e.,

$$(b_1, b_2, \ldots, b_m) = (a_1, a_2, \ldots, a_m)A^{-1},$$

hence,

$$\begin{aligned} \prod_{i=1}^{m} Prob(g_i = a_i) &= \prod_{(b_1, b_2, \ldots, b_m) = (a_1, a_2, \ldots, a_m)A^{-1}} Prob(f_i(x) = b_i) \\ &= Prob((f_1(x), f_2(x), \ldots, f_m(x)) = (a_1, a_2, \ldots, a_m)A^{-1}) \\ &= Prob((g_1(x), g_2(x), \ldots, g_m(x)) = (a_1, a_2, \ldots, a_m)). \end{aligned}$$

By Definition 2.11, $\{g_1(x), g_2(x), \ldots, g_m(x)\}$ forms a statistically independent Boolean function family.

Necessity: Assume the contrary, i.e., assume that A is not invertible. Then there must exist a column of A which is a linear combination of the rest of the columns. Without loss of generality, let $\alpha_1 = c_2\alpha_2 \oplus \ldots \oplus c_m\alpha_m$, where $c_2, \ldots, c_m$ are coefficients in $GF(2)$. Then from $(g_1(x), g_2(x), \ldots, g_m(x)) = (f_1(x), f_2(x), \ldots, f_m(x))A$, we have $g_1(x) = c_2g_2(x) \oplus \cdots \oplus c_mg_m(x)$. This means that $\{g_1(x), g_2(x), \ldots, g_m(x)\}$ cannot form a statistically independent Boolean function family, because $g_1(x)$ is not statistically independent of the other functions, as its values are uniquely determined once the other functions have a fixed value. This contradicts with the preamble of the theorem, which means that the conclusion must be true. □

More generally, we have

Theorem 2.38. *Let $\{f_1(x), f_2(x), \ldots, f_m(x)\}$ be a statistically independent Boolean function family in $\mathcal{F}_n$. Let A be a $m \times k$ matrix over $GF(2)$. Then*

$$(g_1(x), g_2(x), \ldots, g_k(x)) = (f_1(x), f_2(x), \ldots, f_m(x))A$$

forms a statistically independent Boolean function family, if and only if $Rank(A) = k$, where $Rank(A)$ means the rank of matrix A.

Proof: The proof of the necessity of this theorem is very similar to that of Theorem 2.37. Now we only give the proof of sufficiency. Since $Rank(A) = k$, we can add another $m - k$ columns to form a nonsingular matrix A': $A' = [A, A1]$. By Theorem 2.37, we know that

$$\begin{aligned}(g_1, g_2, \ldots, g_m) &= (f_1, f_2, \ldots, f_m)A' \\ &= ((f_1, f_2, \ldots, f_m)A, (f_1, f_2, \ldots, f_m)A1)\end{aligned}$$

forms a statistically independent Boolean function family, and by Theorem 2.36, the subset $(g_1, g_2, \ldots, g_k) = (f_1, f_2, \ldots, f_m)A$ forms a statistically independent Boolean function family; hence, the conclusion of the theorem is true. □

By Theorem 2.38 we have

Corollary 2.5. *Let $\{f_1(x), f_2(x), \ldots, f_m(x)\}$ be a statistically independent Boolean function family in $\mathcal{F}_n$. Then any two different linear combinations of the family members are statistically independent.*

By Corollary 2.5, for a given statistically independent Boolean function family, the linear combinations can produce 2^m Boolean functions that they are pairwise statistically independent. This conclusion gives a clearer picture about why the inverse of Corollary 2.4 is not true.

Theorem 2.39. *The number of nonconstant Boolean functions (members) in a statistically independent Boolean function family in n variables is at most n, and in this case, all the member functions must be balanced.*

Proof: First we prove that the number of member functions in a statistically independent Boolean function family in n variables is at most n. Assume the contrary; by Theorem 2.36, we can assume that there are $n + 1$ Boolean functions $f_1(x), f_2(x), \ldots, f_{n+1}(x) \in \mathcal{F}_{n+1}$ that they form a statistically independent Boolean function family. By Definition 2.11, for any $(a_1, a_2, \ldots, a_{n+1}) \in GF^{n+1}(2)$, we have

$$Prob((f_1(x), f_2(x), \ldots, f_{n+1}(x)) = (a_1, a_2, \ldots, a_{n+1})) = \prod_{i=1}^{n+1} Prob(f_i(x) = a_i).$$

However, since $x \in GF^n(2)$ has only 2^n possible values, and the output of $(f_1(x), f_2(x), \ldots, f_{n+1}(x))$ cannot cover all the vectors in $GF^{n+1}(2)$, hence there must exist $(a_1, a_2, \ldots, a_{n+1}) \in GF^{n+1}(2)$ such that

$$Prob((f_1(x), f_2(x), \ldots, f_{n+1}(x)) = (a_1, a_2, \ldots, a_{n+1})) = 0.$$

However, for any nonconstant member function $f_i(x)$ and any $a_i \in GF(2)$, we have $Prob(f_i(x) = a_i) \neq 0$, and hence

$$\prod_{i=1}^{n+1} Prob(f_i(x) = a_i) \neq 0.$$

This leads to a contradiction. This means that the assumption that $n + 1$ functions form a statistically independent Boolean function family is not true.

Now we prove that if there are n member functions in such a statistically independent Boolean function family, then every member function must be balanced. Assume that $f_1(x), f_2(x), \ldots, f_n(x)$ forms such a statistically independent Boolean function family. Then for any $(a_1, a_2, \ldots, a_n) \in GF^n(2)$, we have

$$Prob((f_1(x), f_2(x), \ldots, f_n(x)) = (a_1, a_2, \ldots, a_n)) = \prod_{i=1}^{n} Prob(f_i(x) = a_i).$$

Since each member function is not a constant, hence for any $a_i \in GF(2)$, $Prob(f_i(x) = a_i) \neq 0$ holds. This means that for any $(a_1, a_2, \ldots, a_n) \in GF^n(2)$, we have

$$Prob((f_1(x), f_2(x), \ldots, f_n(x)) = (a_1, a_2, \ldots, a_n)) \neq 0.$$

Note that $x \in GF^n(2)$ has exactly 2^n possible values; therefore, for any $(a_1, a_2, \ldots, a_n) \in GF^n(2)$, there is exactly one $x \in GF^n(2)$ such that

$$(f_1(x), f_2(x), \ldots, f_n(x)) = (a_1, a_2, \ldots, a_n)$$

holds. Hence, when x goes through all the possible values in $GF^n(2)$,

$$(f_1(x), f_2(x), \ldots, f_n(x)) = (a_1, a_2, \ldots, a_n)$$

will also go through all the possible values in $GF^n(2)$. When x changes its values in a certain order, all the possible values of a_i form the truth table of $f_i(x)$. It is known that, when $(a_1, a_2, \ldots, a_n)$ goes through all the possible values in $GF^n(2)$, each of its coordinates a_i has equal number of 0's and 1's, which means that $f_i(x)$ is balanced. □

When $\{f_1(x), f_2(x), \ldots, f_n(x)\}$ forms a statistically independent Boolean function family in n variables, from the proof of Theorem 2.39 it is seen that different values of input x will result in a different output of $(f_1(x), f_2(x), \ldots, f_n(x))$. In this sense, the Boolean function group $(f_1(x), f_2(x), \ldots, f_n(x))$ is a permutation on $GF^n(2)$. Since each of the coordinate functions is a Boolean function, such a permutation is called a *Boolean permutation* and is denoted as

$$P(x) = [f_1(x), f_2(x), \ldots, f_n(x)].$$

More properties of Boolean permutations will be studied in Chap. 7.

References

1. Daemen, J.: Cipher and hash function design, Strategies based on linear and differential cryptanalysis. PhD Thesis, Leuven (1995)
2. Lai, X.: Higher order derivatives and differential cryptanalysis. In: Blahut, R.E., et al. (eds.) Communications and Cryptography: Two Sides of One Tapestry, pp. 227–233. Kluwer Academic Publishers, Springer (1994)
3. Mitchell, C.: Enumerating Boolean functions of cryptographic significance. J. Cryptol. **2**(3), 155–170 (1990)
4. Wu, C.K.: Boolean functions in cryptology. Ph.D. Thesis, Xidian University, Xian (1993) (in Chinese)
5. Wu, C.K.: On the independence of Boolean functions. Int. J. Comput. Math. **82**(4), 415–420 (2005)
6. Wu, C.K., Dawson, E.: Existence of generalized inverse of linear transformations over finite fields. Finite Fields Appl. **4**(4), 307–315 (1998)
7. Xiao, G.Z., Shen, B.Z., Wu, C.K., Wang, C.C.: Spectral techniques in coding theory. Discret. Math. **87**, 181–186 (1991)

Chapter 3
Nonlinearity Measures of Boolean Functions

Nonlinearity is an important cryptographic measure to cryptographic Boolean functions, and much study can be found from public literatures (see, e.g., [1, 2, 12, 14]). More generalized cryptographic measures about the nonlinear properties of Boolean functions also include algebraic degree, linear structure property, and higher-order nonlinearity [6, 33, 34]. These properties are extensively studied in this chapter.

3.1 Introduction

Linear functions have simple structures and have limited applications. Nonlinear functions are better than linear ones in many cases, and the problem about how to measure the nonlinear property of Boolean functions has been studied from different aspects [32]. One of the nonlinear properties of Boolean functions is their algebraic degree, and another such property is nonlinearity. Among the nonlinear Boolean functions, some have linear structures which behave like linear functions in this aspect.

Algebraic degree is the most intuitive measure of nonlinear property of Boolean functions, because the algebraic degrees of linear Boolean functions (including affine ones) are always 1, except constants 0 and 1 whose algebraic degrees are marked as 0. However, a Boolean function with very high algebraic degree might be very close to a linear function; even if a function has the highest algebraic degree, it may defer with a linear function at only one input out of 2^n total inputs.

Nonlinearity of Boolean functions is one of the fundamental cryptographic properties; it measures the distance from a Boolean function to the nearest affine Boolean function. When a Boolean function is used in a cryptographic algorithm, the Boolean function must have high nonlinearity; otherwise, it may have vulnerability against best affine approximation attack (known as linear cryptanalysis). From this aspect, a cryptographic Boolean function should have as high nonlinearity as

C.-K. Wu, D. Feng, *Boolean Functions and Their Applications in Cryptography*,
Advances in Computer Science and Technology, DOI 10.1007/978-3-662-48865-2_3

possible or even the highest nonlinearity in some cases. For Boolean functions with multiple outputs (vectorial Boolean functions), apart from linear cryptanalysis, there is another attack known as differential cryptanalysis [7]. Resistance against this kind of cryptanalysis seems to be also the nonlinearity.

For a linear (or an affine) Boolean function $f(x)$ in n variables, it has the property that, for any $\alpha \in GF^n(2)$, the equality $f(x \oplus \alpha) \oplus f(x) = c$ always hold, where $c = 0$ or $c = 1$ is a constant depending on α and not depend on the inputs of $f(x)$. A nonlinear Boolean function may have the property that for certain $\alpha \in GF^n(2)$, the equality $f(x \oplus \alpha) \oplus f(x) = c$ always hold, where $c = 0$ or $c = 1$ is a constant. This property coming from linear functions is called having linear structures. This is also a property of nonlinear Boolean function to be avoided when good nonlinear properties are pursued.

This chapter studies the algebraic degree, nonlinearity, and linear structures of Boolean functions.

3.2 Algebraic Degree and Nonlinearity of Boolean Functions

Let us take the nonlinear combiner shown in Fig. 1.4 for consideration. With respect to linear complexity, assume each LFSR$_i$ is of order n_i, which means that the periodic sequences generated by this LFSR$_i$ will have linear complexity n_i. We also assume that the orders of the LFSR's are co-prime of each other. Then the linear complexity of the sum (bitwise Xor) of two of the sequences generated by LFSR$_i$ and LFSR$_j$ will be $n_i + n_j$. The multiplication of the two sequences, however, will have linear complexity $n_i n_j$, which is much larger than $n_i + n_j$. In general, the summation of t of the LFSR sequences will have linear complexity the sum of the orders of those LFSRs, while the multiplication of the t LFSR sequences will have linear complexity the multiplication of the orders of those LFSRs. It is seen that the multiplication of t of the LFSR sequences will result in a sequence of much higher linear complexity than summation can achieve. This corresponds to a multiplicative term of degree t in the algebraic normal form of the combining function $f(x)$. Therefore, if we expect the output sequence of the nonlinear combiner to have high linear complexity, the corresponding nonlinear combining function is expected to be of high algebraic degree. This is why algebraic degree becomes one of the cryptographic measurements.

If the linear complexity of the nonlinear combiner generator sequences is the only cryptographic requirement to pursue, then we can let the nonlinear combining function $f(x)$ to be of the highest algebraic degree n, where n is the number of variables of $f(x)$, which is also the number of the LFSRs as in the nonlinear combiner model. However, practically there are other cryptographic requirements to meet, and some of the requirements may conflict. A simple observation will find that the output of the multiplication of the LFSRs will have most of 0's and very small amount of 1's, which is a very bad behavior in terms of balance. So to achieve a good compromise of all the required cryptographic properties, it has to sacrifice some of the requirements. For example, practically the Boolean functions used in cryptosystems do not reach the highest algebraic degree. However, the algebraic degree of the employed Boolean functions cannot be too low either. In general, the algebraic degree should be larger than $\frac{n}{2}$.

However, even if a nonlinear combining function has very high algebraic degree, say, the highest possible degree n, it may not be good against other attacks. For example, function $f(x) = x_1 + x_1x_2 \cdots x_n$ in n variables has the highest possible algebraic degree n; however, when we use $f'(x) = x_1$ to approximate $f(x)$, then the approximation is so close that the probability of having an error is only one over 2^n. Then we need to define another nonlinear measurement, the nonlinearity of a Boolean function, to be the minimum distance of a given function to all linear functions. Since the set of linear functions and affine functions differ only by a constant, the concept of *nonlinearity* is extended to be the minimum distance of the given function to all the affine functions, as defined in Eq. 1.37.

Although both the algebraic degree and nonlinearity are measures about how different a Boolean function is with all linear and affine Boolean functions as well as constants 0 and 1, these two measures, however, are quite different. A Boolean function with very high algebraic degree may have low nonlinearity. One such an extreme example is $f(x) = x_1x_2 \cdots x_n$, which has the highest degree n, but its nonlinearity is 1, i.e., the distance between $f(x)$ and the constant 0 is just 1, and they differ at the point when all the $x_i = 1$. On the other hand, a Boolean function with a high nonlinearity may not necessarily have a high algebraic degree. For example, when n is even, it can be verified that $f(x) = x_1x_2 \oplus x_3x_4 \oplus \cdots x_{n-1}x_n$ has the highest nonlinearity, but its algebraic degree is 2, the lowest degree among all nonlinear functions in $\mathcal{F}_n$.

3.3 Walsh Spectrum Description of Nonlinearity

Compared with the algebraic degree, the nonlinearity of a Boolean function reflects another angle of the algebraic complexity of the function and has been extensively studied (see, e.g., [31, 46]) due to its cryptographic significance. In some sense, a high nonlinearity is more critical to ensure than a high algebraic degree.

Let $f(x) \in \mathcal{F}_n$. For any affine function $a(x) = a_0 \oplus l(x)$, the distance between $f(x)$ and $a(x)$ is

$$\begin{aligned} d(f(x), a(x)) &= wt(f(x) \oplus a(x)) \\ &= \sum_{x=0}^{2^n-1} (f(x) \oplus a(x)) \\ &= \frac{1}{2} \sum_{x=0}^{2^n-1} (1 - (-1)^{f(x)+a(x)}) \\ &= 2^{n-1} - \frac{1}{2}(-1)^{a_0} \sum_{x=0}^{2^n-1} (-1)^{f(x)+l(x)} \end{aligned}$$

Write $l(x) = \langle w, x\rangle = w_1x_1 \oplus w_2x_2 \oplus \cdots \oplus w_nx_n$, where w is the coefficient vector; then the above can be written as

$$\begin{aligned} d(f(x), a(x)) &= 2^{n-1} - \frac{1}{2}(-1)^{a_0} \sum_{x=0}^{2^n-1} (-1)^{f(x)+\langle w,x\rangle} \\ &= 2^{n-1} - \frac{1}{2}(-1)^{a_0} S_{(f)}(w) \end{aligned}$$

where $S_{(f)}(w)$ is the type II Walsh value of $f(x)$ on w. Note that $a_0 \in \{0, 1\}$, and $\langle w, x\rangle$ can represent all the linear Boolean functions when w goes through all the possible vectors in $GF^n(2)$, so we have [20]:

$$nl(f) = 2^{n-1} - \frac{1}{2} \max_{w \in GF^n(2)} |S_{(f)}(w)|, \tag{3.1}$$

where $|.|$ means the absolute value. If w' is such a value that $|S_{(f)}(w')| = \max_w |S_{(f)}(w)|$, then if $S_{(f)}(w') > 0$, we have that $d(f(x), \langle w, x\rangle) = \min_{a(x)\in \mathcal{A}_n} d(f, a)$, and hence $l(x) = \langle w, x\rangle$ is the *best linear approximation (BLA)* of $f(x)$; if $S_{(f)}(w') < 0$, we have $d(f(x), \langle w, x\rangle \oplus 1) = \min_{a(x)\in \mathcal{A}_n} d(f, a)$, and hence $a(x) = \langle w, x\rangle \oplus 1$ is the *best affine approximation (BAA)* of $f(x)$.

From Eq. 3.1 we can see that the nonlinearity of a Boolean function depends only on the maximum absolute value of its Walsh spectrum. When the maximum absolute value of its Walsh spectrum is small, then the nonlinearity of $f(x)$ is large. By Parseval's theorem (Theorem 1.8),

$$\sum_{w=0}^{2^n-1} S^2_{(f)}(w) = 2^{2n},$$

we have that

$$2^{\frac{n}{2}} \le \max_w |S_{(f)}(w)| \le 2^n$$

and hence we have

$$0 \le nl(f) \le 2^{n-1} - 2^{\frac{n}{2}-1}. \tag{3.2}$$

Definition 3.1. Let $f(x) \in \mathcal{F}_n$. If the nonlinearity of $f(x)$ reaches the upper bound of Eq. 3.2, i.e.,

$$nl(f) = 2^{n-1} - 2^{\frac{n}{2}-1},$$

then $f(x)$ is called a *bent Boolean function*, or a *bent function* for short.

It is easy to see that the smallest value of the maximum absolute value of the Walsh spectrum exists when all the values of $S^2_{(f)}(w)$ are equal; this is the property that a bent function must meet. Equivalently, it can be stated that a Boolean function $f(x)$ in n variables is bent if $|S_{(f)}(w)| = 2^{\frac{n}{2}}$ holds for all $w \in GF^n(2)$. It is obvious that bent functions exist only when n is even.

Bent functions are a very special class of Boolean functions; their studies are extensive and there are numerous related publications in the public literatures (see, e.g., [3–5, 9, 10, 15, 16, 18, 21, 23, 25, 36, 38]).

It is interesting to note that quadratic bent Boolean functions exist for every even n. It can be verified that $f(x) = x_1x_2 \oplus x_3x_4 \oplus \cdots \oplus x_{n-1}x_n$ is a quadratic bent Boolean function, and every permutation on its variables will yield a quadratic bent Boolean function.

3.4 Nonlinearity of Some Basic Operations of Boolean Functions

In this section we will mainly show the nonlinearity of some operations of Boolean functions. By the definition of nonlinearity of Boolean functions, the following basic property of nonlinearity can easily be verified.

Theorem 3.1. *Let $f(x) \in \mathcal{F}_n$, $a(x) \in \mathcal{A}_n$. Set $g(x) = f(x) \oplus a(x)$. Then $nl(g) = nl(f)$.*

Theorem 3.1 means that the minimum distance of a Boolean function is not affected by adding an affine Boolean function. This is obvious, because if $b(x)$ is such an affine function closest to $f(x)$, then $b(x) \oplus a(x)$ must be closest to $f(x) \oplus a(x)$.

Another transformation that does not change the nonlinearity is a kind of affine transformation on the variables as stated below.

Theorem 3.2. *Let $f(x) \in \mathcal{F}_n$, D be an $n \times n$ nonsingular matrix over $GF(2)$ and $b \in \{0, 1\}$. Set $g(x) = f(xD \oplus b)$. Then*

$$nl(g) = nl(f). \tag{3.3}$$

Proof.

$$\begin{aligned} S_{(g)}(w) &= \sum_{x=0}^{2^n-1} (-1)^{g(x)+\langle w,x\rangle} \\ &= \sum_{x=0}^{2^n-1} (-1)^{f(xD\oplus b)+\langle w,x\rangle} \\ &\overset{y=xD\oplus b}{=} \sum_{y=0}^{2^n-1} (-1)^{f(y)+\langle w,(y\oplus b)D^{-1}\rangle} \end{aligned}$$

$$= \sum_{y=0}^{2^n-1}(-1)^{f(y)+\langle w(D^{-1})^T,(y\oplus b)\rangle}$$

$$= (-1)^{\langle w(D^{-1})^T,b\rangle}\sum_{y=0}^{2^n-1}(-1)^{f(y)+\langle w(D^{-1})^T,y\rangle}$$

$$= (-1)^{\langle w(D^{-1})^T,b\rangle}S_{(f)}(w(D^{-1})^T).$$

This means that $\max_{w\in GF^n(2)}|S_{(f)}| = \max_{w\in GF^n(2)}|S_{(g)}|$, and by Eq. 3.1, we have $nl(g) = nl(f)$. □

Construction of nonlinear Boolean functions is often based on some known ones. One of such constructions is by trivial extension, i.e., let $f(x) \in \mathcal{F}_n$ be a Boolean function in n variables, and let $g(x') = f(x) \oplus cx_{n+1}$, where $c \in \{0,1\}$, $x' = (x, x_{n+1})$; then $g(x')$ is a Boolean function in $n+1$ variables. The nonlinearity of $f(x)$ and that of $g(x')$ have the following relationship.

Theorem 3.3. *Let $f(x) \in \mathcal{F}_n$ be a Boolean function in n variables; let $g(x') = f(x) \oplus cx_{n+1}$, where $c \in \{0,1\}$, $x' = (x, x_{n+1})$. Then we have*

$$nl(g) = 2nl(f) \tag{3.4}$$

Proof.

$$S_{(g)}(w') = \sum_{x'=0}^{2^{n+1}-1}(-1)^{g(x')+w'.x'}$$

$$= \sum_{x'=0}^{2^{n+1}-1}(-1)^{(f(x)\oplus cx_{n+1})+\langle (w,w_{n+1}),(x,x_{n+1})\rangle}$$

$$= \sum_{x=0}^{2^n-1}(-1)^{f(x)+\langle w,x\rangle} + \sum_{x=0}^{2^n-1}(-1)^{f(x)+c+\langle w,x\rangle+w_{n+1}}$$

$$= S_{(f)}(w) + (-1)^{c+w_{n+1}}S_{(f)}(w)$$

$$= (1+(-1)^{c+w_{n+1}})S_{(f)}(w)$$

Regardless whether $c = 0$ or $c = 1$, we can always find w_{n+1} such that $(-1)^{c+w_{n+1}} = 1$; hence, $\max_{w'}|S_{(g)}(w')| = 2\max_w|S_{(f)}(w)|$. By Eq. 3.1 we have

$$nl(g) = 2^n - \frac{1}{2}\max_{w'}|S_{(g)}(w')$$

$$= 2(2^{n-1} - \frac{1}{2}\max_{w} |S_{(f)}(w)|)$$
$$= 2nl(f)$$

which proofs the conclusion. □

Another common construction of cryptographic Boolean functions based on some known ones is the cascade construction. Let $f(x), g(x) \in \mathcal{F}_n$ be two Boolean functions in n variables. Then $\phi(x') = (1 \oplus x_{n+1})f(x) \oplus x_{n+1}g(x)$ is a Boolean function in $n+1$ variables, where $x' = (x, x_{n+1})$. Looking at the truth table of $\phi(x')$, it is a concatenation of the truth table of $f(x)$ and that of $g(x)$; hence, it is called a cascade construction. Now we check how the nonlinearity of this new function is related to the nonlinearities of $f(x)$ and $g(x)$.

Theorem 3.4. *Let $f(x), g(x) \in \mathcal{F}_n$. Define the convolutional product of f and g as a function in $n+1$ variables $\phi \in \mathcal{F}_{n+1}$:*

$$\phi = (1 \oplus x_{n+1})f(x) \oplus x_{n+1}g(x).$$

Then we have

$$nl(\phi) \geq nl(f) + nl(g). \tag{3.5}$$

Equality holds if and only if there exists a $w_0 \in GF^n(2)$ such that the following equations hold simultaneously:

$$\begin{cases} |S_{(f)}(w_0)| = \max_w |S_{(f)}(w)| \\ |S_{(g)}(w_0)| = \max_w |S_{(g)}(w)|. \end{cases}$$

Proof. Denote $x' = (x; x_{n+1})$, $w' = (w; w_{n+1})$. Then we have

$$\begin{aligned} &S_{(\phi)}(w') \\ &= \sum_{x' \in GF^{n+1}(2)} (-1)^{\phi(x') \oplus \langle w', x' \rangle} \\ &= \sum_{x_{n+1}=0} \sum_{x \in GF^n(2)} (-1)^{f(x) \oplus \langle w, x \rangle} + \sum_{x_{n+1}=1} \sum_{x \in GF^n(2)} (-1)^{g(x) \oplus \langle w, x \rangle \oplus w_{n+1}} \\ &= S_{(f)}(w) + (-1)^{w_{n+1}} S_{(g)}(w). \end{aligned}$$

It follows that

$$\max_{w'} |S_{(\phi)}(w')| \leq \max_{w} |S_{(f)}(w)| + \max_{w} |S_{(g)}(w)|.$$

Equality holds if and only if the equations

$$\begin{cases} |S_{(f)}(w_0)| = \max_w |S_{(f)}(w)| \\ |S_{(g)}(w_0)| = \max_w |S_{(g)}(w)| \end{cases}$$

hold simultaneously. By Eq. 3.1 the conclusion follows. □

The following result was developed independently in [30] and in [35].

Theorem 3.5. *Let $f(x_1, \ldots, x_n) = f_1(x_1, \ldots, x_{n_1}) \oplus f_2(x_{n_1+1}, .., x_n)$ which is denoted in brief by $f(x) = f_1(\underline{x}_1) \oplus f_2(\underline{x}_2)$. Let $n_2 = n - n_1$. Then we have*

$$nl(f) = 2^{n_2} nl(f_1) + 2^{n_1} nl(f_2) - 2nl(f_1)nl(f_2) > 2nl(f_1)nl(f_2). \tag{3.6}$$

Proof.

$$\begin{aligned} S_{(f)}(w) &= \sum_{x \in GF^n(2)} (-1)^{f(x) \oplus \langle w, x \rangle} \\ &= \sum_{\underline{x}_1 \in GF^{n_1}(2)} \sum_{\underline{x}_2 \in GF^{n_2}(2)} (-1)^{f_1(\underline{x}_1) + f_2(\underline{x}_2) + \underline{w}_1 \cdot \underline{x}_1 + \underline{w}_2 \cdot \underline{x}_2} \\ &= \sum_{\underline{x}_1 \in GF^{n_1}(2)} (-1)^{f_1(\underline{x}_1) + \underline{w}_1 \cdot \underline{x}_1} \sum_{\underline{x}_2 \in GF^{n_2}(2)} (-1)^{f_2(\underline{x}_2) + \underline{w}_2 \cdot \underline{x}_2} \\ &= S_{(f_1)}(\underline{w}_1) \cdot S_{(f_2)}(\underline{w}_2) \end{aligned}$$

Hence, for $w = (\underline{w}_1, \underline{w}_2)$, $|S_{(f)}(w)|$ reaches the maximum value if and only if both $|S_{(f_1)}(\underline{w}_1)|$ and $|S_{(f_2)}(\underline{w}_2)|$ reach the maximum value simultaneously. Let $w \in GF^n(2)$ be such a vector that $|S_{(f)}(w)|$ reaches the maximum Value; then, by Eq. 3.1, we have

$$\begin{aligned} nl(f) &= \frac{1}{2}(2^n - |S_{(f)}(w)|) \\ &= \frac{1}{2}(2^n - |S_{(f_1)}(\underline{w}_1)| \cdot |S_{(f_1)}(\underline{w}_1)|) \\ &= \frac{2^{n_2}}{2}(2^{n_1} - |S_{(f_1)}(\underline{w}_1)|) + \frac{|S_{(f_1)}(\underline{w}_1)|}{2}(2^{n_2} - |S_{(f_2)}(\underline{w}_2)|) \\ &= 2^{n_2} nl(f_1) + |S_{(f_1)}(\underline{w}_1)| nl(f_2) \end{aligned}$$

Since $|S_{(f_1)}(\underline{w}_1)|$ also reaches the maximum value, taking $|S_{(f_1)}(\underline{w}_1)| = 2^{n_1} - 2nl(f_1)$ into the above, we have

$$nl(f) = 2^{n_2} nl(f_1) + 2^{n_1} nl(f_2) - 2nl(f_1)nl(f_2).$$

By Eq. 3.2, the inequality $nl(f) > 2nl(f_1)nl(f_2)$ becomes obvious. □

By Theorem 3.5 we have

Corollary 3.1. *Let* $f(x_1, \ldots, x_n) = f_1(x_1, \ldots, x_{n_1}) \oplus f_2(x_{n_1+1}, \ldots, x_n)$. *Then* f *is bent if and only if both* f_1 *and* f_2 *are bent.*

Theorem 3.6. *Let* $f(x_1, \ldots, x_n) = f_1(x_1, \ldots, x_{n_1}) \cdot f_2(x_{n_1+1}, .., x_n)$ *which is denoted in brief by* $f(x) = f_1(\underline{x}_1) \cdot f_2(\underline{x}_2)$. *Then*

$$nl(f) \geq nl(f_1)nl(f_2). \tag{3.7}$$

The following lemma is required in the proof of Theorem 3.6.

Lemma 3.1. *Let* $f(x) \in \mathcal{F}_n$. *Then*

$$nl(f) \leq wt(f) \leq 2^n - nl(f). \tag{3.8}$$

Proof. By definition, the nonlinearity of $f(x)$ is the minimum Hamming distance of $f(x)$ and all the affine Boolean functions in $\mathcal{F}_n$; hence, we have $nl(f) \leq d(f(x), 0) = wt(f)$, which proves the left part of the inequality (3.8). Again by the definition of nonlinearity, it is obvious that the nonlinearity of a Boolean function $f(x)$ is the same as that of $f(x) \oplus 1$. Hence, the right part of the inequality can be derived by

$$nl(f) = nl(f \oplus 1) \leq wt(f \oplus 1) = 2^n - wt(f).$$

□

Proof of Theorem 3.6: Let $n_2 = n - n_1$ and denote by $\bar{f} = f \oplus 1$. Then

$$\begin{aligned}
S_{(f)}(w) &= \sum_{x \in GF^n(2)} (-1)^{f_1 f_2 \oplus \langle w, x \rangle} \\
&= \sum_{\underline{x}_1 \in GF^{n_1}(2)} \sum_{\underline{x}_2 \in GF^{n_2}(2)} (-1)^{f_1 f_2 \oplus \langle \underline{w}_1, \underline{x}_1 \rangle + \langle \underline{w}_2, \underline{x}_2 \rangle} \\
&= \sum_{f_1 = 1} (-1)^{\langle \underline{w}_1, \underline{x}_1 \rangle} \sum_{\underline{x}_2 \in GF^{n_2}(2)} (-1)^{f_2 \oplus \underline{w}_2 \cdot \underline{x}_2} \\
&\quad + \sum_{f_1 = 0} (-1)^{\langle \underline{w}_1, \underline{x}_1 \rangle} \sum_{\underline{x}_2 \in GF^{n_2}(2)} (-1)^{\langle \underline{w}_2, \underline{x}_2 \rangle} \\
&= \sum_{\underline{x}_1 \in GF^{n_1}(2)} f_1(\underline{x}_1)(-1)^{\langle \underline{w}_1, \underline{x}_1 \rangle} \sum_{\underline{x}_2 \in GF^{n_2}(2)} (-1)^{f_2 \oplus \underline{w}_2 \cdot \underline{x}_2} \\
&\quad + \sum_{\underline{x}_1 \in GF^{n_1}(2)} (f_1(\underline{x}_1) \oplus 1)(-1)^{\langle \underline{w}_1, \underline{x}_1 \rangle} \sum_{\underline{x}_2 \in GF^{n_2}(2)} (-1)^{\langle \underline{w}_2, \underline{x}_2 \rangle} \\
&= \begin{cases} S_{f_1}(\underline{w}_1) S_{(f_2)}(\underline{w}_2) + 2^{n_2} S_{\bar{f}_1}(\underline{w}_1) & \text{if } \underline{w}_2 = 0, \\ S_{f_1}(\underline{w}_1) S_{(f_2)}(\underline{w}_2) & \text{if } \underline{w}_2 \neq 0. \end{cases}
\end{aligned} \tag{3.9}$$

Note that $\bar{f_1} = f_1 \oplus 1$, so we have

$$\begin{aligned} S_{\bar{f_1}}(\underline{w}_1) &= \sum_{\underline{x}_1 \in GF^{n_1}(2)} (f_1(\underline{x}_1) \oplus 1)(-1)^{\langle \underline{w}_1, \underline{x}_1 \rangle} \\ &= \sum_{\underline{x}_1 \in GF^{n_1}(2)} (-1)^{\langle \underline{w}_1, \underline{x}_1 \rangle} - \sum_{\underline{x}_1 \in GF^{n_1}(2)} f_1(\underline{x}_1)(-1)^{\langle \underline{w}_1, \underline{x}_1 \rangle} \\ &= \begin{cases} 2^{n_1} - S_{f_1}(\underline{w}_1) & if\ \underline{w}_1 = 0, \\ -S_{f_1}(\underline{w}_1) & if\ \underline{w}_1 \neq 0. \end{cases} \end{aligned}$$

And by the conversion of the two types of Walsh transforms, we have

$$S_{f_1}(\underline{w}_1) = \begin{cases} \frac{1}{2}(2^{n_1} - S_{(f_1)}(\underline{w}_1)) & if\ \underline{w}_1 = 0, \\ -\frac{1}{2}S_{(f_1)}(\underline{w}_1) & if\ \underline{w}_1 \neq 0. \end{cases}$$

By substituting these relations into Eq. 3.9 and simplifying the expression, we get

$$S_{(f)}(w) = \begin{cases} 2^{n-1} + 2^{n_1-1}S_{(f_2)}(\underline{w}_2) + 2^{n_2-1}S_{(f_1)}(\underline{w}_1) & \\ \qquad - \frac{1}{2}S_{(f_1)}(\underline{w}_1)S_{(f_2)}(\underline{w}_2) & if\ \underline{w}_1 = 0, \underline{w}_2 = 0 \\ 2^{n_2-1}S_{(f_1)}(\underline{w}_1) - \frac{1}{2}S_{(f_1)}(\underline{w}_1)S_{(f_2)}(\underline{w}_2) & if\ \underline{w}_1 \neq 0, \underline{w}_2 = 0 \\ 2^{n_1-1}S_{(f_2)}(\underline{w}_2) - \frac{1}{2}S_{(f_1)}(\underline{w}_1)S_{(f_2)}(\underline{w}_2) & if\underline{w}_1 = 0, \underline{w}_2 \neq 0 \\ -\frac{1}{2}S_{(f_1)}(\underline{w}_1)S_{(f_2)}(\underline{w}_2) & if\ \underline{w}_1 \neq 0, \underline{w}_2 \neq 0. \end{cases} \tag{3.10}$$

Let $|S_{(f)}(w)|$ reach its maximum value on $w = (\underline{w}_1; \underline{w}_2)$. The following cases will be considered:

(1) If $\underline{w}_1 = 0$, $\underline{w}_2 = 0$. Then $S_{(f_1)}(\underline{w}_1) = 2^{n_1} - 2wt(f_1)$, $S_{(f_2)}(\underline{w}_2) = 2^{n_2} - 2wt(f_2)$. By substitution into the first case of Eq. 3.10, we have

$$S_{(f)}(w) = 2^n - 2wt(f_1)wt(f_2).$$

- If $wt(f_1)wt(f_2) \leq 2^{n-1}$, we have $nl(f) = \frac{1}{2}(2^n - |S_{(f)}(w)|) = wt(f_1)wt(f_2)$, and by Lemma 3.1 we have $nl(f) \geq nl(f_1)nl(f_2)$.
- If $wt(f_1)wt(f_2) > 2^{n-1}$, by Lemma 3.1, we have

$$\begin{aligned} nl(f) &= \frac{1}{2}(2^n - |S_{(f)}(w)|) \\ &= 2^n - wt(f_1)wt(f_2) \\ &\geq 2^n - (2^{n_1} - nl(f_1))(2^{n_2} - nl(f_2)) \\ &= 2^{n_2}nl(f_1) + 2^{n_1}nl(f_2) - nl(f_1)nl(f_2). \end{aligned}$$

Since $nl(f_1) < 2^{n_1-1}$, $nl(f_2) < 2^{n_2-1}$, we have $nl(f) \geq 3nl(f_1)nl(f_2)$.

(2) If $\underline{w}_1 \neq 0, \underline{w}_2 = 0$, by Lemma 3.1, Eq. 3.10 becomes

$$S_{(f)}(w) = \frac{S_{(f_1)}(\underline{w}_1)}{2}(2^{n_2} - S_{(f_2)}(\underline{w}_2)) = S_{(f_1)}(\underline{w}_1)wt(f_2).$$

By Lemma 3.1, we have $nl(f_1) < 2^{n_1-1}$ and $wt(f_2) \leq 2^{n_2} - nl(f_2)$. Therefore, we have

$$\begin{aligned} nl(f) &= \tfrac{1}{2}(2^n - |S_{(f)}(w)|) \\ &= \tfrac{1}{2}(2^n - wt(f_2)|S_{(f_1)}(\underline{w}_1)|) \\ &= 2^{n-1} - 2^{n_1-1}wt(f_2) + \tfrac{wt(f_2)}{2}(2^{n_1} - |S_{(f_1)}(\underline{w}_1)|) \\ &\geq 2^{n-1} - 2^{n_1-1}wt(f_2) + wt(f_2)nl(f_1) \\ &= 2^{n-1} - (2^{n_1-1} - nl(f_1))wt(f_2) \\ &\geq 2^{n-1} - (2^{n_1-1} - nl(f_1))(2^{n_2} - nl(f_2)) \\ &= 2^{n_1-1}nl(f_2) + (2^{n_2} - nl(f_2))nl(f_1) \\ &\geq 2nl(f_1)nl(f_2). \end{aligned}$$

(3) If $\underline{w}_1 = 0, \underline{w}_2 \neq 0$, similar to the proof of case (2), we have $nl(f) \geq 2nl(f_1)nl(f_2)$.

(4) If $\underline{w}_1 \neq 0, \underline{w}_2 \neq 0$, then $S_{(f)}(w) = -\frac{1}{2}S_{(f_1)}(\underline{w}_1)S_{(f_2)}(\underline{w}_2)$. In this case it is easy to see that $|S_{(f)}(w)|$ reaches its maximum value if and only if both $|S_{(f_1)}(\underline{w}_1)|$ and $|S_{(f_2)}(\underline{w}_2)|$ reach their maximum value. Similar to the proof of Theorem 3.5 in [35], we have

$$\begin{aligned} nl(f) &= \tfrac{1}{2}(2^n - |S_{(f)}(w)|) \\ &= \tfrac{1}{2}(2^n - \tfrac{1}{2}|S_{(f_1)}(\underline{w}_1)||S_{(f_2)}(\underline{w}_2)|) \\ &= 2^{n-2} + 2^{n_2-2}(2^{n_1} - |S_{(f_1)}(\underline{w}_1)|) + \tfrac{|S_{(f_1)}(\underline{w}_1)|}{4}(2^{n_2} - |S_{(f_2)}(\underline{w}_2)|) \\ &\geq 2^{n-2} + 2^{n_2-2}nl(f_1) + |S_{(f_1)}(\underline{w}_1)|nl(f_2)/4 \end{aligned}$$

Since $|S_{(f_1)}(\underline{w}_1)| = 2^{n_1} - 2nl(f_1)$, taken into the equation above, we have

$$nl(f) \geq 2^{n-2} + 2^{n_2-2}nl(f_1) + 2^{n_1-2}nl(f_2) - \frac{1}{2}nl(f_1)nl(f_2).$$

Recall that $nl(f_1) < 2^{n_1-1}, nl(f_2) < 2^{n_2-1}$, so we get $nl(f) \geq \frac{3}{2}nl(f_1)nl(f_2)$.

Sum up the cases above, and the conclusion of Theorem 3.6 follows. □

Note: It is noticed from the proof above that the value of $nl(f)$ is nearly about $2nl(f_1)nl(f_2)$.

3.5 Upper and Lower Bounds of Nonlinearity of Boolean Functions

Let $f(x) \in \mathcal{F}_n$; by Parseval's equation, we can obtain an upper bound of nonlinearity in general case, that is,

$$nl(f) \leq 2^{n-1} - 2^{\frac{n}{2}-1}.$$

Obviously, the equality holds if and only if n is even, and such a Boolean function is called a bent function. However, bent functions are not balanced; hence, they are usually not directly applied in cryptosystems. So, in practical cryptosystems, nonlinearity of Boolean functions are usually lower than $2^{n-1} - 2^{\frac{n}{2}-1}$. Therefore, it is important to investigate the upper bounds and lower bounds of nonlinearity.

Theorem 3.7 ([19]). *Considering Boolean functions in $\mathcal{F}_n$, the following conclusions hold.*

(1) When $n = 3, 5, 7$, *then* $\max\{nl(f) : f(x) \in \mathcal{F}_n\} = 2^{n-1} - 2^{\frac{n-1}{2}}$;
(2) When $n \geq 9$ *and n is odd, then*

$$2^{n-1} - 2^{\frac{n-1}{2}} \leq \max\{nl(f) : f(x) \in \mathcal{F}_n\} < 2^{n-1} - 2^{\frac{n-1}{2}-1}.$$

It is known that quadratic bent Boolean functions exist for every even n. When n is odd, then there exists a quadratic bent Boolean function $f(x_1, x_2, \ldots, x_{n-1})$ in $n-1$ variables, which has the highest nonlinearity $2^{n-2} - 2^{\frac{n-1}{2}-1}$. Let $g(x) = f(x_1, x_2, \ldots, x_{n-1}) \oplus x_n$, and then by Theorem 3.3, we have $nl(g) = 2nl(f) = 2^{n-1} - 2^{\frac{n-1}{2}}$. This means that the nonlinearity bound $2^{n-1} - 2^{\frac{n-1}{2}}$ is reachable by quadratic Boolean functions; hence, it is called the *quadratic bound*. Theorem 3.7 shows that this bound also holds for all Boolean functions in $\mathcal{F}_n$ for odd $n \leq 7$.

When $n \geq 9$ is odd, existence of Boolean functions with nonlinearity exceeding the quadratic bound remain unknown until 1983 [22], and balanced such examples are found in [31]. Now considering the general case, we have the following results.

Theorem 3.8. *Let $f(x) \in \mathcal{F}_n$ and denote $W = \{w \in GF^n(2) : S_{(f)}(w) \neq 0\}$; then, the nonlinearity $nl(f)$ of $f(x)$ satisfies*

$$nl(f) \leq 2^{n-1} - \frac{2^{n-1}}{\sqrt{|W|}} \leq 2^{n-1} - 2^{\frac{n}{2}-1}. \tag{3.11}$$

Proof. Let $m = \max\{|S_{(f)}(w)| : w \in GF^n(2)\}$, and then $m^2 = \max\{S^2_{(f)}(w) : w \in GF^n(2)\}$. Let w_0 be such a spectral point satisfying $S^2_{(f)}(w_0) = m^2$. By Parseval's equation $\sum_{w=0}^{2^n-1} S^2_{(f)}(w) = 2^{2n}$, we have $\sum_{w \in W} S^2_{(f)}(w) = 2^{2n}$. Note that when

$$\sum_{w \in W} S^2_{(f)}(w) \leq |W| S^2_{(f)}(w_0) = m^2 |W|,$$

we have $m \geq \frac{2^n}{\sqrt{|W|}}$. Therefore,

$$\begin{aligned} nl(f) &= 2^{n-1} - \frac{1}{2} \max_{w \in GF^n(2)} |S_{(f)}(w)| \\ &= 2^{n-1} - \frac{1}{2} m \\ &\leq 2^{n-1} - \frac{2^{n-1}}{\sqrt{|W|}}. \end{aligned}$$

Note that $|W| \leq 2^n$, so we have

$$2^{n-1} - \frac{2^{n-1}}{\sqrt{|W|}} \leq 2^{n-1} - 2^{\frac{n}{2}-1};$$

hence, the theorem follows. □

The following upper bound of the nonlinearity of Boolean functions is related with the autocorrelation function of itself.

Theorem 3.9. *Let $f(x) \in \mathcal{F}_n$, then the nonlinearity of $f(x)$ satisfies*

$$nl(f) \leq 2^{n-1} - 2^{-\frac{n}{2}-1} \sqrt{\sum_{\tau=0}^{2^n-1} R_f^2(\tau)} \leq 2^{n-1} - 2^{\frac{n}{2}-1} \tag{3.12}$$

Proof. Write $|S_{(f)}(w_0)| = \max\{|S_{(f)}(w)| : w \in GF^n(2)\}$. By Eq. 1.24, we have $2^n R_f(\tau) = \sum_{w=0}^{2^n-1} S_{(f)}^2(w)(-1)^{\langle w, \tau \rangle}$. Hence, we get

$$\begin{aligned} 2^{2n} \sum_{\tau=0}^{2^n-1} R_f^2(\tau) &= \sum_{\tau=0}^{2^n-1} \sum_{w=0}^{2^n-1} \sum_{x=0}^{2^n-1} S_{(f)}^2(w) S_{(f)}^2(x)(-1)^{\langle (w \oplus x), \tau \rangle} \\ &= \sum_{w=0}^{2^n-1} \sum_{x=0}^{2^n-1} S_{(f)}^2(w) S_{(f)}^2(x) \sum_{\tau=0}^{2^n-1} (-1)^{\langle (w \oplus x), \tau \rangle} \\ &= 2^n \sum_{w=0}^{2^n-1} S_{(f)}^4(w) \\ &\leq 2^n S_{(f)}^2(w_0) \sum_{w=0}^{2^n-1} S_{(f)}^2(w) \\ &= 2^{3n} S_{(f)}^2(w_0) \text{ (by Parseval's Equation).} \end{aligned}$$

So, we have

$$\sum_{\tau=0}^{2^n-1} R_f^2 \leq 2^n S_{(f)}^2(w_0).$$

Hence, we have

$$S_{(f)}(w_0) \geq 2^{-\frac{n}{2}} \sqrt{\sum_{\tau=0}^{2^n-1} R_f^2(\tau)}.$$

By

$$nl(f) = 2^{n-1} - \frac{1}{2} \max_{w \in GF^n(2)} |S_{(f)}(w)|,$$

we have

$$nl(f) \leq 2^{n-1} - 2^{-\frac{n}{2}-1} \sqrt{\sum_{\tau=0}^{2^n-1} R_f^2(\tau)}.$$

Since $R_f(0) = 2^n$, we have

$$2^{n-1} - 2^{-\frac{n}{2}-1} \sqrt{\sum_{\tau=0}^{2^n-1} R_f^2(\tau)} \leq 2^{n-1} - 2^{\frac{n}{2}-1},$$

and the theorem follows. □

It is also noted that inequality $nl(f) \leq 2^{n-1} - 2^{-\frac{n}{2}-1} \sqrt{\sum_{\tau=0}^{2^n-1} R_f^2(\tau)}$ is also a more tight bound of nonlinearity than the commonly used inequality $nl(f) \leq 2^{n-1} - 2^{\frac{n}{2}-1}$.

3.6 Nonlinearity of Balanced Boolean Functions

In cryptographic applications, it is a common and primary requirement for Boolean functions to be balanced. Note that the Boolean functions with highest nonlinearity are the bent functions, and bent functions are not balanced. It would be interesting to know what the upper bound or lower bound of the nonlinearity of balanced Boolean functions could be. Some interesting results can be found in [30].

Theorem 3.10. *Let $f(x) \in \mathcal{F}_n$ be balanced ($n \geq 3$). Then the nonlinearity of $f(x)$ is given by*

$$nl(f) \leq \begin{cases} 2^{n-1} - 2^{\frac{n}{2}-1} - 2, & n \text{ even} \\ \lfloor\lfloor 2^{n-1} - 2^{\frac{n}{2}-1} \rfloor\rfloor, & n \text{ odd} \end{cases}$$

where $\lfloor\lfloor x \rfloor\rfloor$ denotes the maximum even integer less than or equal to x.

Proof. Since $f(x)$ is a balanced function, we have $wt(f) = 2^{n-1}$. Obviously, for any affine function $l(x) \in \mathcal{A}_n$, we have

$$wt(l) = \begin{cases} 0, & \text{if } l(x) = 0 \\ 2^n, & \text{if } l(x) = 1 \\ 2^{n-1}, & \text{otherwise.} \end{cases}$$

Since $d(f, l) = wt(f \oplus l) = wt(f) + wt(l) - 2wt(f \cdot l)$, it is known that $d(f, l)$ must be even. On the other hand, since $f(x)$ is balanced, $f(x)$ is not bent and so $nl(f) < 2^{n-1} - 2^{\frac{n}{2}-1}$. Since $nl(f) = \min_{l \in \mathcal{A}_n} d(f, l)$, we have

$$nl(f) \leq \begin{cases} 2^{n-1} - 2^{\frac{n}{2}-1} - 2, & n \text{ even} \\ \lfloor\lfloor 2^{n-1} - 2^{\frac{n}{2}-1} \rfloor\rfloor, & n \text{ odd}, \end{cases}$$

which proves the theorem. □

3.7 Higher-Order Nonlinearity of Boolean Functions

Nonlinearity of a Boolean function $f(x) \in \mathcal{F}_n$ measures the distance between $f(x)$ and all affine Boolean functions $l(x) \in \mathcal{A}_n$. For resistance to the linear cryptanalysis, we expect that the nonlinearity is as high as possible, even close to the maximum value $2^{n-1} - 2^{\frac{n}{2}-1}$. Similarly, we can also consider the higher-order nonlinearity of Boolean functions as a measure against higher-order nonlinear approximation cryptanalysis. Naturally, we also expect the higher-order nonlinearity to be as high as possible. However, it is more difficult to calculate the higher-order nonlinearity than the nonlinearity. So far very few methods are available to calculate the higher-order nonlinearity. In the following, we describe two methods which calculate a lower bound of the higher-order nonlinearity.

It is noted that the rth order Reed-Muller code of length n, denoted by $RM(r, n)$, is actually the set of all Boolean functions in n variables with algebraic degree no more than r. When $r = 1$, $RM(r, n)$ is the set of affine Boolean functions $\mathcal{A}_n$. Let $f(x) \in \mathcal{F}_n$, and define the *rth order nonlinearity* of $f(x)$ to be the smallest distance between $f(x)$ and the code words in $RM(r, m)$, denoted by $nl_r(f)$, i.e.,

$$nl_r(f) = \min_{g(x) \in RM(r,n)} \{d(f, g)\}. \tag{3.13}$$

First, we give some simple properties on higher-order nonlinearities. By the definition of higher-order nonlinearity, it is easy to derive the following theorem.

Theorem 3.11. *Let $f(x) \in \mathcal{F}_n$, and then for any $g(x) \in \mathcal{F}_n$ with $deg(g) \leq r$, we have*

$$nl_r(f \oplus g) = nl_r(f).$$

Before considering the rth order nonlinearity, we consider a special case when the algebraic degree of Boolean functions to be considered is $r + 1$. We have

Theorem 3.12. *The tight lower bound of rth order nonlinearity of Boolean functions in $\mathcal{F}_n$ with algebraic degree $r + 1$ is 2^{n-r-1}.*

Proof. Since the minimum distance of $RM(r, n)$ is equal to 2^{n-r}, for every $r < n$, we have $nl_r(f) \geq 2^{n-r-1}$ for every function $f(x)$ of algebraic degree exactly $r + 1 \leq n$. Moreover, if $f_0(x) \in RM(r, n)$ has minimum weight and is of algebraic degree $r+1$, then we have $wt(f_0) = 2^{n-r-1}$; otherwise, the minimum distance of $RM(r, n)$ would be larger than 2^{n-r-1} which is not true. Hence, we have $nl_r(f_0) = 2^{n-r-1}$, because in this case the distance between $f_0(x)$ and the constant 0 is 2^{n-r-1}. This means that 2^{n-r-1} is the tight upper bound of rth order nonlinearity of Boolean functions with algebraic degree $r + 1$. □

It is known that every Boolean function in n variables can be written as a concatenation of two Boolean functions in $n - 1$ variables, i.e., $f(x) \in \mathcal{F}_n$ can be written as

$$f(x) = (x_n \oplus 1)f_0(x') \oplus x_n f_1(x')$$

where $f_0(x'), f_1(x') \in \mathcal{F}_{n-1}$. Then we have the following theorem.

Theorem 3.13. *Let $f(x) = (x_n \oplus 1)f_0(x') \oplus x_n f_1(x')$ be a Boolean function in n variables, where both $f_0(x')$ and $f_1(x')$ are Boolean functions in $n - 1$ variables. Then we have*

$$nl_r(f) \geq nl_r(f_0) + nl_r(f_1). \tag{3.14}$$

Proof. Let $g(x) = (x_n \oplus 1)g_0(x') \oplus x_n g_1(x')$ be a Boolean function in n variables, where both $g_0(x')$ and $g_1(x')$ are Boolean functions in $n-1$ variables. Then we have

$$d(f, g) = d(f_0, g_0) + d(f_1, g_1);$$

this is naturally interpreted as follows: the Hamming distance between two Boolean functions is the sum of the Hamming distance of the first half of the truth tables of these functions and those of the second half. Hence, if $g(x)$ is of algebraic degree r, then we have

$$d(f, g) \geq nl_r(f_0) + nl_r(f_1).$$

When $g(x)$ has the closest Hamming distance with $f(x)$, it gives

$$nl_r(f) \geq nl_r(f_0) + nl_r(f_1).$$

This proves the theorem. □

Now consider a special case. If $f_0(x') = f_1(x')$, then $f(x) = f_0(x') \oplus x_n(f_0(x') \oplus f_0(x')) = f_0(x')$, which means that $f(x)$ is independent of x_n. In this case, if $g(x') \in \mathcal{F}_{n-1}$ with algebraic degree r is the best approximation for $f_0 = f_1$, then $g(x')$ now being viewed as an n-variable Boolean function has distance $2nl_r(f_0)$ from $f(x)$, i.e., $nl_r(f) = 2nl_r(f_0)$.

3.8 Linear Structures of Boolean Functions

Linear structures of block ciphers have been investigated for their cryptographic significance. It has been pointed out in [8, 11, 24] that block ciphers having linear structures are vulnerable to attacks which could be much faster than exhaustive key search. In [17] it was shown that linear structures of a cryptographic function can be used to simplify the expression of the function. In this section linear structures of Boolean functions are studied by two subclasses, invariant linear structures and complementary linear structures. It should be noted that linear structures are normally studied in general rather than being distinguished in those subclasses. By this treatment, some new results are obtained.

Definition 3.2. Let $f(x) \in \mathcal{F}_n$. If for some $\alpha \in GF^n(2)$, $f(x \oplus \alpha) \oplus f(x) = c$ holds for all $x \in GF^n(x)$, where $c \in GF(2)$ is a constant, then α is called a *linear structure* of $f(x)$. More specifically, α is called an *invariant linear structure* of $f(x)$ if $c = 0$ and a *complementary linear structure* of $f(x)$ if $c = 1$.

Denote by $V_L(f)$ the set of all linear structures of $f(x)$, by $V_I(f)$ ($\subseteq V_L(f)$) the subset of invariant linear structures, and by $V_C(f)$ ($\subseteq V_L(f)$) the subset of complementary linear structures. It is easy to verify the following results:

Theorem 3.14. *Let $f(x) \in \mathcal{F}_n$. Then the following is true with respect to invariant and complementary linear structures of $f(x)$:*

- *Both $V_L(f)$ and $V_I(f)$ form a vector subspace of $GF^n(2)$. However, $V_C(f)$ is not a vector subspace of $GF^n(2)$ if it is not empty.*
- *$V_L(f) = V_I(f)$ if and only if $V_C(f)$ is empty.*
- *If $V_C(f)$ is not empty, let $\alpha \in V_C(f)$; then*

$$\phi : x \longrightarrow x \oplus \alpha$$

is a one-to-one mapping from $V_I(f)$ to $V_C(f)$, and in this case we have $dim(V_I(f)) = dim(V_L(f)) - 1$.

Proof. Let $\alpha, \beta \in GF^n(2)$ be two linear structures of $f(x)$, then by Definition 3.2, there exists $c_1, c_2 \in GF(2)$ such that both $f(x \oplus \alpha) = c_1$ and $f(x \oplus \beta) = c_2$ hold for all $x \in GF^n(2)$. So we have

$$\begin{aligned} f(x \oplus \alpha \oplus \beta) \oplus f(x) &= (f(x \oplus \alpha \oplus \beta) \oplus f(x \oplus \alpha)) \oplus (f(x \oplus \alpha) \oplus f(x)) \\ &= c_2 \oplus c_1 \end{aligned}$$

Note that $c_2 \oplus c_1 \in GF(x)$ is a constant, so by Definition 3.2, $\alpha \oplus \beta$ is also a linear structure of $f(x)$. This proves that the linear structures of $f(x)$ form a vector subspace of $GF^n(2)$. Note that when $c_1 = c_2 = 0$, we also have $c_1 \oplus c_2 = 0$, which means that when α and β are two invariant linear structures of $f(x)$, then $\alpha \oplus \beta$ is also an invariant linear structure of $f(x)$. The above discussion includes the case when $\alpha = \beta$, which yields $\alpha \oplus \beta = 0$ to be a trivial invariant linear structure of $f(x)$.

However, when $c_1 = c_2 = 1$, we have $c_1 \oplus c_2 = 0$, which means that when α and β are two complementary linear structures of $f(x)$, then $\alpha \oplus \beta$ is an invariant linear structure of $f(x)$; this proves that $V_C(f)$ is not a vector subspace of $GF^n(2)$.

If $V_L(f)$ is empty, then it is obvious that $V_L(f) = V_I(f)$ holds. On the other hand, if $V_L(f)$ is not empty, since $V_I(f) \subseteq V_L(f)$, it is obvious that $V_L(f) \neq V_I(f)$.

If $\alpha \in V_C(f)$ is not zero, then it is easy to verify that the function $\phi(x) = x \oplus \alpha$ maps an invariant linear structure of $f(x)$ into a complementary linear structure, and in the meantime, it also maps a complementary linear structure into an invariant linear structure; this means that $\phi(x)$ is a one-to-one mapping from $V_I(f)$ to $V_C(f)$; hence, we have $dim(V_I(f)) = dim(V_L(f)) - 1$. □

The following theorem describes in general the structure of Boolean functions which have linear structures.

Theorem 3.15. *Let $f(x) \in \mathcal{F}_n$. Then for all $x \in GF^n(2)$ and all $\alpha \in V_L(f)$, we have*

$$f(x \oplus \alpha) = f(x) \oplus f(\alpha) \oplus f(0),$$

where

$$f(\alpha) = \begin{cases} f(0), & \text{if } \alpha \in V_I(f), \\ f(0) \oplus 1, & \text{if } \alpha \in V_L(f) - V_I(f). \end{cases}$$

Proof. By Definition 3.2, for any $\alpha \in V_L(f)$, $f(x) \oplus f(x \oplus \alpha) = c$ is a constant. Now we see what the constant c equals. Since it is a constant, it should be the same for all the values of x. Let $x = 0$ and we get $c = f(0) \oplus f(\alpha)$, which means that $f(x \oplus \alpha) = f(x) \oplus f(\alpha) \oplus f(0)$ holds. Now we further examine what $f(\alpha)$ is. If $\alpha \in V_I(f)$ is an invariant linear structure of $f(x)$, then by definition, $c = 0$, hence $f(\alpha) = f(0)$. If $\alpha \in V_C(f) = V_L(f) - V_I(f)$ is a complementary linear structure of $f(x)$, then $c = 1$ and hence $f(\alpha) = f(0) \oplus 1$. □

In [17] it was shown how to simplify a function over a finite field having linear structures. For the case of Boolean functions, it can be refined as follows:

Theorem 3.16. *Let* $f(x) \in \mathcal{F}_n$ *and* $dim(V_I(f)) = k$. *Then there exists an invertible matrix A over GF*(2) *such that*

$$g(x_1, \ldots, x_n) = f((x_1, \ldots, x_n)A) = g^*(x_{k+1}, \ldots, x_n),$$

where $g^*(x_{k+1}, \ldots, x_n)$ *has no nonzero invariant linear structure. Moreover,* $g^*(x_{k+1}, \ldots, x_n)$ *has a complementary linear structure, or equivalently it can be written as* $g^*(x_{k+1}, \ldots, x_n) = x_{k+1} \oplus g_1^*(x_{k+2}, \ldots, x_n)$, *if and only if* $f(x)$ *has a complementary linear structure.*

Proof. Choose a nonsingular $n \times n$ matrix in such a way that its first k columns are a basis of $V_I(f)$ and the $(k+1)$th column is possibly a complementary linear structure of f, and the rest of columns are arbitrary vectors. Then it can be verified that the matrix satisfies the requirement. □

From Theorem 3.16, it is easy to deduce the inequality about the degree of a nonlinear Boolean function having linear structures [17]. It is

$$deg(f) + dim(V) \leq n.$$

As well, if $dim(V) = k$, then we have that the nonlinearity of f satisfies $nl(f) \leq 2^{n-1} - 2^{\frac{n+k}{2}-1}$.

Now we turn our attention to the spectral description of linear structures by the treatment with two subclasses.

Theorem 3.17. *Let* $f(x) \in \mathcal{F}_n$. *Then* α *is an invariant linear structure of* $f(x)$, *i.e.,* $f(x) \oplus f(x \oplus \alpha) = 0$, *if and only if the self-correlation of* $f(x)$ *satisfies the following:*

$$R_f(\alpha) = wt(f). \tag{3.15}$$

Proof. $R_f(\alpha) = wt(f)$ if and only if $f(x \oplus \alpha) = 1$ whenever $f(x) = 1$, i.e., if and only if every component in the truth table of $f(x)$ and that of $g(x) = f(x \oplus \alpha)$ are the same. By the definition of self-correlation function (Definition 1.3), the conclusion follows. □

Theorem 3.18. *Let* $f(x) \in \mathcal{F}_n$. *Then* α *is a complementary linear structure of* $f(x)$, *i.e.,* $f(x) \oplus f(x \oplus \alpha) = 1$, *if and only if* $f(x)$ *is balanced and the self-correlation of* $f(x)$ *satisfies the following:*

$$R_f(\alpha) = 0. \tag{3.16}$$

Proof. The proof of the theorem is similar to that of Theorem 3.17 and is omitted here. □

Let $f(x) \oplus f(x \oplus \alpha) = c$. Then

$$\begin{aligned}
&\sum_{x=0}^{2^n-1}(f(x) \oplus f(x \oplus \alpha))(-1)^{\langle \omega, x\rangle} \\
&= S_f(\omega) + (-1)^{\langle \omega, \alpha\rangle} S_f(\omega) - 2\sum_{x=0}^{2^n-1} f(x)f(x \oplus \alpha)(-1)^{\langle \omega, x\rangle} \\
&= (1 + (-1)^{\langle \omega, \alpha\rangle})S_f(\omega) - 2\sum_{x=0}^{2^n-1} f(x)(f(x) \oplus c)(-1)^{\langle \omega, x\rangle} \\
&= (1 + (-1)^{\langle \omega, \alpha\rangle})S_f(\omega) - 2(S_f(\omega) - cS_f(\omega)) \\
&= (2c - 1 + (-1)^{\langle \omega, \alpha\rangle})S_f(\omega).
\end{aligned}$$

On the other hand, since $f(x) \oplus f(x \oplus \alpha) = c$, by Lemma 1.1 we have

$$\sum_{x=0}^{2^n-1}(f(x) \oplus f(x \oplus \alpha))(-1)^{\langle \omega, x\rangle} = \begin{cases} c \cdot 2^n & \text{if } \omega = 0, \\ 0 & \text{else.} \end{cases}$$

By the conversion between the two types of Walsh transforms, we have

$$(2c - 1 + (-1)^{\langle \omega, \alpha\rangle})(-\frac{1}{2}S_{(f)}(\omega)) = 0, \text{ if } \omega \neq 0 \text{ and}$$

$$(2c - 1 + (-1)^{\langle \omega, \alpha\rangle})(2^n - S_{(f)}(\omega))/2 = c \cdot 2^n, \text{ if } \omega = 0$$

i.e.,

$$\begin{cases} (2c - 1 + (-1)^{\langle \omega, \alpha\rangle})S_{(f)}(\omega) = 0 & \text{if } \omega \neq 0, \\ cS_{(f)}(\omega) = 0 & \text{if } \omega = 0. \end{cases}$$

Setting $c = 0$, we have $S_{(f)}(\omega) = 0$ for all ω with $\langle \omega, \alpha\rangle = 1$. By setting $c = 1$, we have $S_{(f)}(\omega) = 0$ for all ω with $\langle \omega, \alpha\rangle = 0$. It will be shown that these conditions are also sufficient for α to be a linear structure of $f(x)$.

Theorem 3.19. *Let $f(x) \in \mathcal{F}_n$. Then α is an invariant linear structure of $f(x)$, i.e., $f(x) \oplus f(x \oplus \alpha) = 0$, if and only if $S_{(f)}(\omega) = 0$ holds for all ω with $\langle \omega, \alpha\rangle = 1$.*

Proof. Necessity has been shown as above. So only the sufficiency needs to be proved. From the conversion of the two types of Walsh transforms, we know that, for $\omega \neq 0$, $S_{(f)}(\omega) = 0$ if and only if $S_f(\omega) = 0$. Since $S_f(\omega) = 0$ for every ω with $\langle \omega, \alpha\rangle = 1$, by Theorem 1.6 we have $wt(f) = 2^{-n}\sum_{\langle \omega, \alpha\rangle=0} S_f^2(\omega)$. By Theorem 1.7,

$$\begin{aligned}
R_f(\alpha) &= 2^{-n}\sum_{\omega} S_f^2(\omega)(-1)^{\langle \omega, \alpha\rangle} \\
&= 2^{-n}\sum_{\langle \omega, \alpha\rangle=0} S_f^2(\omega) \\
&= wt(f).
\end{aligned}$$

By Theorem 3.17 the conclusion then follows. □

Theorem 3.20. *Let $f(x) \in \mathcal{F}_n$. Then α is an invariant linear structure of $f(x)$, i.e., $f(x) \oplus f(x \oplus \alpha) = 1$, if and only if $S_{(f)}(\omega) = 0$ holds for all ω with $\langle \omega, \alpha \rangle = 0$.*

Proof. Necessity is as above and the sufficiency is as follows. Note that $S_{(f)}(0) = 0$ means that $f(x)$ is balanced, and for every $\omega \neq 0$, $S_{(f)}(\omega) = 0$ if and only if $S_f(\omega) = 0$. By Theorem 1.6 we have

$$\sum_{\langle \omega, \alpha \rangle = 1} S_f^2(\omega) + S_f^2(0) = 2^n wt(f).$$

Since $f(x)$ is balanced, we have $S_f(0) = wt(f) = 2^{n-1}$. So $\sum_{\langle \omega, \alpha \rangle = 1} S_f^2(\omega) = 2^{2n-2}$. By Theorem 1.7 we have

$$\begin{aligned} R_f(\alpha) &= 2^{-n} \textstyle\sum_{\omega} S_f^2(\omega)(-1)^{\langle \omega, \alpha \rangle} \\ &= 2^{-n} S_f^2(0) - 2^{-n} \textstyle\sum_{\langle \omega, \alpha \rangle = 1} S_f^2(\omega) \\ &= 0. \end{aligned}$$

And by Theorem 3.18 the conclusion then follows. □

For an arbitrary nonzero vector $\alpha \in GF^n(2)$, define $A_i = \{\omega \in GF^n(2) : \langle \omega, \alpha \rangle = i\}$, $i = 0, 1$. Then it is easy to show by establishing a one-to-one mapping from A_0 to A_1 that $|A_0| = |A_1|$, where $|A|$ means the cardinality of set A. This proves that:

Corollary 3.2. *If $f(x)$ has a nonzero linear structure, then at least half of its Walsh spectrums vanish, i.e., the Walsh transform of $f(x)$ takes value zero on at least half of its inputs.*

Theorem 3.21. *Let $f(x) \in \mathcal{F}_n$ and $V_L(f) = V_I(f) \cup V_C(f)$ be the set of linear structures of the function $f(x)$. If the dimension of $V_L(f)$ is k, then we have*

$$\frac{nl(f)}{2^{n-1}} + \frac{1}{\sqrt{2^{n-k}}} \leq 1.$$

Proof. Since the dimension of $V_L(f) = V_I(f) \cup V_C(f)$ is k, assume that $\beta_1, \beta_2, \cdots, \beta_k \in V_I(f)$ are linearly independent vectors. We also assume that $\beta_1, \beta_2, \cdots, \beta_{k-1}$ are invariant linear structures of $f(x)$ and $f(x \oplus \beta_k) \oplus f(x) = c$. That means that we have

$$f(x) = f(x \oplus \beta_1) = \cdots = f(x \oplus \beta_{k-1}) = f(x \oplus \beta_k) \oplus c,$$

so we have

$$\begin{aligned} S_{(f)}(w) &= (-1)^{\langle \beta_1, w \rangle} S_{(f)}(w) \\ &= \cdots \end{aligned}$$

$$= (-1)^{\langle \beta_{k-1}, w\rangle} S_{(f)}(w)$$
$$= (-1)^{c \oplus \langle \beta_k, w\rangle} S_{(f)}(w).$$

If $S_{(f)}(w) \neq 0$, then we have

$$\langle \beta_1, w\rangle = \cdots = \langle \beta_{k-1}, w\rangle = c \oplus \langle \beta_k, w\rangle = 0. \tag{3.17}$$

Since $\beta_1, \beta_2, \cdots, \beta_{k-1}, \beta_k$ are linearly independent on $GF(2)$, the system of linear equations (3.17) has 2^{n-k} solutions over $GF^n(2)$. Hence, the number of nonzero spectra of $f(x)$ is at most 2^{n-k}, i.e.,

$$|\{w \in GF^n(2)|S_{(f)}(w) \neq 0\}| \leq 2^{n-k}.$$

By Parseval's equation $\sum_{w=0}^{2^n-1} S_{(f)}^2(w) = 2^{2n}$ and

$$\sum_{w=0}^{2^n-1} S_{(f)}^2(w) \leq 2^{n-k} \max_{w \in GF^n(2)} S_{(f)}^2(w),$$

we have

$$\max_{w \in GF^n(2)} |S_{(f)}(w)| \geq 2^{\frac{n+k}{2}}.$$

By Eq. 3.1, $nl(f) = 2^{n-1} - \frac{1}{2} \max_{w \in GF^n(2)} |S_{(f)}(w)|$, we get

$$\frac{nl(f)}{2^{n-1}} = 1 - \frac{1}{2^n} \max_{w \in GF^n(2)} |S_{(f)}(w)| \leq 1 - 2^{-\frac{n-k}{2}};$$

hence, the theorem follows. □

3.9 Remarks

Nonlinearity is only one of the important cryptographic properties of Boolean functions; other cryptographic properties are also needed in practice [26–29, 37]. The same case applies to the other chapters in this book, although each chapter is mainly focused on a specific cryptographic property.

References

1. Beth, T., Ding, C.: On almost perfect nonlinear permutations. In: Advances in Cryptology, Proceedings of Eurocrypt'93. LNCS 765, pp. 65–76. Springer, Berlin/New York (1994)
2. Carlet, C., Ding, C.: Highly nonlinear mappings. J. Complex. **20**, 205–244 (2004)
3. Carlet, C., Dobbertin, H., Leander, G.: Normal extensions of bent functions. IEEE Trans. Inf. Theory **IT-50**(11), 2880–2885 (2004)
4. Carlet, C., Guillot, P.: An alternate characterization of the bentness of binary functions with uniqueness. Des. Codes Cryptogr. **14**, 133–140 (1998)
5. Carlet, C., Gouget, A.: An upper bound on the number of m-resilient bent functions. In: Advances in Cryptology, Proceedings of Asiacrypt 2002. LNCS 2501, pp. 484–496. Springer, Berlin/New York (2002)
6. Carlet, C., Tarannikov, Y.: Covering sequences of Boolean functions and their cryptographic significance. Des. Codes Cryptogr. **25**, 263–279 (2002)
7. Chabaud, F., Vaudenay, S.: Links between differential and linear cryptanalysis. In: Advances in Cryptology, Proceedings of Eurocrypt'94. LNCS 950, pp. 356–365. Springer, Berlin/Heidelberg (1995)
8. Chaum, D., Evertse, J.H.: Cryptanalysis of DES with a reduced number of rounds. In: Advances in Cryptology, Proceedings of Crypto'85. LNCS 218, pp. 192–211. Springer, Berlin (1986)
9. Dobbertin, H.: Construction of bent functions and balanced Boolean functions with high nonlinearity. In: Fast Software Encryption 1994. LNCS 1008, pp. 61–74. Springer, Berlin (1995)
10. Dobbertin, H.: A survey of some recent results on bent functions. In: Sequences and Their Applications – SETA 2004. LNCS 3486, pp. 1–29. Springer, Berlin (2005)
11. Evertse, J.-H.: Linear structures in block ciphers. In: Advances in Cryptology, Proceedings of Eurocrypt'87. LNCS 304, pp. 249–266. Springer, Berlin (1988)
12. Fedorova, M., Tarannikov, Y.: On the construction of highly nonlinear resilient Boolean functions by means of special matrices. In: Proceedings of Indocrypt 2011. LNCS 2247, pp. 254–266. Springer, Heidelberg/New York (2001)
13. Filiol, E., Fontaine, C.: Highly nonlinear balanced Boolean functions with a good correlation immunity. In: Advances in Cryptology, Proceedings Eurocrypt'98. LNCS 1403, pp. 475–488. Springer, Berlin (1998)
14. Fontaine, C.: The nonlinearity of a class of Boolean functions with short representation. In: Proceedings of the 1st International Conference on the Theory and Applications of Cryptology (PRAGOCRYPT'96). CTU Publishing House, Prague, pp. 129–144
15. Kumar, P.V., Scholtz, R.A.: Bounds on the linear span of Bent sequences. IEEE Trans. Inf. Theory **IT-29**(6), 854–862 (1983)
16. Kumar, P.V., Scholtz, R.A.: Generalized bent functions and their properties. J. Combin. Theory (A) **40**, 90–107 (1985)
17. Lai, X.: Additive and linear structures of cryptographic functions. FSE **1994**(1008), 75–85 (1995)
18. Lempel, A., Cohn, M.: Maximal families of Bent sequences. IEEE Trans. Inf. Theory **IT-28**(6), 865–868 (1982)
19. MacWilliams, F.J., Sloane, N.J.A.: The Theory of Error-Correcting Codes. North-Holland, New York (1977)
20. Meier, W., Staffelbach, O.: Nonlinearity criteria for cryptographic functions. In: Advances in Cryptology, Proceedings of Eurocrypt'89. LNCS 434, pp. 549–562. Springer, Berlin (1990)
21. Nyberg, K.: Construction of Bent functions and different sets. In: Advances in Cryptology, Proceedings of Eurocrypt'90. LNCS 473, pp. 151–160. Springer, Berlin (1991)
22. Patterson, N.J., Wiedemann, D.H.: The covering radius of the $[2^{15}, 16]$ Reed-Muller code is at least 16276. IEEE Trans. Inf. Theory **IT-29**(3), 354–356 (1983)

23. Qu, C., Seberry, J., Pieprzyk, J.P.: Homogeneous bent functions. Discret. Appl. Math. **102**, 133–139 (2000)
24. Reeds, J.A., Manferdeli, J.L.: DES has no per round linear factors. In: Advances in Cryptology, Proceedings of Crypto'84. LNCS 196, pp. 377–389. Springer, Berlin/Heidelberg (1985)
25. Rothaus, O.S.: On 'bent' functions. J. Combin. Theory (A) **20**, 300–305 (1976)
26. Sarkar, P., Maitra, S.: Nonlinearity bounds and constructions of resilient Boolean functions. In: Advances in Cryptology, Proceedings of Crypto'2000. LNCS 1880, pp. 515–532. Springer, Berlin (2000)
27. Sarkar, P., Maitra, S.: Construction of nonlinear Boolean functions with important cryptographic properties. In: Advances in Cryptology, Proceedings of Eurocrypt'2000. LNCS 1807, pp. 485–506. Springer, Berlin (2000)
28. Sarkar, P., Maitra, S.: Construction of nonlinear resilient Boolean functions using 'small' affine functions. IEEE Trans. Inf. Theory **IT-50**(1), 2185–2193 (2004)
29. Schnoor, C.P.: The multiplicative complexity of Boolean functions. In: Proceedings of AAECC-6, pp. 45–58. Springer, Berlin (1989)
30. Seberry, J., Zhang, X.M., Zheng, Y.: On construction and nonlinearity of correlation immune functions, (extended abstract). In: Advances in Cryptology, Proceedings of Eurocrypt'93. LNCS 765, pp. 181–199. Springer, Berlin (1994)
31. Seberry, J., Zhang, X.M., Zheng, Y.: Nonlinearly balanced Boolean functions and their propagation characteristics (extended abstract). In: Advances in Cryptology, Proceedings of Crypto'93. LNCS 773, pp. 49–60. Springer, Berlin (1994)
32. Seberry, J., Zhang, X.M., Zheng, Y.: Relationships among nonlinearity criteria (extended abstract). In: Advances in Cryptology, Proceeding of Eurocrypt'94. LNCS 950, pp. 376–388. Springer, Berlin (1995)
33. Sun, G., Wu, C.: The lower bound on the second order nonlinearity of a class of Boolean functions with high nonlinearity. Appl. Algebra Eng. Commun. Comput. (AAECC) **22**(1), 37–45 (2011)
34. Tang, D., Carlet, C., Tang, X.: On the second-order nonlinearities of some bent functions. Inf. Sci. **223**, 322–330 (2013)
35. Wu, C.K.: Boolean functions in cryptology. Ph.D. Thesis, Xidian University, Xian (1993) (in Chinese)
36. Yu, N.Y., Gong, G.: Constructions of quadratic bent functions in polynomial forms. IEEE Trans. Inf. Theory **IT-52**(2), 3291–3299 (2006)
37. Zheng, Y., Xhang, X.M., Imai, H.: Restriction, terms and nonlinearity of Boolean functions. Theor. Comput. Sci. **226**, 207–223 (1999)
38. Zheng, Y., Zhang, X.M.: On plateaued functions. IEEE Trans. Inf. Theory **IT-47**(3), 1215–1223 (2001)

Chapter 4
Correlation Immunity of Boolean Functions

The concept of correlation immunity was proposed by Siegenthaler in 1984. It is a security measure to the correlation attack of nonlinear combiners. This chapter first briefly describes the correlation attack of nonlinear combiners, which gives the rationale about why correlation immunity is a reasonable security measure, and then the correlation immunity of Boolean functions is studied. Different approaches to the constructions of Boolean functions are introduced, which yields a way in theory to exhaustively construct all the correlation immune Boolean functions, and such an example is given for the correlation immune Boolean functions in four variables. Correlation immune Boolean function with some other cryptographic properties are also studied in brief. In the end, the concept of ε-correlation immunity is introduced to reflect the resistance against correlation attack when the Boolean function is not correlation immune in the traditional sense.

4.1 The Correlation Attack of Nonlinear Combiners

Nonlinear combiner is a popular pseudorandom sequence generator for stream ciphers [33]. The basic structure of nonlinear combiners in stream ciphers is shown in Fig. 4.1.

The correlation attack proposed by Siegenthaler [37] makes use of the correlation information between the output sequence (z_k) of the nonlinear combiner and each input sequence (x_k^i) of the combining function $f(x)$ and to use the statistical analysis trying to recover the initial state as well as the feedback function of each LFSR_i individually. This approach is also called *divide and conquer* attack, which significantly reduces the complexity than the brute force attack. Below we will give a brief description about the divide and conquer attack model as described in [37].

In the security analysis, it is always assumed that the structure of the generator is known, i.e., the lengths of each LFSR and the nonlinear combining function

C.-K. Wu, D. Feng, *Boolean Functions and Their Applications in Cryptography*,
Advances in Computer Science and Technology, DOI 10.1007/978-3-662-48865-2_4

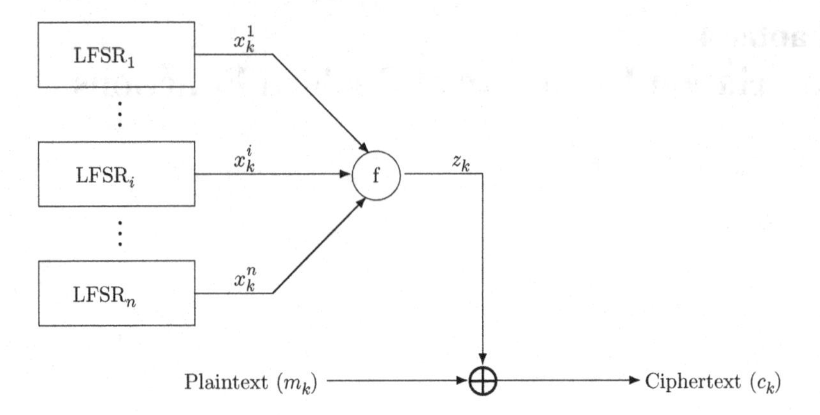

Fig. 4.1 A nonlinear combiner of stream ciphers

$f(x)$. The attack proposed in [37] does not assume the knowledge of the primitive feedback polynomial of each LFSR which is only of certain limited amount to search for.

Assume that all the LFSRs in the combiner of Fig. 4.1 are maximum length sequence generators, i.e., each LFSR$_i$ of order r_i generates an m-sequence of period $p_i = 2^{r_i} - 1$, and there are R_i primitive polynomials of degree r_i (which is the number of different m-sequences of order r_i such that they are not equivalent by cyclic shift). Then under the brute force attack, the number of all the possible keys for the nonlinear combiner (different initial states and different feedback function of each of the LFSR have been taken into account) is

$$K = \prod_{i=1}^{n} R_i(2^{r_i} - 1).$$

With the correlation attack, information about each input sequence (x_k^i) can be extracted from the output sequence (z_k), and hence the attack can concentrate each of the individual LFSR sequences, and the number of trials in the worst case is reduced to approximately

$$K' = \sum_{i=1}^{n} R_i 2^{r_i}.$$

The correlation attack is a probabilistic attack which assumes some statistical properties of the combining function $f(x)$. Assume in the ideal case that each of the LFSRs in Fig. 4.1 produces a pseudorandom sequence with uniform probability distribution, i.e., $Prob(x_k^i = 0) = Prob(x_k^i = 1)$, and assume that $Prob(z_k = 0) = Prob(z_k = 1)$. Let

$$Prob(z_k = x_k^i) = q_i, \tag{4.1}$$

and assume the plaintext comes from a memoryless binary source, which satisfies

$$Prob(y_k = 0) = p_0 \tag{4.2}$$

Then it is easy to compute

$$\begin{aligned} Prob(c_k \oplus x_k^j = 0) &= Prob(z_k = x_k^j) \cdot Prob(y_k = 0) \\ &\quad + Prob(z_k \neq x_k^j) \cdot Prob(y_k = 1) \\ &= 1 - (p_0 + q_j) + 2p_0 q_j \\ &= p_e \end{aligned} \tag{4.3}$$

When $j = 0$, let x_k^0 be an hypothetical random variable which are independent of any x_k^i $(i > 0)$ and with uniform probability distribution. Then compute the correlation of sequences c_k and x_k^j as

$$\alpha = \sum_{k=1}^{N}(1 - 2(c_k \oplus x_k^j)) = N - 2\sum_{k=1}^{N}(c_k \oplus x_k^j), \ \ j \in \{0, 1, \ldots, n\} \tag{4.4}$$

By the central limit theorem, when N is sufficiently large, α approaches to a normal distribution (or Gaussian distribution). In an attack, attackers use hypothetical LFSR of length r_i which produce sequence (x_k^0) for testing. By choosing a nonzero initial state and an arbitrary primitive polynomial as the feedback polynomial, compute the correlation α_0 between N bits of output of the hypothetical LFSR and N bits of the real ciphertext. Then there are two hypotheses to consider:

H_1: There are $N > r_i$ coincidences between the output of the hypothetical LFSR and LFSR$_i$, referring to the above cases, and this is the case when α_0 is the correlation between z_k and x_k^i, $i \in \{1, 2, \ldots, n\}$.

H_0: There are $N > r_i$ disagreement between the output of the hypothetical LFSR and LFSR$_i$, referring to the above cases, and this is the case when α_0 is the correlation between z_k and x_k^0.

In order to make a decision about the two hypotheses, a threshold value T is needed. When $\alpha_0 < T$, then accept the hypothesis H_0, and when $\alpha_0 \geq T$, accept H_1. Let the probability density function of the probabilistic variable α be $P_{\alpha|H_k}(x)$. If $q_i = \frac{1}{2}$ or $p_0 = \frac{1}{2}$, then by Eq. 4.3, we have $p_e = \frac{1}{2}$. In this case no decision can be made, because in this case the probability distribution of α under the two hypotheses is the same. Here the discussed attack depends on the number of wrong decisions, i.e., the number of cases when $\alpha \geq T$. So we define a *false alarm probability* $P_f = Prob(\alpha \geq T|H_0)$. In order to determine an appropriate threshold T, we also need to consider the probability $P_m = Prob(\alpha < T|H_1)$. We have

$$P_f = \int_T^{\infty} P_{\alpha|H_0}(x)dx \tag{4.5}$$

$$P_m = \int_{-\infty}^{T} P_{\alpha|H_1}(x)dx \tag{4.6}$$

With the help of the function

$$Q(x) = \frac{1}{\sqrt{2\pi}} \int_x^{\infty} e^{-\frac{y^2}{2}} dy \tag{4.7}$$

we can get the following expressions:

$$P_f = Q(|\frac{T}{\sqrt{N}}|) \tag{4.8}$$

$$P_m = Q(|\frac{N(2p_e - 1) - T}{2\sqrt{N}\sqrt{p_e(1 - p_e)}}|) \tag{4.9}$$

Denote by

$$\gamma_0 = \frac{N(2p_e - 1) - T}{2\sqrt{N}\sqrt{p_e(1 - p_e)}}, \tag{4.10}$$

then the expression of P_f and P_m can be written as

$$P_f = Q(|\sqrt{N}(2p_e - 1) - 2\gamma_0\sqrt{p_e(1 - p_e)}|), \tag{4.11}$$

$$P_m = Q(|\gamma_0|). \tag{4.12}$$

In order to attack the stream cipher model as in Fig. 4.1, the following process is to be taken: first to determine the probability q_i by $f(x)$, and to determine the probability p_0 according to the coding method of the plaintext, then compute p_e using Eq. 4.3. For any chosen probability P_m, by Eq. 4.12, it is known that γ_0 is a constant, and from Eq. 4.11, it is known that the false alarm probability $P(\alpha \geq T|H_0)$ is a function of N. In order to recover LFSR$_i$, choose an arbitrary primitive polynomial as its feedback polynomial and an arbitrary nonzero state as its initial state, and let it produce a sequence, and then compute the correlation between this sequence and the ciphertext sequence. For any event with $\alpha \geq T$, H_0 is accepted, i.e., the LSFR$_i$ is supposed to have been recovered. However, the probability of event $\alpha \geq T$ is P_f, and our decision may be wrong. So we need to test more ciphertexts for all the events $\alpha \geq T$. If for all the $2^{r_i} - 1$ different states, the decision is always to reject H_1, then change another primitive polynomial and to repeat the test. In the worst case, we need to test for about $R_i 2^{r_i}$ times. The false alarm probability depends on the length of ciphertext N. Choose N_1 such that

$$P_f = \frac{1}{R_i 2^{r_i}} \tag{4.13}$$

and then by Eq. 4.11, we have

$$\frac{1}{R_i 2^{r_i}} = Q(|\sqrt{N_1}(2p_e - 1) - 2\gamma_0\sqrt{p_e(1-p_e)}|). \tag{4.14}$$

Using the inequality

$$Q(x) < \frac{1}{2}2^{-\frac{x^2}{2}}, \quad x > 0 \tag{4.15}$$

we have an upper bound of N_1

$$N_1 < \left[\frac{\frac{1}{\sqrt{2}}\sqrt{\ln(R_i 2^{r_i-1})} + \gamma_0\sqrt{p_e(1-p_e)}}{p_e - \frac{1}{2}}\right]^2. \tag{4.16}$$

The upper bound in Eq. 4.16 gives the length of required ciphertext to enable an attack on the model in Fig. 4.1. If the length of the ciphertext is no less than this upper bound, then when conducting such an attack, the number of tests can be minimized, and when a decision is made, the probability of false alarm is minimized. More detailed description of the correlation attack can be found in [37, 38].

In order to resist the correlation attack as described above, the combining function $f(x)$ needs to have some special properties. Siegenthaler [36] introduced the concept of correlation immunity of Boolean functions. The correlation immunity of Boolean functions has attracted much study (see, e.g., [11–13, 13, 18, 25, 32, 34, 43, 51]). This chapter will show how such functions can have resistance against the correlation attack. Then it will study the construction issues of the correlation immune functions.

4.2 The Correlation Immunity and Correlation Attacks

In most shift register-based key stream generators, the key stream is usually generated by a nonlinear combination of some shift register sequences. These base shift registers are usually linear ones, and sequences generated by them are of very low linear complexity and can easily be retrieved by the Berlekamp-Massey algorithm [26]. The output of the combined sequence, however, should be designed to have good algebraic and statistical properties. A potential attack is to seek the cross-correlation between an input and the output of the combining function known as the correlation attack [37]. So, statistical independence of Boolean functions of their input variables is of practical significance. The concept of correlation immunity of Boolean functions [36], which is designed to resist the correlation attack, is the development of the concept of statistical independence with a stronger constraint.

In Sect. 2.6, the concept of statistical independence was introduced. A Boolean function $f(x) \in \mathcal{F}_n$ being statistically independent of its variables $x_{i_1}, x_{i_2}, \ldots, x_{i_k}$ means that any fixed value of $x_{i_1}, x_{i_2}, \ldots, x_{i_k}$ will not change the probability of $f(x)$ to be 0 or 1. The concept of correlation immunity of Boolean functions is in some sense more restricted than the statistical independence.

Definition 4.1. Let $f(x) \in \mathcal{F}_n$. If for some integer k, and for any set of indices $1 \leq i_1 < i_2 < \cdots < i_k \leq n$, $f(x)$ is always statistically independent of $x_{i_1}, x_{i_2}, \ldots, x_{i_k}$, then $f(x)$ is said to be *correlation immune* (CI) of order k. The value k is called the correlation immunity of $f(x)$ and is denoted by $CI(f) = k$.

Furthermore, if $f(x)$ is balanced, then $f(x)$ is called to be *resilient* of order k or *k-resilient* for short.

From the properties of statistical independence, we know that if $f(x)$ is correlation immune of order k, then it must be correlation immune of any order $m < k$ as well. In this sense, when we talk about the correlation immunity, it is meant the maximum known value of the correlation immunity. If the correlation immunity is 0, then $f(x)$ is rather called to be not correlation immune. If the correlation immunity is n, it is easy to deduce that $f(x)$ must be a constant. Therefore, our discussion about correlation immune functions will assume that the correlation immunity is between 1 and $n - 1$.

Now let us look at the problem about how correlation immune functions can have resistance against the correlation attack. Assume that the nonlinear combining function $f(x)$ as in the nonlinear combiner is correlation immune, and then for any $1 \leq j \leq n$, $f(x)$ is statistically independent of x_j. By Theorem 2.20 we know that for any $a, b \in \{0, 1\}$, we have

$$Prob(x_j = b|f(x) = a) = Prob(x_j = b) = \frac{1}{2}.$$

Hence, we have

$$\begin{aligned}
Prob(f(x) = x_j) &= Prob(f(x) = 0,\ x_j = 0) + Prob(f(x) = 1,\ x_j = 1) \\
&= Prob(f(x) = 0)Prob(x_j = 0|f(x) = 0) \\
&\quad + Prob(f(x) = 1)Prob(x_j = 1|f(x) = 1) \\
&= \frac{1}{2}Prob(f(x) = 0) + \frac{1}{2}Prob(f(x) = 1) \\
&= \frac{1}{2}
\end{aligned}$$

which is the value of q_j as defined in Eq. 4.1, and by Eq. 4.3 we have $p_e = 1 - (p_0 + q_j) - 2p_0q_j = \frac{1}{2}$, and then by Eq. 4.16, we know that the upper bound of N_1 is infinity. This means that one can never get sufficient amount of data (ciphertext) to conduct the correlation attack effectively. This is why correlation immune functions are supposed to be of resistance against correlation attacks.

The objective of correlation attack is to try to recover the LFSRs one by one using the correlation information between the nonlinear combining function $f(x)$ and its variables. It could be modified to recover the linear combinations of those LFSRs by the correlation information of the combining function and the linear combinations of the variables. If n linearly independent linear combinations of the LFSRs can be recovered, then the original LFSRs can be recovered easily. Even if some of the linear combinations of the LFSRs are reconstructed, the computational complexity for computing the rest of the linear combinations can be reduced. In order to enable the nonlinear combining function to be resistant against such a modified correlation attack, the combining function should have high-order correlation immunity. The concept of higher-order correlation immunity is to provide resistance against modified correlation attack so that $Prob(f(x) = x_{i_1} \oplus x_{i_2} \oplus \cdots \oplus x_{i_k}) = \frac{1}{2}$ always holds. However, practically the correlation immunity doesn't need to be very high, as there is much computational overhead to recover the original LFSRs even if some of their linear combinations are found.

4.3 Correlation Immunity of Boolean Functions

In order to study the correlation immunity of Boolean functions, we introduce a new concept of partial correlation immunity.

Definition 4.2. Let $f(x) \in \mathcal{F}_n$. The function $f(x) \in \mathcal{F}_n$ is called to be *correlation immune with respect to the subset* $T \subset \{1, 2, \ldots, n\}$ if the probability for f to take any value from $\{0, 1\}$ remains unchanged when any of the values of $\{x_i, i \in T\}$ are fixed in advance while other variables are free and randomly independent. The correlation immunity with constraint on a particular set T is called a *partial correlation immunity* of $f(x)$.

Note that the partial correlation immunity does not have an order defined but must have a set associated. The function $f(x)$ is said to be *correlation immune (CI) of order k* if for every T of cardinality at most k, f is correlation immune with respect to T. It is noticed that $f(x)$ is correlation immune of order k, implying that it is correlation immune of any order less than k as well. The largest possible value of k is called the *correlation immunity* of f. It is easy to see that this definition of correlation immunity is the same as in Definition 4.1.

Let $z = \bigoplus_{i=1}^{n} c_i x_i$ be a (nonzero) linear combination of the variables, where $c_i \in \{0, 1\}$. Then Boolean function $f(x) \in \mathcal{F}_n$ is said to be *correlation immune in z* if the probability for f to take any value from $\{0, 1\}$ is unchanged given that z is assigned any fixed value in advance. This is actually the case when $T = \{(c_1, c_2, \ldots, c_n)\}$ has a single element, and in this particular case, the partial correlation immunity can be referred to as the correlation immunity in this element (vector). It is noted that the partial correlation immunity with respect to a vector (coefficient vector of a linear combination of the variables) is the same as the statistical independence. However, for the convenience of description, we keep the new term of correlation immunity.

Lemma 4.1. *Let $f(x) \in \mathcal{F}_n$. Then $f(x)$ is correlation immune of order t if and only if for every $\gamma \in GF^n(2)$ with $wt(\gamma) \leq t$, $f(x)$ is correlation immune in combined variable $z = \langle \gamma, x \rangle = \gamma_1 x_1 \oplus \gamma_2 x_2 \oplus \cdots \oplus \gamma_n x_n$*

Proof. It is trivial to prove that $f(x)$ is correlation immune with respect to $T \subset \{1, 2, \ldots, n\}$, if and only if $f(x)$ is correlation immune in $z = \langle \gamma, x \rangle$ for all

$$\gamma \in \{\gamma : \gamma_i = 1 \text{ implies that } i \in T\}.$$

A generalization of this observation is that $f(x)$ is correlation immune with respect to all T of cardinality $\leq t$, if and only if $f(x)$ is correlation immune in every $z = \langle \gamma, x \rangle$ with $wt(\gamma) \leq t$. Therefore, the conclusion of Lemma 4.1 follows. □

It should be noted that $f(x)$ is correlation immune in z_1 and z_2 individually does not imply that it is correlation immune in $z_1 \oplus z_2$. For example, although $f(x_1, x_2, x_3) = x_3 \oplus x_1 x_2 \oplus x_1 x_3 \oplus x_2 x_3$ is a first-order correlation immune function, it is easy to verify that it is not correlation immune in $x_1 \oplus x_2$.

Let $f(x) \in \mathcal{F}_n$, $g(y) \in \mathcal{F}_k$, $D = (d_1^T, d_2^T, \ldots, d_k^T)$ be an $n \times k$ binary matrix with $rank(D) = k$, where $d_i \in GF^n(2)$. Let $f(x) = g(xD) = g(y)$. It is known that each y_i is the linear combination of x_j's with coefficients vector d_i, i.e., $y_i = \langle x, d_i \rangle$. Let $z = \bigoplus_{i=1}^n c_i x_i$ be another variable. Then it is obvious that $f(x)$ is correlation immune in z if and only if $g(y)$ is correlation immune in z. Denote by $\gamma = (c_1, c_2, \ldots, c_n)$. We have

Lemma 4.2. *If $rank[D; \gamma^T] = k + 1$, where $[A; B]$ means the concatenation of matrices A and B, then for any Boolean function $g(y) \in \mathcal{F}_k$, $g(xD)$ is independent of $z = \langle \gamma, x \rangle$ and hence is correlation immune in z.*

Proof. Let $y = (y_1, y_2, \ldots, y_k) = xD$. It is noticed that $rank[D; \gamma^T] = k + 1$ if and only if variables $y_1, y_2, \ldots, y_k$ together with z are all independent, and consequently $g(xD)$ is independent of z. So we have

$$Prob(g(xD) = 1 | z = 1) = Prob(g(y) = 1 | z = 1) = Prob(g(y) = 1).$$

This means that $g(xD)$ is correlation immune in z. □

4.4 Correlation Immune Functions and Error-Correcting Codes

Since the discussion below uses the concept of linear code, here we briefly introduce the relevant concepts.

Definition 4.3. An error-correcting code of length n, denoted by C, is a collection of vectors in $GF^n(2)$ (since we only consider binary codes). And for any $\alpha, \beta \in C$, $\alpha \neq \beta$, we have that the minimum distance of code C is defined as

$$d = \min\{d(\alpha, \beta) : \alpha, \beta \in C\}$$

where $d(\alpha, \beta) = wt(\alpha \oplus \beta)$ is the Hamming distance between α and β.

If C forms a k-dimensional vector subspace of $GF^n(2)$, then C is called a $[n, k, d]$ linear code.

The error-correction capability is $t = \lfloor \frac{d-1}{2} \rfloor$, where $\lfloor X \rfloor$ is the largest integer that is less than or equal to X. The objective of error-correcting codes is to achieve the maximum distance d and hence the maximum error-correction capability when n and the cardinality of the code space $|C|$ are fixed or to find the largest possible $|C|$ when n and d are fixed, where $|C|$ is the cardinality of C. When C forms a linear subspace of $GF^n(2)$, C is called a *linear code*. Since the code words (elements in C) and their linear combinations are all in C, it is possible to select some linearly independent code words from C so that every code word in C can be represented as a linear combination of these code words. Put these linearly independent code words as a matrix G, where the rows of G are code words in C, and any code word in C can be represented as a linear combination of the rows of G, and then G is called a *generating matrix* of C. The number of rows of G determines the size of C. When using a generating matrix G to represent a linear code, the code is often denoted as C_G.

The following theorem reveals a close connection between linear codes and the correlation immunity of Boolean functions.

Theorem 4.1. *If G is a generating matrix of an $[n, k, d]$ linear code, then for any $g(y) \in \mathcal{F}_k$, the correlation immunity of $f(x) = g(xG^T)$ is at least $d - 1$.*

Proof. For any vector $\gamma \in GF^n(2)$ with $wt(\gamma) \leq d-1$, since the minimum distance of C is d, γ cannot be represented by a linear combination of the rows of G; hence, $rank[G^T; \gamma^T] = k + 1$. By Lemma 4.2 we know that $f(x) = g(xG^T)$ is correlation immune in $z = \langle \gamma, x \rangle$. Since γ is an arbitrary vector with Hamming weight less than d, by Lemma 4.1, the correlation immunity of $f(x)$ is at least $d - 1$. □

In order for the function f to have correlation immunity of order larger than $d-1$, by the definition of correlation immunity and Lemmas 4.1 and 4.2, we need to make $g(y)$, or equivalently $f(x) = g(xG^T)$, to be correlation immune in every $z = \langle x, \gamma \rangle$ with $wt(\gamma) = d$. It is obvious that $rank[G^T, \gamma^T] = k$ if and only if γ is a code word of C_G, the linear code generated by G. By Lemma 4.2 we know that for those γ with Hamming weight d which are not code words of C_G, the function f is already correlation immune in $z = \langle x, \gamma \rangle$. So we have

Lemma 4.3. *Let G be a generating matrix of an $[n, k, d]$ linear code, and $f(x) = g(xG^T)$. Then f is correlation immune of order $\geq d$ if and only if for every $\alpha \in GF^k(2)$ with $wt(\alpha G) = d$, $g(y)$ is correlation immune in $z = \langle \alpha, y \rangle$.*

Proof. It can be proven by setting $\gamma = \alpha G$ and consequently we have $\langle \alpha, y \rangle = \langle x, \gamma \rangle$. By Lemma 4.1 the conclusion follows. □

By generalizing Lemma 4.3, we have

Theorem 4.2. *Let G be a generating matrix of an $[n, k, d]$ linear code, and $f(x) = g(xG^T)$. Then a necessary and sufficient condition for the function f to be correlation immune of order m is that for every $\alpha \in GF^k(2)$ with $d \leq wt(\alpha G) \leq m$, $g(y)$ is correlation immune in $z = \langle \alpha, y \rangle$.*

Corollary 4.1. *If the i-th row vector of G is a code word with nonzero minimum Hamming weight d and the function $g(y)$ is not correlation immune in y_i, then the correlation immunity of $f(x) = g(xG^T)$ is exactly $(d - 1)$.*

Now we consider the inverse question for general correlation immune functions. Given an m-th-order correlation immune function $f \in \mathcal{F}_n$, can it be written as $f(x) = g(xD)$, where $g \in \mathcal{F}_k$ is algebraic nondegenerate and D^T is a generating matrix of an $[n, k, d]$ linear code with $k \leq n$ and $d \geq 1$? The following theorem gives a positive answer. Furthermore, it can be shown that the code generated by D^T is unique.

Theorem 4.3. *Let $f(x) \in \mathcal{F}_n$. Then it can be written as $f(x) = g(xD)$, where $g \in \mathcal{F}_k$ is algebraic nondegenerate and D^T is a generating matrix of an $[n, k, d]$ linear code with $k \leq n$ and $d \geq 1$. Moreover, the linear code is unique for any given $f(x)$.*

Proof. From the discussion above, what we need to show is the uniqueness of the code. On the contrary we suppose $f(x) = g_1(xD_1) = g_2(xD_2)$, where $C_{D_1^T} \neq C_{D_2^T}$. Then there must exist a column α of D_1 which is linearly independent of the column vectors of D_2. Without loss of generality, let α be the first column of D_1. Then by Lemma 4.2 we know that $f(x)$ is independent of $\langle \alpha, x \rangle$, and equivalently $g_1(y)$ must be independent of y_1. This is in contradiction with the nondegeneracy assumption of $g(x)$. So the conclusion of Theorem 4.3 is true. □

4.5 Construction of Correlation Immune Boolean Functions

Since correlation immune functions have resistance against correlation attack, in order for the nonlinear combiner or the like to be secure against the correlation attack, the employed nonlinear Boolean function should have certain degree of correlation immunity. It should also be noted that it is misleading to strongly require a high order of correlation immunity of Boolean functions, because there should be a trade-off among the correlation immunity and other algebraic properties of cryptographic Boolean functions. For example, in the extreme case when a Boolean function in n variables is correlation immune of order $n - 1$, the function must be linear or affine due to the relationship between the correlation immunity and the algebraic degree. Nevertheless, the correlation immunity of Boolean functions has been emphasized in public literatures (see, e.g., [1, 15, 16, 35]) as an important security measure. Furthermore, correlation immunity may also lead to other interesting properties of Boolean functions which may be useful in cryptography. For example, in [6] it is shown that the nonlinear points of a correlation immune function form an orthogonal array [2], which is an important tool in the designing of message authentication schemes (see [4]). Therefore, how to construct correlation immune functions becomes practically important.

A necessary step to enable correlation immune functions to be practically useful is to have good methods to efficiently construct such functions. There are some constructions available from the public literatures. Here we explore the constructions based on the availability assumption of good linear error-correcting codes.

4.5.1 Known Constructions of Correlation Immune Boolean Functions

There have been alternative ways for constructing correlation immune functions (see, e.g., [5, 6, 10, 35, 36]). Some of them studied correlation immune functions over a general finite field (e.g., [5, 10]). This book only considers Boolean functions defined over the binary field $GF(2)$.

Lemma 4.4 ([36]). *Let $f_1(x), f_2(x) \in \mathcal{F}_n$ be two k-th-order correlation immune functions with $wt(f_1) = wt(f_2)$. Then*

$$f(x_1, \ldots, x_{n+1}) = x_{n+1} f_1(x) \oplus (1 \oplus x_{n+1}) f_2(x) \tag{4.17}$$

is a k-th-order correlation immune function with $wt(f) = 2wt(f_1)$.

Lemma 4.4 gives a construction of correlation immune functions based on known correlation immune functions in fewer number of variables. This construction works only when correlation immune functions in fewer number of variables are given.

Lemma 4.5 ([6]). *Let $f_1(x) \in \mathcal{F}_n$ be balanced. Write $\overline{x} = (x_1 \oplus 1, \ldots, x_n \oplus 1)$. Then*

1. *$f(x_1, \ldots, x_{n+1}) = f_1(x) \oplus x_{n+1}$ is a $(k+1)$-th-order correlation immune function in $\mathcal{F}_{n+1}$ if and only if $f_1(x)$ is a k-th-order correlation immune function of $\mathcal{F}_n$.*
2. *$f(x_1, \ldots, x_{n+1}) = f_1(x) \oplus x_{n+1}(f_1(x) \oplus f_1(\overline{x}))$ is a $(k+1)$-th-order correlation immune function in $\mathcal{F}_{n+1}$ if and only if $f_1(x)$ is a k-th-order correlation immune function of $\mathcal{F}_n$.*

The two constructions above are both based on known correlation immune functions. In [6] a more direct construction is proposed which can be described as follows:

Lemma 4.6 ([6]). *Let n_1, n_2, n be positive integers with $n_1 + n_2 = n$, $r(y), \phi_i(y) \in \mathcal{F}_{n_2}$, $i = 1, \ldots, n_1$. Let*

$$f(x;y) = \bigoplus_{i=1}^{n_1} x_i \phi_i(y) \oplus r(y). \tag{4.18}$$

Then $f(x;y)$ is a balanced Boolean function in $\mathcal{F}_n$ with correlation immunity of order

$$k \geq \min\{wt(\phi_1(y), \ldots, \phi_{n_1}(y)) : y \in GF^{n_2}(2)\}.$$

4.5.2 Construction of Correlation Immune Boolean Functions Based on A Single Code

Denote by e_i the vector in $GF^n(2)$ with 1 in its i-th coordinate and 0 elsewhere, and we have

Theorem 4.4. *Let $f(x)$ be a Boolean function in n variables and $g(y)$ a Boolean function in $k(k < n)$ variables, and let $f(x) = g(xD) = g(y)$, where D is a binary matrix of order $n \times k$ with $rank(D) = k$. Let $[D : e_i^T]$ be the conjunction by D and e_i^T, which is an $n \times (k+1)$ matrix. If*

$$rank[D : e_i^T] = k + 1, \tag{4.19}$$

where e_i^T is the transposed vector of e_i, then $f(x)$ is statistically independent of x_i.

Proof. Let $A = [D : e_i^T]$. Since $rank(A) = k + 1$, variables $xA = (y_1, y_2, \ldots, y_k, x_i)$ are independent ones. So for any $a, b \in GF(2)$, we have

$$\begin{aligned} &Prob(f(x) = b \mid x_i = a) \\ &= Prob(g(y) = b \mid x_i = a) \\ &= Prob(g(y) = b) \\ &= Prob(f(x) = b) \end{aligned}$$

This implies that $f(x)$ is statistically independent of x_i. □

Theorem 4.5. *Let D be an $n \times k$ binary matrix and rank(D)=k. If*

$$rank[D : e_{i_1}^T, e_{i_2}^T, \ldots, e_{i_m}^T] = k + m \tag{4.20}$$

holds for any indices $i_1, i_2, \ldots, i_m$ with $1 \leq i_1 < i_2 < \ldots < i_m \leq n$, then for any Boolean function $g(y)$ in k variables, $f(x) = g(xD)$ is a Boolean function in n variables which is correlation immune of order $\geq m$.

Proof. By the induction on Theorem 4.4, it is obvious. □

From Theorem 4.5, it can be seen that the problem of constructing correlation immune functions can be converted into the construction of matrices satisfying Eq. 4.20. Once such a matrix is obtained, we can construct 2^{2^k} (the number of Boolean functions in k variables) m-th-order correlation immune functions in n variables (including the two trivial functions $f = 0$ and $f = 1$). Denote by $[n, k, d]$ a linear error-correcting code with length n, dimension k, and minimum distance d. The following theorem yields a way to construct matrices satisfying Eq. 4.20.

Theorem 4.6. *Let D be an $n \times k$ binary matrix with rank(D)=k. Then Eq. 4.20 holds for D if and only if D^T is a generating matrix of an $[n, k, d]$ linear code, where $d > m$.*

Proof. Sufficiency: Denote $D = [d_1^T, d_2^T, \ldots, d_k^T]$, where d_i^T is a vector in $GF^n(2)$. Since D^T is a generating matrix of an $[n, k, d]$ linear code, $d_1, \ldots, d_k, e_{i_1}, \ldots, e_{i_m}$ must be linearly independent vectors. Assume the contrary, and so suppose there exist not-all-zero constants $c_1, c_2, \ldots, c_{k+m} \in GF(2)$ such that

$$c_1 d_1 \oplus \cdots \oplus c_k d_k \oplus c_{k+1} e_{i_1} \oplus \cdots \oplus c_{k+m} e_{i_m} = 0$$

or

$$c_1 d_1 \oplus \cdots \oplus c_k d_k = c_{k+1} e_{i_1} \oplus \cdots \oplus c_{k+m} e_{i_m}.$$

Since there is some c_i to be nonzero, each side of the above equation is a nonzero vector. However, the left side gives a vector with Hamming weight $d > m$ and the right side gives a vector with Hamming weight $\le m$. This is a contradiction.

Necessity: Let the matrix D satisfy Eq. 4.20, i.e., for any $1 \le i_1 < i_2 < \ldots < i_m \le n$, $d_1, \ldots, d_k$ together with $e_{i_1}, \ldots, e_{i_m}$ must be linearly independent vectors. So any linear combination of $d_1, \ldots, d_k$ must be a vector with Hamming weight $> m$ (otherwise it would be equal to a linear combination of some $e_{i_1}, \ldots, e_{i_m}$, contradiction). This proves that D^T is a generating matrix of a linear code with length n, dimension k, and minimum distance $d > m$. □

Corollary 4.2. *Let $g(y) \in \mathcal{F}_k$, D^T be a generating matrix of an $[n, k, d]$ linear code. Then $f(x) = g(xD) \in \mathcal{F}_n$ is a correlation immune function of order $\ge d - 1$.*

By Corollary 4.2 we can form an algorithm for constructing correlation immune functions.

Algorithm 4.1

---

(1) Choose a generating matrix G of an $[n, k, d]$ linear code.
(2) Choose a Boolean function $g(y)$ in k variables.
(3) Then $f(x) = g(xG^T)$ is a Boolean function in n variables with correlation immunity of order $\ge d - 1$.

4.5.3 Preliminary Enumeration of Correlation Immune Boolean Functions

From Algorithm 4.1 we know that given a generating matrix of an $[n, k, d]$ linear code, we can construct 2^{2^k} correlation immune functions in n variables with correlation immunity of order $\geq d-1$. In nonlinear combining generators, the combining functions are usually required to be balanced or have good property of being balanced. How many balanced correlation immune functions can be constructed by Algorithm 4.1? First we give

Lemma 4.7. *Let D be an $n \times k$ binary matrix with rank$(D) = k$, and let $g(y)$ a Boolean function in k variables. Then*

$$Prob(g(xD) = 1) = Prob(g(y) = 1). \tag{4.21}$$

Proof. This is true from the fact that $(y_1, \ldots, y_k) = (x_1, \ldots, x_n)D$ are k independent variables if and only if rank(D)=k. □

By Lemma 4.7 we know that the number of balanced correlation immune functions constructed by Algorithm 4.1 is just the number of balanced Boolean functions in k variables. So we have

Theorem 4.7. *If there exists an $[n, k, d]$ linear code. Then the number of balanced $(d-1)$-th-order correlation immune functions in n variables is at least*

$$\binom{2^k}{2^{k-1}} \approx \frac{1}{\sqrt{\pi}} 2^{2^k - \frac{k-1}{2}} \tag{4.22}$$

Proof. It is known that the number of balanced Boolean functions in k variables is just the left side of Eq. 4.22. From Corollary 4.2 and Lemma 4.7, we have the conclusion. □

4.5.4 Construction of Correlation Immune Boolean Functions Using a Family of Error-Correcting Codes

In order to construct more correlation immune functions, the possibility of using a set of error-correcting codes is explored in this section. Recall that in general case there are many different generating matrices for the same linear code. Theorem 4.8 gives the relationship between the correlation immune functions constructed by Algorithm 4.1 by using different generating matrices.

Theorem 4.8. *Let D be an $n \times k$ binary matrix and H be obtained from D by nonsingular transforms on its columns, i.e., there exists a $k \times k$ binary nonsingular* *matrix P such that $H = DP$. Denote by $S_D = \{g(xD) : g(y) \in \mathcal{F}_k\}$, $S_H = \{g(xH) : g(y) \in \mathcal{F}_k\}$. Then we have $S_D = S_H$.*

Proof. Denote $D = [d_1^T, d_2^T, \ldots, d_k^T]$, $H = [h_1^T, h_2^T, \ldots, h_k^T]$, $P = (p_{ij})_{k\times k}$. Since $H = DP$, we have $h_i^T = \bigoplus_{j=1}^k p_{ji} d_j^T$, $i = 1, 2, \ldots, k$. So for any Boolean function $g(y)$ in k variables, we have

$$\begin{aligned} g(xH) &= g(xh_1^T, xh_2^T, \ldots, xh_k^T) \\ &= g\left[\bigoplus_{j=1}^k p_{j1}(xd_j^T), \bigoplus_{j=1}^k p_{j2}(xd_j^T), \ldots, \bigoplus_{j=1}^k p_{jk}(xd_j^T)\right] \\ &= g_1(xd_1^T, xd_2^T, \ldots, xd_k^T) \\ &= g_1(xD) \end{aligned}$$

where g_1 is another Boolean function in k variables. This shows that $S_H \subseteq S_D$. Since $D = HP^{-1}$, it can be shown that $S_D \subseteq S_H$. So we have $S_D = S_H$. □

Theorem 4.8 shows that those correlation immune functions constructed by Algorithm 4.1 do not rely on the choice of the generating matrices of a linear code. If we apply a permutation to the generating matrix of some $[n, k, d]$ code, it results in an $[n, k, d]$ code. What is the relationship between the sets of correlation immune functions generated by Algorithm 4.1 based on those two different linear codes? More generally we have

Theorem 4.9. *Let C_1 and C_2 be $[n, k_1, d]$ and $[n, k_2, d]$ linear codes and G_1 and G_2 be their generating matrices, respectively. Let the dimension of the subcode $C_1 \cap C_2$ be $k(k \leq \min\{k_1, k_2\})$. Then*

$$\left|S_{G_1^T} \cap S_{G_2^T}\right| = 2^{2^k} \tag{4.23}$$

Proof. Let $G = \begin{bmatrix} \alpha_1 \\ \alpha_2 \\ \vdots \\ \alpha_k \end{bmatrix}$ be the generating matrix of $C_1 \cap C_2$. Then C_1 and C_2 must have generating matrices of the form

$$G_1 = \begin{bmatrix} \alpha_1 \\ \vdots \\ \alpha_k \\ \beta_1 \\ \vdots \\ \beta_{k_1-k} \end{bmatrix}, \quad G_2 = \begin{bmatrix} \alpha_1 \\ \vdots \\ \alpha_k \\ \gamma_1 \\ \vdots \\ \gamma_{k_2-k} \end{bmatrix}$$

respectively. It is easy to verify that for any Boolean function $f_1 \in \mathcal{F}_{k_1}$ and $f_2 \in \mathcal{F}_{k_2}$, $f_1(xG_1^T) \in S_{G_2^T}$ if and only if f_1 depends only on the first k variables, and $f_2(xG_2^T) \in S_{G_1^T}$ if and only if f_2 depends only on the first k variables. So the number of functions in $S_{G_1^T} \cap S_{G_2^T}$ is equal to the number of functions in k variables, i.e.,

$$\left|S_{G_1^T} \cap S_{G_2^T}\right| = 2^{2^k}.$$

□

The following two theorems can easily be derived from Theorem 4.9:

Theorem 4.10. *Let C_i be an $[n, k_i, d_i]$ linear code having a generating matrix G_i, i=1,2,.... If*

$$\max\{dim(C_i \cap C_j) : i \neq j\} = k. \tag{4.24}$$

Then any function in the form $f(xG_i^T)$ with degree $\geq k+1$ is not included in the set $\bigcup_{j \neq i} S_{G_j^T}$.

Theorem 4.11. *Let C_i be an $[n, k_i, d_i]$ linear code with generating matrix $G_i, i = 1, 2, \cdots$. If $dim[C_i \cap (\cup_{j \neq i} C_j)] = t$, then any function in the form $f(xG_i^T)$ with degree $\geq t+1$ should not be included in the set $\bigcup_{j \neq i} S_{G_j^T}$.*

Based on Theorems 4.10 and 4.11, we can develop another algorithm as follows which is the generalization of Algorithm 4.1.

Algorithm 4.2

(1) Choose $[n, k_i, d_i]$ linear codes C_i with generating matrices G_i, $i = 1, 2, \cdots$.
(2) Compute $k = max\{dim(C_i \bigcap C_j) : i \neq j\}$, $d = min\{d_i : i = 1, 2, \ldots\}$.
(3) Let $k_m = max\{k_i : i = 1, 2, \ldots\}$. Choose a Boolean function g_m in k_m variables, then $f(x) = g_m(xG_m^T)$ is a correlation immune function in n variables with correlation immunity of order $\geq d-1$.
(4) Execute the following two steps:

- For each $i \neq m$, choose a Boolean function g_i in k_i variables with degree $\geq k+1$, and then $f_i(x) = g_i(xG_i^T)$ is a correlation immune function in n variables with correlation immunity $\geq d-1$ and with degree $\geq k+1$.
- For each $i \neq m$, compute $t_i = dim[C_i \cap (\cup_{j \neq i} C_j)]$, $T_i = min\{k, t_i\}$, and choose a Boolean function g_i in k_i variables with degree $\geq T_i + 1$. Then $f_i(x) = g_i(xG_i^T)$ is a correlation immune function in n variables with correlation immunity of order $\geq d-1$ and with degree $\geq T_i + 1$.

It should be noted that functions generated by the two steps in step 4 may be in duplication, but they are not covered by the former three steps. Meanwhile, this algorithm is by no means optimum; it only gives a way to generate some (not the largest number) of the distinct correlation immune functions.

4.6 Lower Bounds on Enumeration of the Correlation Immune Functions Constructible from the Error-Correcting Code Construction

In the following discussion, we shall use symbol $\sigma_n(t)$ to denote the number of correlation immune functions in n variables with correlation immunity of order $\geq t$ and $\sigma'_n(t)$ to denote the number of such nontrivial functions. Now some lower bounds can be derived.

Theorem 4.12.

$$\sigma_n(1) \geq 2^{2^{n-1}} + 2 \sum_{i=1}^{\lfloor (n-1)/2 \rfloor} \binom{n}{2i+1} \tag{4.25}$$

where $\lfloor a \rfloor$ is the largest integer $\leq a$.

Proof. By Algorithm 4.2 we know that, with an $[n, n-1, 2]$ even weight linear code, there are $2^{2^{n-1}}$ Boolean functions in n variables with correlation immunity of odd order (including the trivial ones $f = 0$ and $f = 1$) which are constructible. For each vector of length n and Hamming weight odd (≥ 3), two more nontrivial functions with correlation immunity of order even can be constructed. Take all these cases into account, and we have the desired conclusion. □

Theorem 4.13. *If $m \geq n/2$, then*

$$\sigma'_n(m) \geq \sum_{k=m+1}^{n} \binom{n}{k} \tag{4.26}$$

Proof. Directly from Theorem 4.10. □

With the help of the following lemmas, we can discuss the case when $m < n/2$.

Lemma 4.8 ([40]). *If integers n, k, d satisfy*

$$V(n, d-1) < 2^{n-k+1}, \tag{4.27}$$

where $V(n, d-1) = \sum_{i=0}^{d-1} \binom{n}{i}$, then there must exist a binary $[n, k, d]$ linear code.

Lemma 4.9 (Estimation for a sum of binomial coefficients [22]). *For integers n, m, if $m < n/2$, then*

$$\sum_{i=0}^{m} \binom{n}{i} \leq 2^{nH_2(m/n)} \tag{4.28}$$

where $H_2(x) = -x\log_2 x - (1-x)\log_2(1-x)$.

By Lemmas 4.8 and 4.9, we have

Theorem 4.14. *For any integers m, n, if* $m < n/2$, *then we have*

$$\sigma_n(m) \geq 2^{2^n(\frac{m}{n})^m(1-\frac{m}{n})^{n-m}} \tag{4.29}$$

Proof. By Lemmas 4.8 and 4.9, if $2^{nH_2(m/n)} < 2^{n-k+1}$, i.e., $k < n[1-H_2(m/n)]+1$, then an $[n,k,m+1]$ binary linear code exists, and by Theorem 4.5, there are 2^{2^k} Boolean functions in n variables with correlation immunity of order $\geq m$ which are constructible. Let $k = \lceil n - nH_2(m/n)\rceil$, where $\lceil a\rceil$ denotes the smallest integer $\geq a$. Then we have

$$2^k \geq 2^n/2^{nH_2(m/n)} = 2^n(m/n)^m(1-m/n)^{n-m}$$

Then the conclusion of Theorem 4.14 follows. □

By the theory of error-correcting codes (e.g., [22]), we know that each irreducible polynomial of degree t over $GF(2^m)$ corresponds to a binary, irreducible Goppa code of length $n = 2^m$, dimension $k \geq n - tm$, and minimum distance $d \geq 2t+1$. The theory of finite fields (see [21]) also shows that the number of irreducible polynomials of degree t over $GF(2^m)$, I_t satisfies

$$I_t \geq \frac{n^t}{t}[1 - \frac{1}{n^{t/2-1}}] \tag{4.30}$$

So we have I_t irreducible Goppa codes of length $n = 2^m$, dimension $k \geq n - tm$, and minimum distance $d \geq 2t+1$. Let their generating matrices be $G_1, G_2, \ldots, G_{I_t}$, respectively. Then with G_1 we can construct 2^{2^k} Boolean functions in n variables with correlation immunity of order $\geq 2t$. By Theorem 4.9, with each $G_i, i = 2, 3, \ldots, I_t$, we can construct at least $2^{2^k}/2$ more correlation immune functions of degree k. This can be summarized as

Theorem 4.15. *Let* $n = 2^m$, *m an integer, and* $t < n/m$ *be an arbitrary integer. Then*

$$\sigma_n(2t) \geq \frac{I_t+1}{2}\cdot 2^{2^k} \geq \frac{n^t}{2t}(1 - \frac{1}{n^{t/2-1}})\cdot 2^{2^n/n^t} + 2^{2^n/n^t-1} \tag{4.31}$$

4.7 Examples

Here we list some examples to show the procedure of construction with corresponding enumeration of correlation immune functions by a family of error-correcting codes.

Example 1. For $n = 3$, with a generating matrix $\begin{bmatrix} 1 & 1 & 0 \\ 0 & 1 & 1 \end{bmatrix}$ of a [3,2,2] linear code, we can construct $2^{2^2} = 16$ first-order correlation immune functions. With a generating matrix (1 1 1) of [3,1,3] linear code, two more nontrivial second-order correlation immune functions can be constructed. These are all the correlation immune functions in three variables.

Example 2. For $n = 4$, with the matrix

$$\begin{bmatrix} 1\,1\,0\,0 \\ 0\,1\,1\,0 \\ 0\,0\,1\,1 \end{bmatrix},$$

one can construct $2^{2^3} = 256$ functions with correlation immunity of odd order. With the matrices (1 1 1 0), (1 1 0 1), (1 0 1 1), (0 1 1), there are $4 \times 2 = 8$ nontrivial functions with correlation immunity of order 2.

Example 3. For $n = 5$, with the generating matrix of a [5, 4, 2] linear code, we can construct $2^{2^4} = 65{,}536$ Boolean functions with correlation immunity of odd order. With matrices

$$\begin{bmatrix} 00111 \\ 11010 \end{bmatrix}, \begin{bmatrix} 01011 \\ 10101 \end{bmatrix}, \begin{bmatrix} 01101 \\ 10110 \end{bmatrix}, \begin{bmatrix} 01110 \\ 11001 \end{bmatrix}, \begin{bmatrix} 10011 \\ 11100 \end{bmatrix},$$

$(2^{2^2} - 4) \times 5 = 60$ nontrivial Boolean functions with correlation immunity of order ≥ 2 can be constructed. Note that the correlation immune functions in the form $g(xD)$ based on these matrices have been covered by the former ones if and only if $g(y_1, y_2) = y_1 \oplus y_2 \oplus c$ with c being 0 or 1. And with matrices

$$\begin{bmatrix} 00111 \\ 11001 \end{bmatrix}, \begin{bmatrix} 00111 \\ 11100 \end{bmatrix}, \begin{bmatrix} 01011 \\ 10110 \end{bmatrix}, \begin{bmatrix} 01011 \\ 11100 \end{bmatrix}, \begin{bmatrix} 01101 \\ 10011 \end{bmatrix},$$

$$\begin{bmatrix} 01101 \\ 11010 \end{bmatrix}, \begin{bmatrix} 01110 \\ 10011 \end{bmatrix}, \begin{bmatrix} 01110 \\ 10101 \end{bmatrix}, \begin{bmatrix} 10101 \\ 11010 \end{bmatrix}, \begin{bmatrix} 10110 \\ 11001 \end{bmatrix},$$

$2^{2^2-1} \times 10 = 80$ more functions of degree 2 and with correlation immunity of order ≥ 2 can be constructed. There are only 10 nontrivial functions with correlation immunity of order 3 (which can be generated by using the matrices above). There are only two nontrivial fourth-order correlation immune functions. The number of correlation immune functions which can be constructed by Algorithm 4.2 is shown in Table 4.1.

Table 4.1 Number of constructible correlation immune functions for $n = 5$

m	1	2	3	4	5
$\sigma_5(m)$	65,676	160	14	2	2

Example 4. For $n = 6$, with the generating matrix of a [6, 5, 2] even weight linear code, we can construct $2^{2^5} = 4{,}294{,}967{,}296$ correlation immune functions with odd order correlation immunity. With matrices

$$G_1 = \begin{bmatrix} 100011 \\ 101100 \\ 010110 \end{bmatrix}, \quad G_2 = \begin{bmatrix} 100110 \\ 111000 \\ 010101 \end{bmatrix}$$

we can construct $2^{2^3} = 256$ nontrivial correlation immune functions with correlation immunity of order ≥ 2. Among them $2^{3+1} = 16$ affine functions (including linear and trivial ones) have been covered by the former step. With matrices

$$G_3 = \begin{bmatrix} 101001 \\ 110100 \\ 001110 \end{bmatrix}, \quad G_4 = \begin{bmatrix} 101010 \\ 110001 \\ 000111 \end{bmatrix},$$

since

$$C_{G_3} \cap (C_{G_1} \cup C_{G_2}) = \{(000000), (111010)\},$$
$$C_{G_4} \cap (C_{G_1} \cup C_{G_2} \cup C_{G_3}) = \{(000000), (101101)\},$$

where C_G is the linear code with G as a generating matrix, then by Theorem 4.9, there are $(2^{2^3} - 4) \times 2 = 504$ more correlation immune functions of order ≥ 2 that are constructible. With matrices

$$\begin{bmatrix} 100011 \\ 101100 \\ 010101 \end{bmatrix}, \begin{bmatrix} 100011 \\ 110100 \\ 001101 \end{bmatrix}, \begin{bmatrix} 100011 \\ 110100 \\ 001110 \end{bmatrix}, \begin{bmatrix} 100011 \\ 111000 \\ 001101 \end{bmatrix},$$

$$\begin{bmatrix} 100011 \\ 111000 \\ 001110 \end{bmatrix}, \begin{bmatrix} 100101 \\ 101010 \\ 010011 \end{bmatrix}, \begin{bmatrix} 100101 \\ 101010 \\ 010110 \end{bmatrix}, \begin{bmatrix} 100101 \\ 110010 \\ 001110 \end{bmatrix},$$

$$\begin{bmatrix} 100101 \\ 110010 \\ 011100 \end{bmatrix}, \begin{bmatrix} 100101 \\ 111000 \\ 001110 \end{bmatrix}, \begin{bmatrix} 100101 \\ 111000 \\ 010110 \end{bmatrix}, \begin{bmatrix} 100110 \\ 101001 \\ 010011 \end{bmatrix},$$

$$\begin{bmatrix} 100110 \\ 101001 \\ 010101 \end{bmatrix}, \begin{bmatrix} 100110 \\ 110001 \\ 001101 \end{bmatrix}, \begin{bmatrix} 100110 \\ 110001 \\ 011100 \end{bmatrix}, \begin{bmatrix} 100110 \\ 111000 \\ 001101 \end{bmatrix},$$

$$\begin{bmatrix} 101001 \\ 110010 \\ 000111 \end{bmatrix}, \begin{bmatrix} 101001 \\ 110010 \\ 001110 \end{bmatrix}, \begin{bmatrix} 101001 \\ 110100 \\ 000111 \end{bmatrix}, \begin{bmatrix} 101010 \\ 110001 \\ 001101 \end{bmatrix},$$

Table 4.2 Number of constructible correlation immune functions for $n = 6$

m	1	2	3	4	5	6
$\sigma_6(m)$	4,294,968,172	880	136	16	4	2

$$\begin{bmatrix}101010\\110100\\000111\end{bmatrix}, \begin{bmatrix}101010\\110100\\001101\end{bmatrix}, \begin{bmatrix}101100\\110001\\000111\end{bmatrix}, \begin{bmatrix}101100\\110001\\010110\end{bmatrix},$$

$$\begin{bmatrix}101100\\110010\\000111\end{bmatrix}, \begin{bmatrix}101100\\110010\\010101\end{bmatrix},$$

we can construct $2^{2^3-1} \times 26 = 3328$ correlation immune functions of degree 3 and with correlation immunity of order ≥ 2. For the construction of correlation immune functions of order ≥ 3, we can use the following matrices, which can yield $2^{2^2-1} \times 15 = 120$ functions of degree 2 (it can easily be verified that they yield functions with correlation immunity of order exactly 3):

$$\begin{bmatrix}001111\\110011\end{bmatrix}, \begin{bmatrix}011011\\101101\end{bmatrix}, \begin{bmatrix}011101\\100111\end{bmatrix}, \begin{bmatrix}010111\\101110\end{bmatrix}, \begin{bmatrix}011110\\101011\end{bmatrix},$$

$$\begin{bmatrix}001111\\110100\end{bmatrix}, \begin{bmatrix}001111\\110110\end{bmatrix}, \begin{bmatrix}010111\\101011\end{bmatrix}, \begin{bmatrix}010111\\101101\end{bmatrix}, \begin{bmatrix}011011\\100111\end{bmatrix},$$

$$\begin{bmatrix}011011\\101110\end{bmatrix}, \begin{bmatrix}011101\\101011\end{bmatrix}, \begin{bmatrix}011101\\101110\end{bmatrix}, \begin{bmatrix}011110\\100111\end{bmatrix}, \begin{bmatrix}011110\\101101\end{bmatrix}.$$

The number of nontrivial fourth-order correlation immune functions is 12, which can be constructed based on the following matrices and have not been covered by the above procedures: $(011111), (101111), (110111), (111011), (111101), (111110)$. Only two nontrivial correlation immune functions can be constructed based on the matrix (111111). Table 4.2 shows the number of such functions for $n = 6$.

4.8 Exhaustive Construction of Correlation Immune Boolean Functions

Theoretically by using Theorems 4.1 and 4.2, the complete set of correlation immune functions can be constructed. By applying Theorem 4.2, we can see when the correlation immunity is larger than or equal to the minimum distance of the code. In order to do this, we need to construct Boolean functions which are correlation immune in some of their variables and/or their linear combinations. Denote by $\hat{x}_i = (x_1, \ldots, x_{i-1}, x_{i+1}, \ldots, x_n)$. Then we have

Lemma 4.10. *Let* $f(x) = x_i f_1(\hat{x}_i) \oplus f_2(\hat{x}_i)$. *Then* $f(x)$ *is correlation immune in* x_i *if and only if*

$$wt(f_1 \oplus f_2) = wt(f_2). \tag{4.32}$$

Proof. By writing $f(x) = x_i(f_1(\hat{x}_i) \oplus f_2(\hat{x}_i)) \oplus (1 \oplus x_i)f_2(\hat{x}_i)$, it can be seen that $f(x)$ is correlation immune in x_i if and only if $wt(f_1 \oplus f_2) = wt(f_2) = \frac{1}{2}wt(f)$. □

Lemma 4.11. *Let* $f(x) \in \mathcal{F}_n$. *Then* $deg(f) < n$ *if and only if the Hamming weight of* $f(x)$ *is an even number.*

In [36] it was shown that if $f(x) \in \mathcal{F}_n$ is correlation immune (of order ≥ 1), then $deg(f) \leq n - 1$. We further prove that

Lemma 4.12. *Let* $f(x) \in \mathcal{F}_n$. *If* $deg(f) = n$ *then* $f(x)$ *is not correlation immune in any linear combination of its variables.*

Proof. Assume the contrary, $f(x)$ is correlation immune in $\langle \alpha, x \rangle$, and without loss of generality the first coordinate of α is assumed to be not zero. Denote by δ_i the vector in $GF^n(2)$ with i consecutive ones followed by zeros. Let $D = [\alpha^T, \delta_2^T, \ldots, \delta_n^T]$. Then $g(x) = f(xD^{-1})$ is correlation immune in x_1 and hence can be written as $g(x) = x_1 g_1(\hat{x_1}) \oplus g_2(\hat{x_1})$. By Lemma 4.10, we know that

$$\begin{aligned} wt(g_1) &= wt((g_1 \oplus g_2) \oplus g_2) \\ &= wt(g_1 \oplus g_2) + wt(g_2) - 2wt((g_1 \oplus g_2) \cdot g_2) \\ &= 2wt(g_2) - 2wt((g_1 \oplus g_2) \cdot g_2) \end{aligned}$$

is an even number, and by Lemma 4.11 we have $deg(f) = deg(g) = deg(g_1) + 1 < (n-1) + 1 = n$. This is a contradiction. So the conclusion of Lemma 4.12 must be true. □

Let $f(x) = g(xG^T)$ be a Boolean function in $\mathcal{F}_n$, where g is algebraically nondegenerate, and G is a generating matrix of an $[n, k, d]$ linear code. It is easy to see that by a linear transform on the rows of G, we can always make the row vectors of G satisfy

$$wt(g_1) \leq wt(g_2) \leq \cdots \leq wt(g_k),$$

and there does not exist another basis $\beta_1, \beta_2, \ldots, \beta_k$ of C_G with $wt(\beta_1) \leq wt(\beta_2) \leq \cdots \leq wt(\beta_k)$ such that $wt(\beta_i) < wt(g_i)$ for some $1 \leq i \leq k$. Constructions can always be based on this assumption. Such a matrix will be called a *minimum weight generating matrix.*

It is noticed that under a permutation on the variables of a Boolean function, the correlation immunity of the function is an invariant. To simplify the problem, we will treat two correlation immune functions as equivalent if they are equivalent by a permutation on the variables. For the function $f(x) = g(xG^T)$, a permutation on x is equivalent to the same permutation on the column vectors of G. Complements of correlation immune functions can be left out in the early steps and then added at last. So the exhaustive construction can be outlined as follows:

Algorithm 4.3 For all integers $k \in \{1, 2, \ldots, n\}$, conduct the following steps:

(1) Search the minimum weight generating matrices G_i, $i \in I$, of $[n, k]$ codes such that they are not column equivalent, where I is the set of complete index.
(2) List all nontrivial Boolean functions $g(y) \in \mathcal{F}_k$ such that $g(\mathbf{0}) = 0$.
(3) Match each $g(y)$ with every G_i to see if $f_i(x) = g(xG_i^T)$ is correlation immune of any order according to Theorem 4.2.
(4) For those $f_i(x)$ with a certain order of correlation immunity, permute their variables to get an equivalent class of correlation immune functions.
(5) Complement every correlation immune function obtained above.

Theoretically the above steps can exhaustively generate all the correlation immune functions. However, because of the large number of correlation immune functions in n variables when n is sufficiently large, it is not surprising to see that the above steps are not practically efficient in terms of computational complexity (such as step 3). So more efficient constructions of specific correlation immune functions are still required.

4.9 An Example of Exhaustive Construction of Correlation Immune Functions

It is not surprising that to accomplish an exhaustive construction of correlation immune functions in n variables is not practical when n is fairly large, even if the method described in Sect. 4.5 is used. However, as an interesting practice, we show here a small example of how all the correlation immune functions are constructed.

We consider the correlation immunity of Boolean functions in $n = 4$ variables. All correlation immune functions will be presented by means of representatives, i.e., their complements and/or variable-permutation equivalences. First of all we know that

$$f(x_1, x_2, x_3, x_4) = c_1x_1 \oplus c_2x_2 \oplus c_3x_3 \oplus c_4x_4$$

is correlation immune of order $wt(\gamma) - 1$ if $\gamma = (c_1, c_2, c_3, c_4) \neq \mathbf{0}$ or 4 if $\gamma = \mathbf{0}$. Then we consider functions in the form $g(xG^T)$, where g is an algebraic nondegenerate Boolean function in two variables and G is a generating matrix of [4, 2] code. Since the function $g(y)$ has only two variables, it is easy to see that g is algebraically nondegenerate if and only if $deg(g) = 2$, and by Lemma 4.12 such a function is not correlation immune in any linear combination of its variables. All possible representatives of such functions are as follows:

$$y_1y_2,$$
$$y_1y_2 \oplus y_1,$$
$$y_1y_2 \oplus y_2,$$
$$y_1y_2 \oplus y_1 \oplus y_2.$$

In order for the constructed function to be correlation immune of order at least one, the only possible codes useful are [4, 2, 2] codes. Recall that a permutation on the column vectors of matrix G is equivalent to the same permutation performed on the variables of the constructed correlation immune functions. So under column permutation equivalence, we have three different linear codes with matrices

$$\begin{bmatrix} 1\,1\,0\,0 \\ 1\,0\,1\,0 \end{bmatrix}, \begin{bmatrix} 1\,1\,0\,0 \\ 0\,0\,1\,1 \end{bmatrix}, \begin{bmatrix} 1\,1\,0\,0 \\ 1\,0\,1\,1 \end{bmatrix}.$$

By Corollary 4.1 we know that all the constructed functions (with 12 representatives) are exactly first-order correlation immune. All these functions also have the properties that *algebraic degree* = *2, nonlinearity* = *4, number of invariant linear structures* = *4, number of complementary linear structures* = *0.*

Now we consider algebraic nondegenerate functions in three variables and the family of [4, 3] linear codes. It is known that there are totally $2^{2^3} = 256$ Boolean functions in three variables. Among them half are of degree 3 which are algebraically nondegenerate according to Corollary 2.1 (they are useless in constructing correlation immune functions according to Corollary 4.1 because every [4, 3] linear code has a code word with Hamming weight one), and $2^{3+1} = 16$ are affine ones. So only 112 functions are of degree 2 with half being complements of the other. It can be verified that those algebraically degenerate functions can always be written as $y_1y_2, y_1y_2 \oplus y_1, y_1y_2 \oplus y_2$, and $y_1y_2 \oplus y_1 \oplus y_2$ and their complements. When y_1 and y_2 are as follows (order is ignored):

$$\begin{cases} y_1 = x_1 \oplus x_2 \\ y_2 = x_3 \end{cases}, \begin{cases} y_1 = x_1 \oplus x_3 \\ y_2 = x_2 \end{cases}, \begin{cases} y_1 = x_2 \oplus x_3 \\ y_2 = x_1 \end{cases}, \begin{cases} y_1 = x_1 \oplus x_2 \\ y_2 = x_2 \oplus x_3 \end{cases},$$

they form 16 algebraically degenerate functions of degree 2. When $y_1 = 1$ while y_2 is any Boolean function in two variables from x_1, x_2, x_3 with degree 2, y_1y_2 has 12 different forms. Altogether we have 28 algebraically degenerate functions of degree 2 and with constant term 0. So there are 28 algebraically nondegenerate Boolean functions of degree 2 which have constant term 0, namely,

$$\begin{aligned}
&x_1x_2 \oplus \{x_3,\ x_1 \oplus x_3,\ x_2 \oplus x_3,\ x_1 \oplus x_2 \oplus x_3\},\\
&x_1x_3 \oplus \{x_2,\ x_1 \oplus x_2,\ x_2 \oplus x_3,\ x_1 \oplus x_2 \oplus x_3\},\\
&x_2x_3 \oplus \{x_1,\ x_1 \oplus x_2,\ x_1 \oplus x_3,\ x_1 \oplus x_2 \oplus x_3\},\\
&x_1x_2 \oplus x_1x_3 \oplus \{x_2,\ x_3,\ x_1 \oplus x_2,\ x_1 \oplus x_3\},\\
&x_1x_2 \oplus x_2x_3 \oplus \{x_1,\ x_3,\ x_1 \oplus x_2,\ x_2 \oplus x_3\},\\
&x_1x_3 \oplus x_2x_3 \oplus \{x_1,\ x_2,\ x_1 \oplus x_3,\ x_2 \oplus x_3\},\\
&x_1x_2 \oplus x_1x_3 \oplus x_2x_3 \oplus \{0,\ x_1 \oplus x_2,\ x_1 \oplus x_3,\ x_2 \oplus x_3\}
\end{aligned}$$

It is easy to verify that no function above is correlation immune. So by Theorem 4.2, in order for the function $g(xG^T)$ to be correlation immune, there are at most two linearly independent code words with Hamming weight one. Therefore, only the following minimum weight generating matrices of [4, 3] linear codes need to be considered (without being column permutation equivalent):

$$G_1 = \begin{bmatrix} 1\,1\,0\,0 \\ 0\,1\,1\,0 \\ 0\,0\,1\,1 \end{bmatrix},\ G_2 = \begin{bmatrix} 1\,0\,0\,0 \\ 0\,1\,1\,0 \\ 0\,0\,1\,1 \end{bmatrix} \text{ and } G_3 = \begin{bmatrix} 1\,0\,0\,0 \\ 0\,1\,0\,0 \\ 0\,0\,1\,1 \end{bmatrix}.$$

Matching the 28 functions above with G_1, we can construct 28 first-order correlation immune functions. These functions are actually constructed based on Theorem 4.1 and have been discussed in [42]. By Theorem 4.2, if $g(y)$ is correlation immune in y_1, then $g(xG_1^T)$ is correlation immune of order ≥ 1. Among the above algebraically nondegenerate functions, only the following ones are correlation immune in x_1:

$$\begin{array}{c} x_1x_2 \oplus \{x_3,\ x_1 \oplus x_3,\ x_2 \oplus x_3,\ x_1 \oplus x_2 \oplus x_3\} \\ x_1x_3 \oplus \{x_2,\ x_1 \oplus x_2,\ x_2 \oplus x_3,\ x_1 \oplus x_2 \oplus x_3\} \\ x_1x_2 \oplus x_1x_3 \oplus \{x_2,\ x_3,\ x_1 \oplus x_2,\ x_1 \oplus x_3\}. \end{array}$$

Matching them with G_2, we can generate 12 first-order correlation immune functions in four variables. By permutations on the variables, more correlation immune functions can be generated. Note that all these functions are not constructible by the methods in [42].

It can also be verified that functions

$$x_1x_2 \oplus \{x_3,\ x_1 \oplus x_3,\ x_2 \oplus x_3,\ x_1 \oplus x_2 \oplus x_3\}$$

are also correlation immune in x_2 as well. Matching with G_3, we can get four more first-order correlation immune functions in four variables which are not constructible by the methods in [42] either. In addition, all of the above-constructed functions also have the properties that *algebraic degree* = *2, nonlinearity* = *4, number of invariant linear structures* = *2, and number of complementary linear structures* = *2.*

By computing search, we found that there are 192 functions in $\mathcal{F}_4$ which are algebraically nondegenerate and with first-order correlation immunity. They also have the properties that *algebraic degree* = *3, nonlinearity* = *4, number of invariant linear structures* = *1, and number of complementary linear structures* = *0, and propagation criterion order* = *0.* Among them 96 are listed below by truth table expression, and the other 96 are just the complements of those in the list.

0001011010011000 0001011010100100 0001011011000010
0001100101101000 0001100110100100 0001100111000010
0001101001100100 0001101010010100 0001101011000001
0001110001100010 0001110010010010 0001110010100001
0010010101101000 0010010110011000 0010010111000010
0010011001011000 0010011010010100 0010011011000001
0010100101011000 0010100101100100 0010100111000001

0010110001010010 0010110001100001 0010110010010001
0011010001001010 0011010010000110 0011010010001001
0011100001000110 0011100001001001 0011100010000101
0011110111011010 0011110111100110 0011110111101001
0011111011010110 0011111011011001 0011111011100101
0100001101101000 0100001110011000 0100001110100100
0100011000111000 0100011010010010 0100011010100001
0100100100111000 0100100101100010 0100100110100001
0100101000110100 0100101001100001 0100101010010001
0101001000101100 0101001010000110 0101001010001001
0101100000100110 0101100000101001 0101100010000011
0101101110111100 0101101111100110 0101101111101001
0101111010110110 0101111010111001 0101111011100011
0110000100101100 0110000101001010 0110000110001001
0110001000011100 0110001001001001 0110001010000101
0110010000011010 0110010000101001 0110010010000011
0110011110111100 0110011111011010 0110011111101001
0110100000011001 0110100000100101 0110100001000011
0110101101111100 0110101111011001 0110101111100101
0110110101111010 0110110110111001 0110110111100011
0110111001111001 0110111010110101 0110111011010011
0111011010011110 0111011010101101 0111011011001011
0111100101101110 0111100110101101 0111100111001011
0111101001101101 0111101010011101 0111101011000111
0111110001101011 0111110010011011 0111110010100111

All the correlation immune functions in four variables can be obtained by a permutation on the variables and by counting the complements of the above-constructed functions.

4.10 Construction of High-Order Correlation Immune Boolean Functions

From the above, every correlation immune function can be written as $g(xD)$, where g is an algebraic nondegenerate function and D^T is a minimum weight generating matrix of an $[n, k, d]$ linear code. In this section, we will concentrate mainly on the construction of those functions whose correlation immunity is no less than d.

For any Boolean function $f(x) \in \mathcal{F}_n$, set

$$\Delta_f = \{w \in GF^n(2) : \ f(x) \text{ is correlation immune in } \langle w, x\rangle\}. \tag{4.33}$$

Then by Theorem 4.2, we have

Theorem 4.16. *Let $g(y) \in \mathcal{F}_k$ and G be a generating matrix of an $[n, k, d]$ linear code. Set $f(x) = g(xG^T)$. Then the correlation immunity of $f(x)$ is*

$$\min_{\alpha \notin \Delta_g} wt(\alpha G) - 1. \tag{4.34}$$

Moreover we have

$$AD(f) = n - k + AD(g). \tag{4.35}$$

where $AD(f)$ is the degeneracy of $f(x)$ as defined in Definition 2.4.

Proof. The former part (Eq. 4.34) comes directly from Theorem 4.2. So we only need to prove the latter part. Assume $AD(g) = t$, i.e., there exists an algebraic nondegenerate function $g_1 \in \mathcal{F}_{k-t}$ and a $k \times (k-t)$ matrix D such that $g(y) = g_1(yD)$. So $f(x) = g_1(xG^T D)$, and $AD(f) \geq n - (k - t) = n - k + AD(g)$.

On the other hand, since $rank(G) = k$, we can assume, without loss of generality, that the first k columns of G are linearly independent, and we write $G = [G_1; G_2]$. Then $g(y) = f(yG_1^{-1}, 0, \cdots, 0)$. This means that if f can be algebraically degenerated to a function in r variables, then g can be algebraically degenerated to a function in no more than r variables, i.e., $k - AD(g) \leq n - AD(f)$ or $AD(f) \leq n - k + AD(g)$.

In light of the above discussion, the conclusion follows. □

In order to determine Δ_f for a general Boolean function $f(x) \in \mathcal{F}_n$, we have

Theorem 4.17. *Let $f(x) \in \mathcal{F}_n$ and $w \in GF^n(2)$. Then $w \in \Delta_f$ if and only if*

$$S_f(w) = 0. \tag{4.36}$$

Proof. It is easy to see that

$$\begin{aligned}
w \in \Delta_f &\Longleftrightarrow f(x) \text{ is correlation immune in } \langle w, x \rangle \\
&\Longleftrightarrow Prob(f(x) = 1 | \langle w, x \rangle = 0) = Prob(f(x) = 1 | \langle w, x \rangle = 1) \\
&\Longleftrightarrow \sum_{\langle w, x \rangle = 0} f(x) - \sum_{\langle w, x \rangle = 1} f(x) = 0 \\
&\Longleftrightarrow S_f(w) = \sum_x f(x)(-1)^{\langle w, x \rangle} = \sum_{\langle w, x \rangle = 0} f(x) - \sum_{\langle w, x \rangle = 1} f(x) = 0.
\end{aligned}$$

Hence, the conclusion of the theorem holds. □

By Theorem 4.17, Eq. 4.34 can be rewritten as

$$\min_{\alpha:\ S_g(\alpha) \neq 0} wt(\alpha G) - 1. \tag{4.37}$$

It is seen that using the techniques of Walsh transforms, the correlation immunity of $f(x) = g(xG^T)$ can easily be determined by Eq. 4.37.

Note that $g(y)$ can always be chosen as algebraically nondegenerate which enables us to construct correlation immune functions with least possible algebraic degeneration. When we use Theorem 4.16 to construct correlation immune functions, it is noticed that an $[n, k, d]$ linear code normally has several code words of Hamming weight d. So in general it is hard to find a Boolean function which can match a generating matrix of this linear code to generate correlation immune functions of order $\geq d$. However, it is easy to find Boolean functions which are correlation immune in part of their variables and their linear combinations as shown in the following:

Corollary 4.3. *Let $g(y) \in \mathcal{F}_k$ be correlation immune in its first t variables and their nonzero linear combinations. Let G be a generating matrix of an $[n-t, k-t, d]$ linear code. Then the correlation immunity of function $f(x) = g(x\hat{G}^T)$ is at least $d-1$, where*

$$\hat{G} = \begin{bmatrix} D & 0 \\ 0 & G \end{bmatrix},$$

and D is an arbitrary nonsingular binary matrix of order $t \times t$.

We note that when Corollary 4.3 is used to construct correlation immune functions, the size of D is normally small. For special cases, we have

Corollary 4.4. *If G is a generating matrix of an $[n, k, d]$ linear code and the row vectors of G include all the code words with Hamming weight d, then for any algebraic nondegenerate Boolean function $g(y)$ in k variables with correlation immunity of order t, $f(x) = g(xG^T)$ is a correlation immune function of order $t+1$.*

4.11 Construction of Correlation Immune Boolean Functions with Other Cryptographic Properties

The correlation immunity, being another cryptographic requirement, has some conflict with the algebraic degree. Siegenthaler [36] has proved the relationship between the correlation immunity m and the algebraic degree d of a Boolean function, which says

Theorem 4.18 ([36]). *Let $f(x) \in \mathcal{F}_n$ has algebraic degree d and correlation immunity m. Then*

$$m + d \leq n. \tag{4.38}$$

Furthermore, if $f(x)$ is balanced, then

$$m + d \leq n - 1. \tag{4.39}$$

One remarkable result on the spectral description of correlation immunity of Boolean functions is in [44], which says:

Lemma 4.13 (Xiao-Massey). *Let $f(x) \in \mathcal{F}_n$. Then $f(x)$ is correlation immune of order k if and only if $S_f(w) = 0$ for every w with $1 \leq wt(w) \leq k$.*

It is noted that the Xiao-Massey theorem is a direct corollary of Theorem 2.22.

Since correlation immunity is not the only cryptographic measure, and other cryptographic requirements may have conflicts with the correlation immunity, so when constructing Boolean functions with multiple cryptographic properties, one or more of the properties have to sacrifice to certain degree. From the discussion above, we see that correlation immune functions usually have a good structure, so we will use this structure as a basis to add on more other cryptographic properties.

4.11.1 Correlation Immune Functions with Good Balance

From the viewpoint of cryptographic applications, we aim to construct correlation immune functions with as good balance as possible. Define the *bias* of a Boolean function $f(x) \in \mathcal{F}_n$ to be

$$\delta(f) = \sum_{x=0}^{2^n-1} (-1)^{f(x)}.$$

Then the bias of correlation immune functions given in the form $f(x) = g(xG^T)$ can easily be controlled by choosing $g(y)$ to have a good balance property.

Lemma 4.14. *Let $f(x) = g(xD)$, where g is an algebraically nondegenerate Boolean function in k variables and D^T is a generating matrix of an $[n, k, d]$ linear code. Then we have*

$$\delta(g) = \delta(f),$$

In particular, $f(x)$ is balanced (when $\delta(f) = 0$) if and only if $g(y)$ is balanced as well.

Proof. Denote by $KerD = \{x : xD = 0\}$. For any $y \in GF^k(2)$, since $rank(D) = k$, there must exist an $x \in GF^n(2)$ such that $y = xD$. So $x + KerD$ is the set of all solutions of equation $xD = y$. This means that when there exists an y such that $g(y) = 1$, there will exist 2^{n-k} vectors x such that $xD = y$ and $f(x) = 1$. So we have that $wt(f) = 2^{n-k} \cdot wt(g)$. By the definition of bias, the conclusion of Lemma 4.14 holds. □

4.11.2 Correlation Immune Functions with High Algebraic Degree

Algebraic degree is one criterion to measure the nonlinearity of Boolean functions. In practical applications, a correlation immune function is required to have as high algebraic degree as possible. Otherwise there may be a risk in decreasing its security when the low-order approximation technique [31] is applied.

Siegenthaler proved in [36] that for any Boolean function $f \in \mathcal{F}_n$ which is correlation immune of order m, its algebraic degree $deg(f)$ satisfies $m + deg(f) \leq n$. With the theory of error-correcting codes, we have the following conclusion:

Theorem 4.19. *If there exists an $[n, k, d]$ maximum distance separable (MDS) code, then the number of $(d-1)$-th-order correlation immune functions satisfying $(d-1) + deg(f) = n$ is at least 2^{2^k-1}.*

In order to prove Theorem 4.19, the following lemmas are needed.

Lemma 4.15. *Let $f(x) \in \mathcal{F}_n$ and A be an $n \times n$ invertible binary matrix. Then $deg(f(xA)) = deg(f(x))$.*

Proof. Denote $f_1(x) = f(xA)$. It is obvious that the expansion of $f(xA)$ does not generate a term with degree $> deg(f(x))$, so we have $deg(f_1(x)) \leq deg(f(x))$. On the other hand, from the invertibility of A, we have $f(x) = f_1(xA^{-1})$ and hence $deg(f(x)) \leq deg(f_1(x))$. Therefore, $deg(f_1(x)) = deg(f(x))$. □

Lemma 4.16. *Let D be an $n \times k$ $(k \leq n)$ binary matrix and let $f(x) = g(xD)$, where $g \in \mathcal{F}_k$. Then $deg(f) = deg(g)$ holds for any g if and only if $rank(D) = k$.*

Proof. By row transformation, matrix D can be written as

$$D = A \begin{pmatrix} I_r & 0 \\ 0 & 0 \end{pmatrix} P$$

where A is an $n \times n$ invertible matrix, I_r is an $r \times r$ $(r \leq k)$ identity matrix, and P is a $k \times k$ permutation matrix. Then

$$f(x) = g(xD) = g(xA \begin{pmatrix} I_r & 0 \\ 0 & 0 \end{pmatrix} P)$$

Denote $f_1(x) = f(xA^{-1})$, $g_1(y) = g(yP)$, where $x \in GF^n(2)$ and $y \in GF^k(2)$. Then

$$\begin{aligned} f_1(x) &= f(xA^{-1}) \\ &= g(xA^{-1}D) \\ &= g(x \begin{pmatrix} I_r & 0 \\ 0 & 0 \end{pmatrix} P) \end{aligned}$$

$$= g_1(x \begin{pmatrix} I_r & 0 \\ 0 & 0 \end{pmatrix})$$

$$= g_1(x_1, \ldots, x_r, 0, \ldots, 0).$$

From the equation above, it can be seen that

$$deg(f_1) = deg(g_1(x_1, \ldots, x_r, 0, \ldots, 0)) = deg(g_1(y))$$

holds for any $g_1(y) \in \mathcal{F}_k$ if and only if $r = k$, i.e., if and only if $rank(D) = k$. Notice that by Lemma 4.15, we have $deg(g_1) = deg(g)$ and $deg(f_1) = deg(f)$. So we get that $deg(f) = deg(g)$ holds for any $g(y) \in \mathcal{F}_k$ if and only if $rank(D) = k$. □

From Lemma 4.16 we see that the maximum algebraic degree of the function written as $f(x) = g(xD)$ is k. In this case by Corollary 4.1 and Lemma 4.12, the correlation immunity of $f(x)$ is exactly $d - 1$, where D^T is the generating matrix of an $[n, k, d]$ linear code. This is consistent with Siegenthaler's inequality [36]. The discussion above also shows that we can construct correlation immune functions which meet the equality (maximum correlation immunity/algebraic degree) of Siegenthaler's inequality.

Proof of Theorem 4.19: Given an $[n, k, d]$ MDS code with generating matrix G, we have $d = n - k + 1$ or $k + d - 1 = n$. For any Boolean function in k variables, by Algorithm 4.1, $f(x) = g(xG^T)$ is a Boolean function in n variables with correlation immunity at least $d - 1$. By Lemma 4.16, we have $deg(f) = deg(g)$, and it is easy to verify that $g_1(xG^T) \neq g_2(xG^T)$ if $g_1 \neq g_2$, so the number of Boolean functions in k variables with degree k is no more than the number of Boolean functions in n variables of degree k and with correlation immunity at least $d - 1$. It is easy to verify that the number of Boolean functions in k variables with degree k is 2^{2^k-1}, so the conclusion of Theorem 4.19 holds. □

4.11.3 Correlation Immune Functions with High Nonlinearity

Nonlinearity of Boolean functions is a measure of the distance of Boolean functions to the nearest affine ones [29]. If the nonlinearity of a Boolean function is very low, then it can be approximated by an affine Boolean function with high correlation with the affine function [14] and hence is cryptographically insecure. By the Walsh spectrum representation of nonlinearity as in Eq. 3.1 and the conversion between the two types of Walsh transforms, it is easy to deduce that

Lemma 4.17.

$$nl(f) = \min\{wt(f),\ 2^n - wt(f),\ 2^{n-1} - \max\{|S_f(\omega)| : \omega \neq 0\}. \tag{4.40}$$

Lemma 4.18. *Let $f(x) = g(xG^T)$, where g is an algebraic nondegenerate Boolean function in k variables and G is a generating matrix of an $[n, k, d]$ linear code. Then*

$$nl(f) \leq 2^{n-k} nl(g).$$

Proof. By the definition of nonlinearity, there exists an affine function $l(y)$ in k variables such that $wt(g(y) \oplus l(y)) = nl(g)$. Hence, we have $wt(g(xG^T) \oplus l(xG^T)) = 2^{n-k} nl(g)$ and again by the definition we have $nl(f) \leq 2^{n-k} nl(g)$. □

Furthermore, we can prove

Theorem 4.20. *Let D be an $n \times k$ $(k \leq n)$ binary matrix. Then $rank(D) = k$ if and only if for any Boolean function $g(y) \in \mathcal{F}_k$ and $f(x) = g(xD)$, and we have*

$$nl(f) = 2^{n-k} nl(g). \tag{4.41}$$

In order to prove Theorem 4.20, the following lemma will be used.

Lemma 4.19. *Let $D = \begin{bmatrix} D_1 \\ 0 \end{bmatrix}$ be an $n \times k$ binary matrix, where D_1 is a $k \times k$ nonsingular matrix. Let $f(x) = g(xD)$. Then we have*

$$nl(f) = 2^{n-k} nl(g).$$

Proof. For any vector $\alpha \in GF^n(2)$, we will write $\underline{\alpha}_1 = (\alpha_1, \cdots, \alpha_k)$. It is easy to see that

$$\begin{aligned} KerD &= \{(0, \ldots, 0, x_{k+1}, \ldots, x_n) : x_i \in GF(2)\}, \\ (KerD)^{\perp} &= \{(x_1, \ldots, x_k, 0, \ldots, 0) : x_i \in GF(2)\}. \end{aligned}$$

Noticing that $GF^n(2) = (KerD)^{\perp} \oplus KerD$, we have

$$\begin{aligned} S_f(\omega) &= \textstyle\sum_x f(x)(-1)^{\langle \omega, x \rangle} \\ &= \textstyle\sum_x g(xD)(-1)^{\langle \omega, x \rangle} \\ &= \textstyle\sum_{x \in (KerD)^{\perp}} \sum_{y \in KerD} g((x \oplus y)D)(-1)^{\langle \omega, (x \oplus y) \rangle} \\ &= \textstyle\sum_{x \in (KerD)^{\perp}} g(xD)(-1)^{\langle \omega, x \rangle} \sum_{y \in KerD} (-1)^{\langle \omega, y \rangle}. \end{aligned}$$

By Lemma 1.1 we know that $S_f(\omega) = 0$ if $\omega \notin (KerD)^{\perp}$. If $\omega \in (KerD)^{\perp}$ we have

$$\begin{aligned} S_f(\omega) &= 2^{n-k} \textstyle\sum_{x \in (KerD)^{\perp}} g(xD)(-1)^{\langle \omega, x \rangle} \\ &= 2^{n-k} \textstyle\sum_{x \in (KerD)^{\perp}} g(\underline{x}_1 D_1)(-1)^{\langle \underline{\omega}_1, \underline{x}_1 \rangle} \quad \text{(by Theorem 1.5)} \\ &= 2^{n-k} S_g(\underline{\omega}_1 (D_1^{-1})^T). \end{aligned}$$

This means that

$$\max\{|S_f(\omega)| : \omega \neq 0\} = \max\{2^{n-k}|S_g(\underline{\omega}_1)| : \underline{\omega}_1 \neq 0\}$$

Notice that $wt(f) = 2^{n-k}wt(g)$. By Lemma 4.17 we have $nl(f) = 2^{n-k}nl(g)$. □

Proof of Theorem 4.20:
Necessity: Since $rank(D) = k$, there must exist a nonsingular $n \times n$ matrix R such that $RD = D' = [\begin{smallmatrix} D_1 \\ 0 \end{smallmatrix}]$. Write

$$f_1(x) = f(xR) = g(xRD) = g(xD').$$

Then by Lemma 4.19 we have $nl(f_1) = 2^{n-k}nl(g)$, and by Theorem 3.2 we have $nl(f) = nl(f_1)$. So the conclusion follows.

Sufficiency: On the contrary we assume that $rank(D) < k$. Then the columns of $D = [d_1^T, \cdots, d_k^T]$ are linearly dependent, i.e., for some i-th column of D, there must exist $a_j \in GF(2)$ such that

$$d_i = a_1d_1 \oplus \cdots \oplus a_{i-1}d_{i-1} \oplus a_{i+1}d_{i+1} \oplus \cdots \oplus a_kd_k.$$

If d_i is an all-zero vector, then for any $j \neq i$, set $g(y) = y_iy_j$ to be a quadratic function which has nonzero nonlinearity, $f(x) = g(xD) = (xd_i^T)(xd_j^T) = 0$ has zero nonlinearity. If d_i is a nonzero vector, then set $g(y) = y_i(a_1y_1 \oplus \cdots \oplus a_{i-1}y_{i-1} \oplus a_{i+1}y_{i+1} \oplus \cdots \oplus a_ky_k)$ to be a quadratic function which has nonzero nonlinearity. Then

$$\begin{aligned} f(x) &= g(xD) \\ &= (xd_i^T)(a_1xd_1^T \oplus \cdots \oplus a_{i-1}xd_{i-1}^T \oplus a_{i+1}xd_{i+1}^T \oplus \cdots \oplus a_kxd_k^T) \\ &= (xd_i^T)(xd_i^T) \\ &= xd_i^T \end{aligned}$$

is a linear function which has zero nonlinearity. This is a contradiction with Eq. 4.41, and hence the conclusion of Theorem 4.20 is true. □

From Theorem 4.20 we know that if a correlation immune function is constructed in the form $f(x) = g(xD)$, where D is an $n \times k$ matrix with $rank(D) = k$, then $f(x)$ has maximum possible nonlinearity if and only if $g(x)$ has the maximum possible nonlinearity as well. There have been alternative methods for constructing Boolean functions with high nonlinearity (refer to [7–9, 35, 46]). With Boolean functions having high-order nonlinearity, correlation immune functions having high nonlinearity can be constructed according to Theorem 4.20.

4.11.4 Correlation Immune Functions with Propagation Criterion

Unlike other properties, the propagation criterion is not inheritable from g to f for the expression $f(x) = g(xD)$, i.e., g satisfying propagation criterion does not guarantee that f does. For example, let

$$D = \begin{bmatrix} 1\,0\,0\,0\,0 \\ 0\,1\,0\,0\,0 \\ 0\,0\,1\,0\,0 \\ 0\,0\,0\,1\,0 \\ 0\,0\,0\,0\,1 \\ 1\,1\,1\,1\,1 \end{bmatrix}.$$

Although $g(y) = y_1y_2 \oplus y_2y_3 \oplus y_3y_4 \oplus y_4y_5 \oplus y_1y_5$ satisfies the propagation criterion of order 4, $f(x) = g(xD) = x_1x_2 \oplus x_2x_3 \oplus x_3x_4 \oplus x_4x_5 \oplus x_1x_5 \oplus x_6$ does not satisfy the propagation criterion of order 1. In order to study the way that the propagation property of f relates to that of g more precisely, for $f(x) \in \mathcal{F}_n$, we denote by

$$NP(f) = \{\alpha \in GF^n(2) : f(x) \oplus f(x \oplus \alpha) \text{ is not balanced}\}.$$

Theorem 4.21. *Let $f(x) = g(xD)$, where $g(y) \in \mathcal{F}_k$ and D is an $n \times k$ binary matrix with $rank(D) = k$. Then the propagation criterion order of $f(x)$ is*

$$PC(f) = \min_{\alpha D \in NP(g)} wt(\alpha) - 1.$$

Proof. We first prove that $\alpha \in NP(f)$ if and only if $\alpha D \in NP(g)$. It is easy to verify (refer the proof of Lemma 4.14) that when $rank(D) = k$, xD forms k random variables provided that x is a collection in n random variables with uniform probability distribution. So $g(xD) \oplus g(xD \oplus \beta)$ is unbalanced if and only if $\beta \in NP(g)$. So $\alpha \in NP(f) \Longleftrightarrow f(x) \oplus f(x \oplus \alpha)$ is unbalanced $\Longleftrightarrow g(xD) \oplus g(xD \oplus \alpha D)$ is unbalanced $\Longleftrightarrow \alpha D \in NP(g)$. By the definition that

$$PC(f) = \min_{\alpha \in NP(f)} wt(\alpha) - 1$$

the conclusion follows. □

Particularly, when g satisfies the propagation criterion of the maximum order k, i.e., g is a bent function (or g is perfect nonlinear and k is even in this case), we have

Corollary 4.5. *Let $g \in \mathcal{F}_k$ and g satisfy the propagation criterion of order k, i.e., g is perfect nonlinear, and let D be an $n \times k$ matrix with $rank(D) = k$. Then $f(x) = g(xD)$ satisfies the propagation criterion of order k.*

Proof. Note that $g(y) \in \mathcal{F}_k$ satisfies the propagation criterion of order k if and only if $NP(g) = \{0\}$. Since $rank(D) = k$, it is obvious that $\alpha D \in NP(g)$ or equivalently $\alpha D = 0$ only if $wt(\alpha) \geq k + 1$. We can also find an α with $wt(\alpha) = k + 1$ such that $\alpha D = 0$. So by Theorem 4.21, the conclusion of Corollary 4.5 is true. □

In the case of Corollary 4.5, function $f(x)$ has the same propagation criterion order as that of $g(y)$. Is it possible that $f(x)$ has a higher propagation criterion order than that of $g(y)$? The answer is yes as demonstrated by the following example. It can be verified that $g(x_1, \ldots, x_5) = x_1x_2 \oplus x_3x_4 \oplus x_5$ satisfies propagation criterion of order 0. Let

$$A = \begin{bmatrix} 1\,1\,1\,0\,0 \\ 0\,0\,1\,1\,1 \\ 1\,0\,0\,1\,0 \\ 0\,1\,0\,0\,1 \\ 1\,0\,0\,0\,0 \\ 1\,1\,1\,1\,1 \end{bmatrix}.$$

Then $f(x_1, \ldots, x_6) = g((x_1, \ldots, x_6)A) = x_1 \oplus x_1x_2 \oplus x_2x_3 \oplus x_4 \oplus x_1x_4 \oplus x_3x_4 \oplus x_1x_5 \oplus x_4x_5 \oplus x_6 \oplus x_1x_6 \oplus x_4x_6 \oplus x_5x_6$ satisfies the propagation criterion of order 3. We can easily find more such examples. However, as the propagation criterion characteristics of different functions are very different, and the choice of the matrices can be variant, we do not have a systematic way for constructing correlation immune functions in the form $f(x) = g(xD)$ enabling the propagation criterion order of f to be higher than that of g. We leave this as an open problem.

4.11.5 Linear Structure Characteristics of Correlation Immune Functions

From the definition of linear structures, it can be seen that the more linear structures a Boolean function has, the closer the function is related to an affine function. In the extreme case when every vector is a linear structure of a Boolean function, it must be an affine one. From a cryptographic point of view, a Boolean function is required to have as few linear structures as possible. However, when a Boolean function can be written as $f(x) = g(xD)$, it definitely has linear structures if $k < n$. The relationship between the linear structures of f and that of g can be described as follows:

Theorem 4.22. *Let $f(x) = g(xD)$, where D is an $n \times k$ ($k \leq n$) matrix with $rank(D) = k$. Then α is an invariant (a complementary) linear structure of $f(x)$ if and only if αD is an invariant (a complementary) linear structure of $g(x)$.*

Proof. The sufficiency is obvious. So we only need to present the proof of the necessity. Assume the contrary, i.e., there exists a vector $\alpha \in GF^n(2)$ such that $f(x) \oplus f(x \oplus \alpha) \equiv c$ and $g(y) \oplus g(y \oplus \alpha D) \not\equiv c$. Let $g(y') \oplus g(y' \oplus \alpha D) \neq c$. Since $rank(D) = k$, there must exist an $x' \in GF^n(2)$ such that $y' = x'D$. So we have

$$f(x') \oplus f(x' \oplus \alpha) = g(x'D) \oplus g((x' \oplus \alpha)D) = g(y') \oplus g(y' \oplus \alpha D) \neq c.$$

This is a contradiction of the assumption. So the conclusion is true. □

Corollary 4.6. *Let $f(x) = g(xD)$, where D is an $n \times k$ ($k \leq n$) matrix with $rank(D) = k$. Denote by V_f and V_g the set of linear structures of f and g, respectively. Then $dim(V_f) = (n-k) + dim(V_g)$, where $dim(.)$ means the dimension of a vector space.*

It can be seen from Corollary 4.6 that even if g has no nonzero linear structures, f may have because the all-zero vector is an invariant linear structure (trivial) of every function. It also implies that a Boolean function may have many invariant linear structures but no complementary ones.

The above shows that if a function is algebraically degenerate, it must have nonzero invariant linear structures. Is this also a sufficient condition for a Boolean function to be algebraically degenerate? The following gives a positive answer:

Theorem 4.23. *Let $f(x) \in \mathcal{F}_n$, $V_I(f)$ be the linear space of all the invariant linear structures of $f(x)$ and $dim(V_I(f)) = k$. Then there must exist a nonsingular matrix A over $GF(2)$ such that*

$$g(x_1, \ldots, x_n) = f((x_1, \ldots, x_n)A) = g_1(x_{k+1}, \ldots, x_n),$$

where $g_1(x_{k+1}, \ldots, x_n)$ has no nonzero invariant linear structures. Moreover, $g_1(x_{k+1}, \ldots, x_n)$ has a complementary linear structure, or equivalently it can be written as $g_1(x_{k+1}, \ldots, x_n) = x_{k+1} \oplus g_2(x_{k+2}, \ldots, x_n)$, if and only if f has a complementary linear structure.

Proof. Let A be such an $n \times n$ binary matrix that, the first k rows of A, $\alpha_1, \ldots, \alpha_k$, form a basis of $V_I(f)$. Let $e_i \in GF^n(2)$ be the vector with the i-th coordinate being one and zero elsewhere. Set $g(x) = f(xA)$. It is easy to verify that $e_1, \ldots, e_k$ form a basis of $V_I(g)$. This means that $g(x)$ is independent of $x_1, \ldots, x_k$ and hence can be written as $g(x) = g_1(x_{k+1}, \ldots, x_n)$. Also note that α is a complementary linear structure of $f(x)$ if and only if αA^{-1} is a complementary linear structure of $g(x)$. So the conclusion follows. □

Note that this result is similar to the one in [19]. However, here we precisely describe the value of k which is the dimension of $V_I(f)$. The proof here is also simpler.

From Theorem 4.23 we have

Corollary 4.7. *Let $f(x) \in \mathcal{F}_n$, $V_I(f)$ be the linear space of all the invariant linear structures of $f(x)$. Then $AD(f) = dim(V_I(f))$. Particularly, $f(x)$ is algebraically nondegenerate if and only if it has no nonzero invariant linear structures.*

Corollary 4.7 gives a relationship between the algebraic degeneration and linear structure characteristics of Boolean functions. We further know that an algebraically nondegenerate function can have at most one complementary linear structure.

Theorem 4.24. *Let $f(x) \in \mathcal{F}_n$, where α is a complementary linear structure of $f(x)$. Then there exists an $n \times n$ nonsingular matrix D such that $g(x) = f(xD) = x_1 \oplus g_1(x_2, \ldots, x_n)$, where g_1 has no linear structures. In this case, $f(x)$ is balanced.*

Proof. Let $D = \begin{bmatrix} \alpha \\ D_1 \end{bmatrix}$ be a nonsingular matrix. Then e_1 is a complementary linear structure of $g(x)$, and by Theorem 4.23, $g(x)$ can be written as $x_1 \oplus g_1(x_2, \ldots, x_n)$. It is easy to verify that $\beta = (0, b_2, \ldots, b_n)$ is an invariant linear structure of $f(x)$ if and only if $\beta_1 = (b_2, \ldots, b_n)$ is an invariant linear structure of g_1, and $\beta = (1, b_2, \ldots, b_n)$ is an invariant linear structure of $f(x)$ if and only if $\beta_1 = (b_2, \ldots, b_n)$ is a complementary linear structure of g_1. Since $f(x)$ has no invariant linear structures, g_1 must have no linear structures. □

Considering the correlation immune functions without linear structures, from the discussion above, it is known that they are algebraically nondegenerate functions which do not have a complementary linear structure. From Lemma 4.24, it is known that those unbalanced correlation immune functions which are algebraically nondegenerate satisfy the requirement, i.e., they do not have linear structures. In the next section, we give constructions of algebraically nondegenerate correlation immune functions which can be formulated by the constructions for correlation immune functions having no linear structures.

4.12 Construction of Algebraically Nondegenerate Correlation Immune Functions

It is noted that the constructions of correlation immune functions described above are based on linear codes, and the constructed correlation immune function must be of the format $f(x) = g(yD)$, where y is a vector variable of less dimension than x. This means that $f(x)$ must be algebraic degenerate. In the applications where algebraic degeneracy is to be avoided, nondegenerate correlation immune functions are more useful. It is noted that correlation immune functions from other constructions are also algebraically degenerate.

4.12.1 On the Algebraic Degeneration of Correlation Immune Functions

Since constructions described in Lemmas 4.4 and 4.5 are based on known correlation immune functions, initial correlation immune functions are required before executing the construction. In [36] it is suggested that $f_1(x)$ can be a linear function with $m+1$ terms and $f_2(x)$ is the one obtained from $f_1(x)$ by permuting the variables. In this case we have

Theorem 4.25. *Let* $f(x_1,\ldots,x_{n+1}) = x_{n+1}l_1(x) \oplus (1 \oplus x_{n+1})l_2(x)$, *where* $l_1(x), l_2(x) \in \mathcal{L}_n$. *Then* $f(x_1,\ldots,x_{n+1})$ *can be algebraically degenerated to a function in no more than three variables.*

Proof. It is known that for a linear function, there is only one nonzero Walsh spectrum. It can be verified that the dimension of the vector space linearly spanned by the nonzero Walsh spectrums of $f(x)$ is at most three, and by Theorem 2.10, the conclusion follows. □

More generally we have

Theorem 4.26. *Let* $f(x_1,\ldots,x_{n+1}) = x_{n+1}f_1(x) \oplus (1 \oplus x_{n+1})f_2(x)$, *where* $f_1(x), f_2(x) \in \mathcal{F}_n$. *Then*

$$AD(f) \geq AD(f_1) + AD(f_2) - n. \tag{4.42}$$

Proof. By Theorem 2.10, $dim(\langle\{w : S_{f_i}(w) \neq 0\}\rangle) = n - AD(f_i),\ i = 1,2$. Then for every $w' = (w; w_{n+1})$, $S_f(w') = 0$ if both $S_{f_1}(w)$ and $S_{f_2}(w)$ vanish, i.e., if $S_{f_1}(w) = S_{f_2}(w) = 0$. This means that $\{w' : w' \neq 0 \text{ and } S_f(w') \neq 0\}$ is a subset of

$$\{(w;0) : w \neq 0 \text{ and } S_{f_1}(w) \neq 0\} \bigcup \{(w;0) : w \neq 0 \text{ and } S_{f_2}(w) \neq 0\} \bigcup \{(0;1)\}.$$

This directly results in the following inequality:

$$\begin{aligned}&dim(\langle\{w' : S_f(w') \neq 0\}\rangle)\\ &\leq dim(\langle\{w : S_{f_1}(w) \neq 0\}\rangle) + dim(\langle\{w : S_{f_2}(w) \neq 0\}\rangle) + 1.\end{aligned}$$

Note by Theorem 2.10 we have that $dim(\langle\{w : S_{f_i}(w) \neq 0\}\rangle) = n - AD(f_i),\ i = 1,2$. So it follows that

$$\begin{aligned}AD(f) &= n + 1 - dim(\langle\{w' : S_f(w') \neq 0\}\rangle)\\ &\geq AD(f_1) + AD(f_2) - n.\end{aligned}$$

□

For the construction of Lemma 4.6, although the algebraic degeneration of the constructed functions can hardly be determined, it is most likely to be degenerate for the ones with higher-order correlation immunity. This is the case since some of the $\phi_i(y)$ should be constant with a value of one in order for $f(x;y)$ to be guaranteed to have higher-order correlation immunity. More precisely we have

Theorem 4.27. *Let $f(x;y) \in \mathcal{F}_n$ be the function of Eq. 4.18. Let the number of constants of $\{\phi_1(y), \ldots, \phi_{n_1}(y)\}$ be t. Then the degree of degeneracy of $f(x)$ satisfies that*

$$AD(f) \geq t - 1. \tag{4.43}$$

Proof. Without loss of generality, we assume that $\phi_1(y), \ldots, \phi_t(y)$ are constants. Then $f(x;y)$ can be expressed as $f(x;y) = c_1x_1 \oplus \cdots \oplus c_tx_t \oplus \bigoplus_{i=t+1}^{n_1} x_i\phi_i(y) \oplus r(y)$, where $c_i \in \{0,1\}$. By a linear transformation, the first part can be changed from the linear combination of $x_1, \ldots, x_t$ to one variable and the others unchanged. This yields a function with at most $n - t + 1$ variables. If $c_1 = \cdots = c_t = 0$, then $f(x;y)$ is actually a function in $n - t$ variables. In both of the cases, Eq. 4.43 always holds. □

The algebraic degeneration of the correlation immune functions constructed in Theorem 4.1 is clearly not zero because it coincides with the definition of algebraic degeneration. More precisely we have that $AD(f) \geq n-k$ holds for every correlation immune function constructed by Theorem 4.1. By the theory of error-correcting codes (see [22]), we know that $n - k \geq d - 1$. So the algebraic degeneration of the correlation immune functions constructed by Theorem 4.1 is larger than or equal to the designed order of correlation immunity.

4.12.2 Construction of Algebraically Nondegenerate Correlation Immune Functions

Correlation immunity also has a strong connection with orthogonal arrays [3]. An *orthogonal array* $OA_\lambda(t,d,v)$ is a $\lambda v^t \times d$ array of v symbols, such that in any t columns of the array, every one of the possible v^t ordered pairs of symbols occurs in exactly λ rows (see [22] for this definition). The following conclusion which has been proved in [6] is a direct deduction of Theorem 2.20.

Theorem 4.28. *Let $f(x) \in \mathcal{F}_n$. Treat the $supp(f) = \{x : f(x) = 1\}$ as a $wt(f) \times n$ matrix with its row vectors being the x's on which $f(x)$ takes value 1, and denote the matrix as T_f. Then $f(x)$ is correlation immune of order k if and only if T_f is an orthogonal array $OA_\lambda(k,n,2)$, where $\lambda = wt(f)/2^k$.*

Let T_f be the $wt(f) \times n$ matrix as defined in Theorem 4.28 and $e_i \in GF^n(2)$ be such a vector with 1 in its i-th coordinate and 0 elsewhere. It is easy to verify that the following conclusion holds:

Lemma 4.20. *$S_f(e_i) = 0$ if and only if the i-th column of T_f is balanced, i.e., in the i-th column of T_f, the number of zeros and that of ones are equal.*

Now a necessary and sufficient condition for judging a Boolean function to be algebraically nondegenerate in terms of its truth table representation can be described as follows:

Theorem 4.29. *Let $f(x) \in \mathcal{F}_n$. Then $f(x)$ is algebraically nondegenerate if and only if there exists an $n \times n$ invertible matrix G such that every column of T_fG is not balanced.*

Proof. Necessity: Let $f(x)$ be an algebraically nondegenerate function. By Theorem 2.10 we know that there are n linearly independent vectors $w_1, \ldots, w_n \in GF^n(2)$ such that $S_f(w_i) \neq 0$, $i = 1, \ldots, n$. Write $G = [w_1^T, \ldots, w_n^T]$. Then G is an $n \times n$ invertible matrix and $G^{-1}G = I = [e_1^T, \ldots, e_n^T]$, or $G^{-1}w_i^T = e_i^T$. Let $f(x) = g(xG)$. By Theorem 1.5 we have $S_f(w) = S_g(w(G^{-1})^T)$. Since $S_f(w_i) \neq 0$, and it should be noticed that $w_i(G^{-1})^T = e_i$, we have $S_g(e_i) \neq 0$. It is easy to verify that $T_g = T_fG$. By Lemma 4.20 it is known that every column of T_g is not balanced.

Sufficiency: Suppose that T_fG has the property that every column is not balanced. Let $g(x)$ be the function with $T_g = T_fG$, then $g(x) = f(xG^{-1})$. By Theorem 1.5 we have $S_g(w) = S_f(wG^T)$. Note that by Lemma 4.20, we have $S_g(e_i) \neq 0$, then $S_f(e_iG^T) \neq 0$, $i = 1, 2, \ldots, n$. This shows that on every row vector of G^T, the Walsh transform of $f(x)$ takes a nonzero value. Recall that G is invertible, its columns can generate the whole space $GF^n(2)$. By Theorems 2.9 and 2.10, we know that $f(x)$ is algebraically nondegenerate. □

By Theorems 4.28 and 4.29, we have

Corollary 4.8. *There exists an algebraically nondegenerate k-th-order correlation immune function in $\mathcal{F}_n$ if and only if there exists an orthogonal array $OA_\lambda(k, n, 2)$ and an $n \times n$ binary invertible matrix G such that every column of $OA_\lambda(k, n, 2)G$ is not balanced.*

The following examples are found based on Corollary 4.8, where each A_i is an orthogonal array, each G_i is an invertible matrix, and A_iG_i has the property that every column is not balanced.

$$A_1 = \begin{bmatrix} 0011 \\ 0010 \\ 0100 \\ 1000 \\ 1101 \\ 1111 \end{bmatrix}, G_1 = \begin{bmatrix} 0011 \\ 0101 \\ 1001 \\ 1110 \end{bmatrix}, A_1G_1 = \begin{bmatrix} 0111 \\ 1001 \\ 0101 \\ 0011 \\ 1000 \\ 0001 \end{bmatrix},$$

$$A_2 = \begin{bmatrix} 0000 \\ 0110 \\ 1011 \\ 1101 \\ 1000 \\ 0111 \end{bmatrix}, G_2 = \begin{bmatrix} 1101 \\ 0111 \\ 0011 \\ 1000 \end{bmatrix}, A_2G_2 = \begin{bmatrix} 0000 \\ 0100 \\ 0110 \\ 0010 \\ 1101 \\ 1100 \end{bmatrix},$$

$$A_3 = \begin{bmatrix} 10000 \\ 10011 \\ 10010 \\ 11111 \\ 11110 \\ 11100 \\ 11010 \\ 11000 \\ 01101 \\ 01100 \\ 01001 \\ 00111 \\ 00110 \\ 00101 \\ 00011 \\ 00001 \end{bmatrix}, G_3 = \begin{bmatrix} 11111 \\ 10100 \\ 01100 \\ 00010 \\ 00001 \end{bmatrix}, A_3G_3 = \begin{bmatrix} 11111 \\ 11100 \\ 11101 \\ 00100 \\ 00101 \\ 00111 \\ 01001 \\ 01011 \\ 11001 \\ 11000 \\ 10101 \\ 01111 \\ 01110 \\ 01101 \\ 00011 \\ 00001 \end{bmatrix}.$$

From the orthogonal arrays above, we get three algebraically nondegenerate functions:

$$f_1(x_1, x_2, x_3, x_4) = x_1 \oplus x_2 \oplus x_3 \oplus x_1x_4 \oplus x_2x_4 \oplus x_1x_2x_3 \oplus x_1x_2x_4 \\ \oplus x_1x_3x_4 \oplus x_2x_3x_4$$

$$f_2(x_1, x_2, x_3, x_4) = 1 \oplus x_2 \oplus x_3 \oplus x_4 \oplus x_2x_4 \oplus x_3x_4 \oplus x_1x_2x_3 \\ \oplus x_1x_2x_4 \oplus x_1x_3x_4 \oplus x_2x_3x_4$$

$$f_3(x_1, x_2, x_3, x_4, x_5) = x_1 \oplus x_5 \oplus x_1x_3 \oplus x_2x_3 \oplus x_3x_4 \oplus x_1x_3x_4 \oplus x_1x_3x_5 \\ \oplus x_1x_4x_5 \oplus x_2x_3x_5 \oplus x_2x_4x_5 \oplus x_3x_4x_5$$

where $f_1, f_2 \in \mathcal{F}_4$, and $f_3 \in \mathcal{F}_5$ is a balanced function. They are all first-order correlation immune functions.

For $f(x) \in \mathcal{F}_n$, denote $S(f) = \{w : S_f(w) \neq 0\}$. By Theorem 1.5 we have $S_f(w) = S_g(wD^T)$, and consequently we have

Theorem 4.30. *Let $f(x) \in \mathcal{F}_n$. If there is an $n \times n$ invertible matrix G such that for every w with $1 \le wt(w) \le k$, we have*

$$w \in S(f)G = \{wG : w \in S(f)\}.$$

Then $f(xG^T)$ is a k-th-order correlation immune function. Moreover, $f(xG^T)$ is algebraic nondegenerate if and only if $f(x)$ is algebraically nondegenerate.

As for the iterative construction described in Lemma 4.4, we have

Theorem 4.31. *Let $f_1(x), f_2(x) \in \mathcal{F}_n$ be two k-th-order correlation immune functions with $wt(f_1) = wt(f_2)$. If $\langle\{w : S_{f_1}(w) + S_{f_2}(w) \neq 0\}\rangle = GF^n(2)$, then $f(x_1, \dots, x_{n+1}) = x_{n+1}f_1(x) \oplus (1 \oplus x_{n+1})f_2(x)$ is an algebraically nondegenerate k-th-order correlation immune function in $n+1$ variables.*

Proof. Denote by $\overline{\omega} = (\omega, \omega_{n+1})$ and $\overline{x} = (x, x_{n+1})$. Then for the functions of Eq. 4.17, we have

$$\begin{aligned} S_f(\overline{\omega}) &= \sum_{\overline{x}} f(\overline{x})(-1)^{\overline{\omega}\cdot\overline{x}} \\ &= \sum_{x_{n+1}=1} \sum_{x} f_1(x)(-1)^{\omega\cdot x \oplus \omega_{n+1}} + \sum_{x_{n+1}=0} \sum_{x} f_2(x)(-1)^{\omega\cdot x} \\ &= (-1)^{\omega_{n+1}} S_{f_1}(\omega) + S_{f_2}(\omega). \end{aligned} \tag{4.44}$$

It is easy to verify that when the dimension of the linear span of $\{\omega : S_{f_1}(\omega) + S_{f_2}(\omega) \neq 0\}$ is n, the dimension of the linear span of $\{\overline{\omega} : S_f(\overline{\omega}) \neq 0\}$ is $n+1$, and hence f is algebraic nondegenerate. □

Theorem 4.31 gives a sufficient condition for the Boolean function f defined by Eq. 4.17 to be algebraically nondegenerate. When the condition of Theorem 4.31 can be satisfied is still not clear. It is anticipated that when one or both of f_1 and f_2 are algebraically nondegenerate, f is likely to be so. It is noticed that Sect. 4.9 listed 96 algebraically nondegenerate correlation immune functions; among them half have Hamming weight 6 and another half have Hamming weight 10. By checking every pair of them with the same Hamming weight, we found that among $2 \times \binom{96}{2} = 9120$ pairs, there are 7680 pairs which can form an algebraic nondegenerate correlation immune function in five variables according to Eq. 4.17, while another 1440 pairs cannot.

In practice it is suggested to use the definition to verify whether the constructed correlation immune function according to Lemma 4.4 is algebraically nondegenerate. Notice in the proof of Theorem 4.31 that for every $\overline{\omega} = (\omega, \omega_{n+1})$, $(-1)^{\omega_{n+1}} S_{f_1}(\omega) + S_{f_2}(\omega) = 0$ if and only if $S_{f_1}(\omega) + (-1)^{\omega_{n+1}} S_{f_2}(\omega) = 0$. So we have

Corollary 4.9. *Let* $f_1(x), f_2(x) \in \mathcal{F}_n$. *Then* $x_{n+1}f_1(x) \oplus (1 \oplus x_{n+1})f_2(x)$ *is algebraically nondegenerate if and only if* $(1 \oplus x_{n+1})f_1(x) \oplus x_{n+1}f_2(x)$ *is algebraically nondegenerate.*

Let $f(x) \in \mathcal{F}_n$. Now we consider the function $F(\overline{x}) = F(x_1, \ldots, x_{n+1}) = x_{n+1} \oplus f(x)$. It is easy to verify that $AD(F) \leq AD(f) + 1$. So $F(\overline{x})$ is algebraic degenerate if $f(x)$ is such. When $f(x)$ is algebraically nondegenerate, the algebraic degeneration of $F(\overline{x})$ is at most one. It is interesting to know when $F(\overline{x})$ is algebraically nondegenerate as well. We have

Theorem 4.32. *Let* $f(x) \in \mathcal{F}_n$ *be an algebraic nondegenerate function and* $F(\overline{x}) = x_{n+1} \oplus f(x)$. *Then* $F(\overline{x})$ *is algebraically nondegenerate if and only if* $f(x)$ *has no complementary linear structures.*

Proof. Necessity: Assume that $f(x)$ has a complementary linear structure α, then $(\alpha, 1)$ is an invariant linear structure of $F(\overline{x})$. By Theorem 4.23, $F(\overline{x})$ is algebraic degenerate.

Sufficiency: If $x_{n+1} \oplus f(x)$ is algebraically degenerate, then by Corollary 4.7, $x_{n+1} \oplus f(x)$ must have an invariant linear structure $(a_1, \ldots, a_{n+1})$. It can easily be verified in this case that $(a_1, \ldots, a_n)$ is an invariant linear structure of $f(x)$ if $a_{n+1} = 0$ and is a complementary linear structure of $f(x)$ if $a_{n+1} = 1$. □

By Theorem 4.32 and Lemma 4.4, we know that if $f(x)$ is a balanced algebraic nondegenerate m-th-order correlation immune function and has no complementary linear structures, then $x_{n+1} \oplus f(x)$ is a balanced algebraically nondegenerate $(m+1)$-th-order correlation immune function in $n+1$ variables. Note that this construction cannot be preceded further as $x_{n+1} \oplus f(x)$ has at least one complementary linear structure. As an example of this construction, we found that the function

$$f(x_1, \ldots, x_5) = x_1 \oplus x_5 \oplus x_2x_3 \oplus x_3x_4 \oplus x_3x_5 \oplus x_1x_2x_3 \oplus x_1x_2x_4$$
$$\oplus x_1x_2x_5 \oplus x_1x_3x_4 \oplus x_1x_3x_5 \oplus x_2x_3x_5$$

is balanced, algebraically nondegenerate, and first-order correlation immune and has no complementary linear structures. Then by Theorem 4.32 and Lemma 4.4, we can construct a Boolean function $x_6 \oplus f(x)$ which is balanced, algebraically nondegenerate, and second-order correlation immune and has only one complementary linear structure (000001).

4.13 The ε-Correlation Immunity of Boolean Functions

It is noted that the correlation immunity is a cryptographic measure about the resistance against correlation attack, and there can be cases where although a combining function is not correlation immune, however the correlation attack still consumes large amount of computation due to the function being "near" to correlation immune. We hereby define a measure about how "near" a function is to being correlation immune. This only makes sense for the functions that are not correlation immune at all. Consider the balancedness of the i-th coordinate of all the vectors in $supp(f)$. If it has a good balance, then $f(x)$ has small correlation with x_i. If it is balanced, then $f(x)$ has no correlation with x_i. If for all $i \in \{1, 2, \ldots, n\}$, $f(x)$ has no correlation with x_i, then $f(x)$ is correlation immune (of order at least 1). Not expecting the correlation immunity of $f(x)$, we define the relative correlation of $f(x)$ with x_i as the difference between the number of 0's and that of 1's in the i-th coordinates of vectors x in $supp(f)$, i.e.,

$$\varepsilon^{(i)}(f) = \left| \sum_{x \in supp(f)} (-1)^{x_i} \right| = \left| wt(f) - 2 \sum_{x \in supp(f)} x_i \right|.$$

By this definition, it is easy to see that the idea of correlation immunity is to find the maximum value of these relative correlations. If the maximum value is 0, then $f(x)$ must be correlation immune (of order 1 or larger); otherwise $f(x)$ is not correlation immune. However, in the case where $f(x)$ is not correlation immune, the value of $\varepsilon^{(i)}(f)$ varies which indicates the different degrees that $f(x)$ has correlation with x_i. The correlation of $f(x)$ with any variables is hence defined as

$$\varepsilon(f) = \max\{\varepsilon^{(i)}(f) : i \in \{1, 2, \ldots, n\}\}.$$

For this consideration, we define the ε-*correlation immunity* of $f(x)$ as

$$\begin{aligned} CI_\varepsilon(f) &= 1 - \frac{\varepsilon(f)}{wt(f)} \\ &= 1 - \frac{1}{wt(f)} \max_{i \in \{1,2,\ldots,n\}} \left| wt(f) - 2 \sum_{x \in supp(f)} x_i \right| \end{aligned} \tag{4.45}$$

It is seen from Eq. 4.45 that $0 \le CI_\varepsilon(f) \le 1$. When $CI_\varepsilon(f) = 1$, it means that $f(x)$ is correlation immune (of at least order 1). Another extreme case is when $CI_\varepsilon(f) = 0$; this means that there exists i such that $x_i = 0$ (or $x_i = 1$) always holds for all $x \in supp(f)$, which means that the correlation between $f(x)$ and this x_i is high (the highest possible case). The ε here means that the indexed correlation immunity is a fractional value between 0 and 1, instead of integral value as the traditional meaning of correlation immunity.

Now we take a look at what the ε-correlation immunity has to do with the correlation attacks proposed by Siegenthaler. Let i be such an index satisfying that

$$CI_\varepsilon(f) = 1 - \frac{1}{wt(f)} \left| wt(f) - 2 \sum_{x \in supp(f)} x_i \right| \triangleq \varepsilon.$$

Then we have

$$Prob(x_i = 1|f(x) = 1) = \frac{\sum_{x \in supp(f)} x_i}{|supp(f)|} = \frac{\sum_{x \in supp(f)} x_i}{wt(f)}$$

and

$$\begin{aligned} Prob(x_i = 0|f(x) = 0) &= \frac{|\overline{supp}(f)| - \sum_{x \in \overline{supp}(f)} x_i}{|\overline{supp}(f)|} \\ &= \frac{2^n - wt(f) - (\sum_{x \in GF^n(2)} x_i - \sum_{x \in supp(f)} x_i)}{2^n - wt(f)} \\ &= \frac{2^{n-1} - wt(f) + \sum_{x \in supp(f)} x_i}{2^n - wt(f)}. \end{aligned}$$

Hence, we have

$$\begin{aligned} q_i &= Prob(f(x) = x_i) \\ &= Prob(f(x) = 1)Prob(x_i = 1|f(x) = 1) \\ &\quad + Prob(f(x) = 0)Prob(x_i = 0|f(x) = 0) \\ &= \frac{wt(f)}{2^n} \cdot \frac{\sum_{x \in supp(f)} x_i}{wt(f)} + \frac{2^n - wt(f)}{2^n} \cdot \frac{2^{n-1} - wt(f) + \sum_{x \in supp(f)} x_i}{2^n - wt(f)} \\ &= \frac{1}{2^n}(2^{n-1} - wt(f) - 2\sum_{x \in supp(f)} x_i) \\ &= \frac{1}{2} - \frac{wt(f)}{2^n}(1 - \varepsilon) \end{aligned} \tag{4.46}$$

If ε is very close to 0, then q_i is very different from $\frac{1}{2}$. Particularly when $f(x)$ is balanced which is often practically required, then q_i is very close to 1 or 0; in which case, we have high confidence to have either $f(x) = x_i$ or $f(x) = x_i \oplus 1$. Consequently by Eq. 4.3, we get that $p_e \approx p_0$ or $p_e \approx 1 - p_0$. It is assumed that $p_0 \neq \frac{1}{2}$; otherwise, we would always have $p_e = \frac{1}{2}$ and hence the correlation attack does not work. It is also easy to verify that these are the cases when $|p_e - \frac{1}{2}|$ reaches the maximum value, and by Eq. 4.16 we know that the minimum amount of data is needed to conduct a correlation attack.

If ε is very close to 1, then q_i is very close to $\frac{1}{2}$, and by Eq. 4.3, p_e is also very close to $\frac{1}{2}$, and consequently large amount of ciphertext is required to conduct a correlation attack. Although such an attack is possible, however, when ε is so close to 1 that results in the bound of Eq. 4.16 to be too large to reach in practice, then the correlation attack becomes practically infeasible.

The concept of higher-order ε-correlation immunity has similar motivation to that of higher-order correlation immunity, and it is to measure the probability of event $(f(x) = x_{i_1} \oplus x_{i_2} \oplus x_{i_k})$ and a corresponding modified correlation attack, for any possible $1 \le i_1 < i_2 < \cdots < i_k \le n$.

In order to compute the ε-correlation immunity of a Boolean function, motivated by Lemma 4.13, we will seek a Walsh spectrum description. It is easy to deduce that the Walsh spectrum of $f(x)$ on e_i (as defined before, e_i is such a vector in $GF^n(2)$ that its i-th coordinate is 1 and 0 elsewhere) is

$$\begin{aligned} S_f(e_i) &= \sum_{x \in GF^n(2)} f(x)(-1)^{e_i \cdot x} \\ &= \sum_{x \in supp(f)} (-1)^{e_i \cdot x} \\ &= \sum_{x \in supp(f)} (-1)^{x_i} \\ &= \sum_{x \in supp(f)} (1 - 2x_i) \\ &= wt(f) - 2 \sum_{x \in supp(f)} x_i \end{aligned}$$

Therefore, we have

$$CI_\varepsilon(f) = 1 - \frac{1}{wt(f)} \max_i |S_f(e_i)| \tag{4.47}$$

Given the relationship of the two types of Walsh spectrums, we have $S_f(e_i) = -\frac{1}{2} S_{(f)}(e_i)$, and hence the ε-correlation immunity can be represented as

$$CI_\varepsilon(f) = 1 - \frac{1}{2wt(f)} \max_i |S_{(f)}(e_i)| \tag{4.48}$$

We will use the concept of ε-correlation immunity to study the majority functions in Chap. 6 and will see that although the majority functions are not correlation immune at all, their ε-correlation immunity however tends to approach 1 with the increase of the number of variables n, which means that they also have a good resistance against correlation attack.

It is also easy to extend the concept of ε-correlation immunity of Boolean functions to an order higher than 1. Note that the basic correlation attack considers $q_i = Prob(f(x) = x_i)$; in general case, we may consider a nonzero linear combination of the LFSR sequences, and the linear combination can be written as $\langle w, x \rangle$, where $w \in GF^n(2)$ is the coefficient vector. Now the probability

$$q_w = Prob(f(x) = \langle w, x \rangle)$$

needs to be considered. When $w = e_i$, we have $\langle w, x \rangle = x_i$ which is the special case. However, even in the general case, one cannot afford to count all the possible linear combinations in a practical attack. So we can restrict that there are at most k LFSR sequences involved in the linear combination, where k is a security parameter. So we need to consider all the linear combinations $\langle w, x \rangle$ with $1 \leq wt(w) \leq k$. Similar to the analysis of how the ε-correlation immunity is related to the basic correlation attack, we define the *k-order-ε-correlation immunity* of $f(x) \in \mathcal{F}_n$ to be

$$CI_\varepsilon^k(f) = 1 - \frac{\max\limits_{w:\ 1 \leq wt(w) \leq k} |S_f(w)|}{wt(f)}. \tag{4.49}$$

When $k = 1$, it becomes the ε-correlation immunity as defined above.

4.14 Remarks

Correlation immunity is an interesting cryptographic property, which is to measure the level of resistance (subject to the order of correlation immunity) against correlation attacks. Since practically used functions are often required to be balanced, so resilient functions are of particular interest. Constructions of correlation immune function with other cryptographic properties are of practical significance; much related work can be found in public literatures, for example, [23, 49, 50]; and much research has been devoted to the constructions of resilient function with other crytographic properties; see, for example, [24, 39, 41, 45, 47, 48]. Based on the correlation attack, some other related attacks are developed, for example, conditional correlation attack [20], fast correlation attack [27, 28, 30], and edit distance correlation attack [17]. Those new attacks indicate new cryptographic measures on the nonlinear functions.

References

1. Anderson, R.J.: Searching for the optimum correlation attacks. In: Proceedings of K.U.Leuven Workshop on Cryptographic Algorithms, Leuven, pp. 56–62 (1994)
2. Beth, T., Jungnickel, D., Lenz, H.: Design Theory. Bibliographisches Institute, Zürich (1986)
3. Bierbrauer, J., Gopalakrishnan, K., Stinson, D.R.: Bounds on resilient functions and orthogonal arrays. In: Advances in Cryptology, Proceedings of Crypto'94. LNCS 839, pp. 247–256. Springer, Berlin/Heidelberg (1994)
4. Brickell, E.: A few results in message authentication. Congr. Numer. **43**, 141–154 (1984)
5. Camion, P., Canteaut, A.: Construction of t-resilient functions over a finite alphabet. In: Advances in Cryptology, Proceedings of Eurocrypt'96. LNCS 1070, pp. 283–293. Springer, Berlin/Heidelberg (1996)

6. Camion, P., Carlet, C., Charpin, P., Sendrier, N.: On correlation-immune functions. In: Advances in Cryptology, Proceedings of Crypto'91. LNCS 576, pp. 86–100. Springer, Berlin/Heidelberg/New York (1992)
7. Carlet, C.: Partially-bent functions. Des. Codes Cryptogr. **3**, 135–145 (1993)
8. Carlet, C.: Two new classes of Bent functions. In: Advances in Cryptology, Proceedings of Eurocrypt'93. LNCS 765, pp. 77–101. Springer, Berlin/Heidelberg (1994)
9. Carlet, C.: Generalized partial spreads. IEEE Trans. Inf. Theory **IT-41**(5), 1482–1487 (1995)
10. Carlet, C.: More correlation-immune and resilient functions over Galois fields and Galois rings. In: Advances in Cryptology, Proceedings of Eurocrypt'97. LNCS 1233, pp. 422–433. Springer, Berlin/Heidelberg (1997)
11. Carlet, C., Sarkar, P.: Spectral domain analysis of correlation immune and resilient boolean functions. Finite Fields Appl. **8**(1), 120–130 (2002)
12. Chee, S., Lee, S., Lee, D.: On the correlation immune functions and their nonlinearity. In: Advances in Cryptoloty, Proceedings of Asiacrypt'96. LNCS 1163, pp. 232–243. Springer, Berlin/Heidelberg (1996)
13. Denisov, O.V.: An asymptotic formula for the number of correlation immune of order *k* boolean functions. Discret. Math. Appl. **2**, 407–426 (1992)
14. Ding, C., Shan, W., Xiao, G.: The Stability Theory of Stream Ciphers. LNCS 561. Springer, Berlin/Heidelberg (1991)
15. Golic, J.D.: On the security of shift register based keystream generators. In: Fast Software Encryption 1993. LNCS 809, pp. 90–100. Springer, Berlin/Heidelberg (1994)
16. Golic, J.D.: Correlation properties of a general binary combiner with memory. J. Cryptol. **9**(2), 111–126 (1996)
17. Golic, J.D., Menicocci, R.: Edit distance correlation attack on the alternating step generator. In: Advances in Cryptology, Proceedings of Crypto'97. LNCS 1294, pp. 499–512. Springer, Berlin (1997)
18. Gopalakrishnan, K., Stinson, D.R.: Three characterizations of non-binary correlation-immune and resilient functions. Des. Codes Cryptogr. **5**(3), 241–251 (1995)
19. Lai, X.: Additive and linear structures of cryptographic functions. In: Fast Software Encryption 1994. LNCS 1008, pp. 75–85. Springer, Berlin/Heidelberg (1995)
20. Lee, S., Chee, S., Park, S., Park, S.: Conditional correlation attack on nonlinear filter generators. In: Advances in Cryptology, Proceedings of Asiacrypt 1996. LNCS 1163, pp. 360–367. Springer, Berlin/Heidelberg (1996)
21. Lidl, R., Niederreiter, H.: Finite Fields. Encyclopedia of Mathematics and Applications, vol. 20. Addison-Wesley, Reading (1983)
22. MacWilliams, F.J., Sloane, N.J.A.: The Theory of Error-Correcting Codes. North-Holland (1977)
23. Maitra, S.: On nonlinearity and autocorrelation properties of correlation immune boolean functions. J. Inf. Sci. Eng. **20**, 305–323 (2004)
24. Maitra, S., Passalic, E.: Further constructions of resilient boolean functions with very high nonlinearity. IEEE Trans. Inf. Theory **IT-48**(7), 1825–1834 (2002)
25. Maitra, S., Sarkar, P.: Hamming weights of correlation immune boolean functions. Inf. Process. Lett. **71**, 149–153 (1999)
26. Massey, J.L.: Shift-register synthesis and BCH decoding. IEEE Trans. Inf. Theory **IT-15**(1), 122–127 (1969)
27. Meier, W., Staffelbach, O.: Fast correlation attacks on stream ciphers. In: Advances in Cryptology, Proceedings of Eurocrypt'88. LNCS 330, pp. 301–314. Springer, New York (1988)
28. Meier, W., Staffelbach, O.: Fast correlation attacks on certain stream ciphers. J. Cryptol. **1**, 159–176 (1989)
29. Meier, W., Staffelbach, O.: Nonlinearity criteria for cryptographic functions. In: Advances in Cryptology, Proceedings of Eurocrypt'89. LNCS 434, pp. 549–562. Springer, Berlin/Heidelberg (1990)
30. Meier, W., Staffelbach, O.: Correlation properties of combiners with memory in stream ciphers. J. Cryptol. **5**(1), 67–86 (1992)

31. Millan, W.: Low order approximation of cipher functions. In: Cryptography: Policy and Algorithms, pp. 144–155. Springer, Berlin/Heidelberg (1996)
32. Rueppel, R.A.: Correlation-immunity and the summation generator. In: Advances in Cryptology, Proceedings of Crypto'85. LNCS 218, pp. 260–272. Springer, Berlin/Heidelberg (1986)
33. Rueppel, R.A.: Analysis and Design of Stream Ciphers. Springer, Berlin/Heidelberg (1986)
34. Schneider, M.: On the construction and upper bounds of balanced and correlation-immune functions. Sel. Areas Cryptogr. Kluwer Academic Publishers, **6544**(3), 73–87 (1997)
35. Seberry, J., Zhang, X.M., Zheng, Y.: On construction and nonlinearity of correlation immune functions, (extended abstract). In: Advances in Cryptology, Proceedings of Eurocrypt'93. LNCS 765, pp. 181–199. Springer, Berlin/Heidelberg/New York (1994)
36. Siegenthaler, T.: Correlation-immunity of nonlinear combining functions for cryptographic applications. IEEE Trans. Inf. Theory **IT-30**(5), 776–780 (1984)
37. Siegenthaler, T.: Decrypting a class of stream ciphers using ciphertext only. IEEE Trans. Comput. **C-34**(1), 81–85 (1985)
38. Siegenthaler, T.: Cryptanalysts' representation of nonlinearly filtered m-sequences. In: Advances in Cryptology, Proceedings of Eurocrypt'85. LNCS 219, pp. 103–110. Springer, Berlin (1986)
39. Stinson, D.R.: Resilient functions and large sets of orthogonal arrays. Congr. Numer. **92**, 105–110 (1993)
40. van Lint, J.H.: Introduction to Coding Theory. Springer, Berlin/Heidelberg (1982)
41. Wu, C.K., Dawson, E.: On construction of resilient functions. In: Information Security and Privacy, Proceedings of First Australasian Conference. LNCS 1172, pp. 79–86. Springer, Berlin/Heidelberg (1996)
42. Wu, C.K., Wang, X.M., Dawson, E.: Construction of correlation immune functions based on the theory of error-correcting codes. In: Proceedings of ISITA96, Victoria, pp. 167–170 (1996)
43. Xian, Y.: Correlation immunity of boolean functions. Electron. Lett. **23**, 1335–1336 (1987)
44. Xiao, G.Z., Massey, J.L.: A spectral characterization of correlation-immune combining functions. IEEE Trans. Inf. Theory **IT-34**(3), 569–571 (1988)
45. Zhang, X.M., Zheng, Y.: On nonlinear resilient functions (extended abstract). In: Advances in Cryptology, Proceedings of Eurocrypt'95. LNCS 921, pp. 274–288. Springer, Berlin/Heidelberg (1995)
46. Zhang, X.M., Zheng, Y.: Auto-correlations and new bounds on the nonlinearity of boolean functions. In: Advances in Cryptology, Proceedings of Eurocrypt'96. LNCS 1070, pp. 294–306. Springer, Berlin/Heidelberg (1996)
47. Zhang, X.M., Zheng, Y.: Cryptographically resilient functions. IEEE Trans. Inf. Theory **IT-43**(5), 1740–1747 (1997)
48. Zhang, F., Hu, Y., Xie, M., Wei, Y.: Constructions of 1-resilient boolean functions on odd number of variables with a high nonlinearity. Secur. Commun. Netw. **5**(6), 614–624 (2011)
49. Zheng, Y., Zhang, X.M.: On relationships among avalanche, nonlinearity and corrlation immunity. In: Advances in Cryptology, Proceedings of Asiacrypt 2000. LNCS 1976, pp. 470–482. Springer, Berlin/Heidelberg (2000)
50. Zheng, Y., Zhang, X.M.: Improved upper bound on the nonlinearity of high order correlation immune functions. In: Selected Areas in Cryptography. LNCS 2012, pp. 262–274. Springer, Berlin/Heidelberg (2001)
51. Zheng, Y., Zhang, X.M.: New results on correlation immune functions. In: Proceedings of 3-rd International Conference on Information Security and Cryptology. LNCS 2015, pp. 49–63. Springer, Berlin/Heidelberg (2001)

Chapter 5
Algebraic Immunity of Boolean Functions

Algebraic immunity is a cryptographic measure about the resistance against algebraic attack which was first proposed by Courtois in 2003 for stream ciphers. This chapter studies some basic properties of algebraic immunity of Boolean functions, including the construction of annihilators of Boolean functions, upper and lower bounds of algebraic immunity, and an approach toward computing the annihilators of Boolean functions.

5.1 Algebraic Attacks on Stream Ciphers

Any cryptographic property comes from a concrete attack or a potential security threat to cryptosystems. The concept of algebraic immunity of Boolean functions comes from the algebraic attack on stream ciphers proposed by Courtois and Meier [9], which has proven to be a very effective attack not only on stream ciphers [13] but also on block ciphers [10]. It is difficult to construct general Boolean function reaching the best algebraic immunity [6]; many of such functions are from a very specific class of Boolean functions or their modifications [18]. However, construction of Boolean functions with high-order algebraic immunity is possible [11, 12], and some even have other cryptographic properties [14].

In order to study the algebraic immunity of Boolean functions, we first give a brief description of the principle of the algebraic attacks on stream ciphers. Our description will be based on the nonlinear feedforward generator as depicted in Fig. 5.1, and the method applies to other models as well with a suitable modification.

In a nonlinear feedforward generator-based stream cipher, the security of the stream cipher is mainly based on the security (randomness, nonlinearity, correlation immunity, etc.) of the output of the nonlinear function $f(x)$. Therefore, the nonlinear feedforward function $f(x)$ plays an essential role in the model. When attacking such a model, it is assumed that the attacker is given sufficiently long sequence of the

C.-K. Wu, D. Feng, *Boolean Functions and Their Applications in Cryptography*,
Advances in Computer Science and Technology, DOI 10.1007/978-3-662-48865-2_5

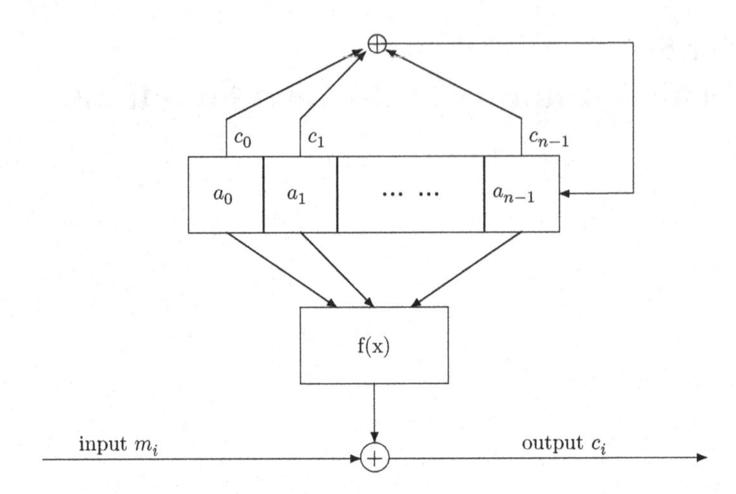

Fig. 5.1 Nonlinear feedforward generator in stream ciphers

output of the generator, not only the ciphertext but the actual output of $f(x)$ in the generator, and the objective of the attack is to reconstruct the initial state of the LFSR in the generator, so that the output of the generator from an arbitrary state can be reconstructed. If some output of the LFSR are known, by using the famous Berlekamp-Massey algorithm (BM-algorithm) [4, 15], only $2n$ consecutive bits of the LFSR are needed to fully reconstruct the LFSR of order n. So the objective of the algebraic attack is to recover $2n$ consecutive bits of the LFSR in the nonlinear feedforward generator.

Let $L(x)$ be the state-updating function, where $L(x)$ is an (n, n) Boolean function, i.e., it is a mapping from $GF^n(2)$ into itself. We will see that such a Boolean function can be represented by a vector Boolean function. More precisely, $L(x)$ can be represented as

$$L(x) = L(x_1, x_2, \ldots, x_n) = (x_2, x_3, \ldots, x_n, c_0x_1 \oplus c_1x_2 \oplus \cdots \oplus c_{n-1}x_n).$$

Let $S = (s_0, s_1, \ldots, s_{n-1})$ denote a state of the LFSR in the feedforward generator. Then $L(x)$, being an operator on the state, when applied to S, outputs

$$S_1 = L(S) = L(s_0, s_1, \ldots, s_{n-1}) = (s_1, s_2, \ldots, s_{n-1}, s_n),$$

where $s_n = c_0s_0 \oplus c_1s_1 \oplus \cdots \oplus c_{n-1}s_{n-1}$. When the operator $L(x)$ applies on S_1, we call that the operator applies on S twice and denote it as $S_2 = L(S_1) = L^2(S)$. In general, when the operator $L(x)$ applies on the state S for k times, it outputs $S_k = L^k(S) = L^k(s_0, s_1, \ldots, s_{n-1})$. The advantage of this notation is that when we treat $s_0, s_1, \ldots, s_{n-1}$ as unknowns, then the output of every $L^k(S)$ is a function of those n unknowns.

Let $f(x) \in \mathcal{F}_n$ be the nonlinear feedforward function in the generator of Fig. 5.1, and then its outputs $b_0, b_1, b_2, \ldots$ can be written as

$$\begin{cases} b_0 = f(s_0, s_1, \ldots, s_{n-1}) \\ b_1 = f(L(s_0, s_1, \ldots, s_{n-1})) \\ b_2 = f(L^2(s_0, s_1, \ldots, s_{n-1})) \\ \qquad \cdots\cdots \end{cases} \tag{5.1}$$

Here $(s_0, s_1, \ldots, s_{n-1})$ are the unknowns which form the initial state of the LFSR, and the output of $f(x)$ is actually the key stream for the stream cipher. Denoting the unknowns as $x = (s_0, s_1, \ldots, s_{n-1})$, then every output bit of $f(x)$ is a function of the unknown x, i.e., $b_i = f(L^i(x))$. Given a sufficient number of b_i, then we have sufficient number of equations of unknown x, and it may be possible (and very likely) to determine the value of x (and hence the values of $s_0, s_1, \ldots, s_{n-1}$) by solving the system of these equations. For traditional notation, we will denote the variable as $x = (x_1, x_2, \ldots, x_n)$.

There are improvements on the algebraic attacks (see, e.g., [1–3]); one of such improvements is called *fast algebraic attack*.

5.2 A Small Example of Algebraic Attack

In order to demonstrate how the algebraic attack works, here we give a toy example. First a concept is introduced.

Definition 5.1. Treating each single term of $f(x_1, x_2, \cdots, x_n)$ in its ANF as a new variable, then the equation $f(x) = 0$ is called a *multivariate equation*.

As an example, the Boolean function in four variables

$$f(x) = x_1 \oplus x_2 \oplus x_3 \oplus x_3x_4 \oplus x_1x_2x_4 \oplus x_1x_3x_4 \oplus x_2x_3x_4$$

has seven single terms in its ANF representation, and the corresponding multivariate equation of $f(x) = 0$ becomes something like $x_1 \oplus x_2 \oplus x_3 \oplus y_1 \oplus y_2 \oplus y_3 \oplus y_4 = 0$, an equation with seven variables (unknowns).

When all the functions in Eq. 5.1 become a multivariate equation, then Eq. 5.1 becomes a system of equations of many variables. Let the degree of $f(x)$ be k, since the degree of $f(x)$ is the same as that of $f(L^k(x))$ for any k, where $L(x)$ is the state-updating function as described above; it is seen that the maximum possible number of variables (or the number of different single terms) of Eq. 5.1 is at most

$$\binom{n}{1} + \binom{n}{2} + \cdots + \binom{n}{k}.$$

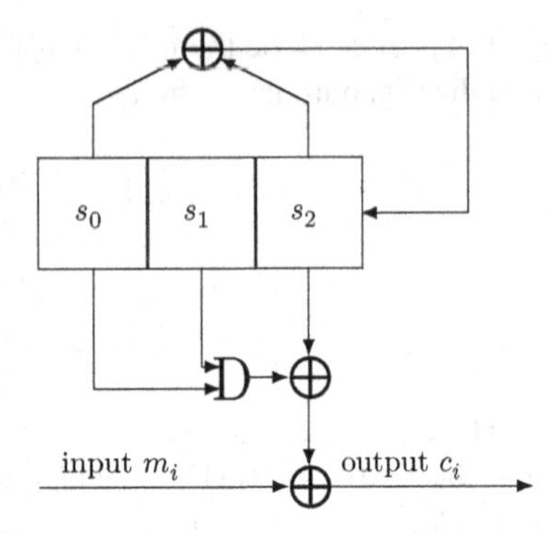

Fig. 5.2 A nonlinear feedforward generator of order 3

If there are more functions in Eq. 5.1 than the number of variables, then Eq. 5.1 is called an *over-defined system*. The process of an algebraic attack is actually to solve such an over- defined system [10]. Now we'll see how to perform an algebraic attack by solving such an over- defined system of equations.

Consider the model as depicted in Fig. 5.2.

In Fig. 5.2, the notation "D" means the AND operator, which is equivalent to the multiplication of Boolean variables. It is clear that the feedback iteration of the LFSR is $a_{i+3} = a_i \oplus a_{i+2}$ for $i = 0, 1, 2, \ldots$ and the feedforward function is $f(x) = x_1x_2 \oplus x_3$.

Let the initial state of the LFSR be $S = (s_0, s_1, s_2)$. Then the relationship between the outputs b_i of $f(x)$ and the LFSR initial state can be represented as follows:

$$\begin{cases} b_0 = f(s_0, s_1, s_2) = s_0s_1 \oplus s_2 \\ b_1 = f(L(s_0, s_1, s_2)) = f(s_1, s_2, s_0 \oplus s_2) = s_1s_2 \oplus s_0 \oplus s_2 \\ b_2 = f(L^2(s_0, s_1, s_2)) = f(s_2, s_0 \oplus s_2, s_0 \oplus s_1 \oplus s_2) = s_0s_2 \oplus s_0 \oplus s_1 \\ b_3 = f(L^3(s_0, s_1, s_2)) = f(s_0 \oplus s_2, s_0 \oplus s_1 \oplus s_2, s_0 \oplus s_1) = s_0s_1 \oplus s_1s_2 \oplus s_1 \oplus s_2 \\ b_4 = f(L^4(s_0, s_1, s_2)) = f(s_0 \oplus s_1 \oplus s_2, s_0 \oplus s_1, s_1 \oplus s_2) = s_0s_2 \oplus s_1s_2 \oplus s_0 \oplus s_2 \\ b_5 = f(L^5(s_0, s_1, s_2)) = f(s_0 \oplus s_1, s_1 \oplus s_2, s_0) = s_0s_1 \oplus s_0s_2 \oplus s_1s_2 \oplus s_0 \oplus s_1 \\ b_6 = f(L^6(s_0, s_1, s_2)) = f(s_1 \oplus s_2, s_0, s_1) = s_0s_1 \oplus s_0s_2 \oplus s_1 \\ \quad \cdots\cdots \end{cases} \tag{5.2}$$

Note that Eq. 5.2 is effectively a system of multivariate equations in six variables s_0, s_1, s_2 and the generated variables s_0s_1, s_0s_2, and s_1s_2. If 6 of such functions are given, the value of these six variables may be determined. When more than six equations are given, the system is an over-defined system. Here we assume that we are given such an over-defined system of equations, i.e., the functions of $b_0, b_1, \ldots, b_6$. Now rewrite this equation as

$$\begin{pmatrix} 0\,0\,1\,1\,0\,0 \\ 1\,0\,1\,0\,0\,1 \\ 1\,1\,0\,0\,1\,0 \\ 0\,1\,1\,1\,0\,1 \\ 1\,0\,1\,0\,1\,1 \\ 1\,1\,0\,1\,1\,1 \\ 0\,1\,0\,1\,1\,0 \end{pmatrix} \begin{pmatrix} s_0 \\ s_1 \\ s_2 \\ s_0s_1 \\ s_0s_2 \\ s_1s_2 \end{pmatrix} = \begin{pmatrix} b_0 \\ b_1 \\ b_2 \\ b_3 \\ b_4 \\ b_5 \\ b_6 \end{pmatrix}.$$

Since the last row of the coefficient matrix is the exclusive-or (XOR) of the other six rows, it is redundant and can be removed. In fact, it is equivalent to remove any row since the remaining coefficient matrix still forms a full rank matrix. For example, when we remove the six-th row instead of the seven-th, the equation then becomes

$$\begin{pmatrix} 0\,0\,1\,1\,0\,0 \\ 1\,0\,1\,0\,0\,1 \\ 1\,1\,0\,0\,1\,0 \\ 0\,1\,1\,1\,0\,1 \\ 1\,0\,1\,0\,1\,1 \\ 0\,1\,0\,1\,1\,0 \end{pmatrix} \begin{pmatrix} s_0 \\ s_1 \\ s_2 \\ s_0s_1 \\ s_0s_2 \\ s_1s_2 \end{pmatrix} = \begin{pmatrix} b_0 \\ b_1 \\ b_2 \\ b_3 \\ b_4 \\ b_6 \end{pmatrix}.$$

Since the coefficient matrix is a full rank one, there is a unique solution for the variables s_0, s_1, s_2 and s_0s_1, s_0s_2, and s_1s_2. Since our target is to recover the initial state of the LFSR, it is not necessary to find the values of s_0s_1, s_0s_2, and s_1s_2. Given any set of specific values of b_0, b_1, b_2, b_3, b_4 and b_6, it is easy to compute the values of s_0, s_1, and s_2.

For a relatively large system, even if we get an over-defined system of equations, it may not form a full rank coefficient matrix, and hence it has many solutions. Among those possibilities, some incorrect ones can be filtered by checking the relationships between single variables $x_1, x_2, \ldots, x_n$ and their products, where the latter ones have been treated as new variables in the process of equation solving. So to determine a unique solution is not always an easy task.

5.3 Annihilators and Algebraic Immunity of Boolean Functions

The above analysis shows that, to break the stream cipher as shown in Fig. 5.1, it needs to solve the system of nonlinear Eq. 5.1, where the unknowns are the coefficients of $f(x)$. However, even if we know that such a function exists and Eq. 5.1 can uniquely determine the values of $x_1, x_2, \ldots, x_n$, to find the actual values for the function is equivalent to solving the system of equations (5.1) which is nonlinear, and the problem may become computationally infeasible when n is large. So far there is no efficient algorithm to solve such a system of nonlinear equations, so we must find a practical way to solve such a system of nonlinear equations; at least that works when n is reasonably small.

One such approach is to replace every nonlinear term by a new variable. Then apart from possible variables $x_1, x_2, \ldots, x_n$, all their products, when they appear in the expression of Eq. 5.1, will become new variables. In general, the number of variables can be up to the number of all the products of $x_1, x_2, \ldots, x_n$, which equals

$$\binom{n}{1} + \binom{n}{2} + \cdots + \binom{n}{n} = 2^n - 1.$$

Apparently when n is large, the above is a very large number, although the actual number of variables is less than $2^n - 1$; for a random Boolean function $f(x)$, the number of terms of $f(x)$ is about half of that number, and the number of terms in Eq. 5.1 is likely to be more than the number of terms of $f(x)$ itself. So after variable replacement, although Eq. 5.1 becomes a system of linear equations, due to the large number of variables, it is computationally infeasible to solve. However, noting that the algebraic degree of $f(x)$ may be far smaller than n, say $deg(f) = t < n$, and noticing that the degree of $f(L^k(x))$ is a function of degree at most t, then the number of products (including single terms $x_1, x_2, \ldots, x_n$) of Eq. 5.1 is

$$\binom{n}{1} + \binom{n}{2} + \cdots + \binom{n}{t}.$$

If t is small enough, then $\binom{n}{1} + \binom{n}{2} + \cdots + \binom{n}{t}$ is not terribly large, and hence it is possible to solve Eq. 5.1 using the technique of variable replacement.

Experiments show that at present people can only handle very little t for $n = 128$ or similar scale. For example, when $t \le 5$, it seems to be possible to solve Eq. 5.1; however when $t > 6$, to solve Eq. 5.1 is infeasible using normal computers.

However, practically the feedforward function is likely to be of an algebraic degree higher than 6; hence the attack via solving a system of nonlinear equations does not work. Meier and Courtois [8, 16] considered the following cases and concluded that under certain circumstances, even if the algebraic degree of $f(x)$ is high, it may be possible to solve the unknowns of Eq. 5.1 with a transformation:

[A1] There exists a Boolean function $g(x) \in \mathcal{F}_n$ of low algebraic degree such that $g(x)f(x) = 0$ holds. In this case, $g(x)$ is called an *annihilator* of $f(x)$. Note that $g(x) = 0$ is a trivial annihilator of any function $f(x)$; so in general, we only consider the nonzero annihilators, unless specified otherwise.

If there exists an annihilator $g(x)$ of $f(x)$ with $deg(g) < deg(f)$, then given a sufficient number of outputs $b_i = f(L^i(x))$, we have

$$f(L^i(x))g(L^i(x)) = b_i g(L^i(x)) = 0.$$

When $b_i = 1$, we must have $g(L^i(x)) = 0$. When there are a sufficient number of such b_i, we have a sufficient number of equations $g(L^i(x)) = 0$. Since the algebraic degree of g is lower than that of $f(x)$, by solving the system of equations $g(L^i(x)) = 0$, we can hopefully find a solution of x which is also the solution for Eq. 5.1.

[A2] There exists a Boolean function $g(x) \in \mathcal{F}_n$ with $deg(g) < deg(f)$ such that $g(x)(1 \oplus f(x)) = 0$ holds. In this case, from $b_i = f(L^i(x))$, we have $(f(L^i(x)) \oplus 1)g(L^i(x)) = (b_i \oplus 1)g(L^i(x)) = 0$. When $b_i = 0$, we must have $g(L^i(x)) = 0$, and the work for finding a solution of Eq. 5.1 becomes that of finding a solution of the system of equations $g(L^i(x)) = 0$, which is easier due to the degree of $g(x)$ being lower than that of $f(x)$.

It is noted that conditions **[A1]** and **[A2]** are independent of each other, i.e., one condition does not affect the other. For example, $f(x) = x_1x_2x_3$ in 3 variables satisfies condition **[A1]**, as $g(x) = x_1 \oplus x_2$ is an annihilator, but not condition **[A2]**, and $f(x) = x_1 \oplus x_1x_2 \oplus x_3 \oplus x_1x_3 \oplus x_4 \oplus x_1x_4 \oplus x_2x_4 \oplus x_1x_2x_4 \oplus x_3x_4 \oplus x_1x_3x_4 \oplus x_2x_3x_4 \oplus x_1x_2x_3x_4$ in four variables has a low-degree annihilator $g(x) = x_1x_2 \in \mathcal{F}_4$, and $f(x) \oplus 1$ also has a low-degree annihilator $h(x) = x_3 \oplus x_1x_3$ (or $h(x) = x_1 \oplus x_1x_2$).

[A3] There exist Boolean functions $g(x) \in \mathcal{F}_n$ and $h(x) \in \mathcal{F}_n$ such that $g(x)f(x) = h(x)$ holds, where $deg(h) < deg(f)$.

It can be shown that this case is equivalent to case **[A2]**. Assume that condition **[A3]** holds. Multiply both sides of $g(x)f(x) = h(x)$ by $f(x)$, and we have $f(x)g(x) = f(x)h(x) = h(x)$, or $h(x)(1 \oplus f(x)) = 0$. So this leads to case **[A2]**.

[A4] There exist low-degree factors of $f(x)$, i.e., there exists a low-degree Boolean function $g(x)$ such that $f(x) = g(x)h(x)$, where $h(x)$ is another Boolean function which may have high or low degree.

For this case, multiply both sides of $f(x) = g(x)h(x)$ by $g(x) \oplus 1$, and we have $f(x)g(x) = g(x)h(x) = f(x)$ or $f(x)(1 \oplus g(x)) = 0$. Since by assumption that $1 \oplus g(x)$ is of low degree, this leads to the case **[A1]**.

By the above discussion, we see that many different cases can be converted into either case **[A1]** or case **[A2]**. This means that, even if the algebraic degree of $f(x)$ is high, as long as there exists a low-degree annihilator of $f(x)$ or of $f(x) \oplus 1$, then the algebraic attack would work. The minimum of algebraic degree of the annihilators of $f(x)$ and that of $1 \oplus f(x)$ is called the *algebraic immunity* of $f(x)$ and is denoted by

$$AI(f) = \min\{deg(g) : f(x)g(x) = 0 \text{ or } (f(x) \oplus 1)g(x) = 0\} \quad (5.3)$$

By this notion, the problem of improving the efficiency of algebraic attack is converted into finding the low-degree annihilator of $f(x)$ or of $1 \oplus f(x)$, and the problem of resisting algebraic attack becomes the problem of finding Boolean functions with high algebraic immunity. It should be noted that the algebraic immunity is just one of many important cryptographic properties.

5.4 Construction of Annihilators of Boolean Functions

Construction of Boolean function with high-order algebraic immunity is not an easy task; that of the highest order of algebraic immunity is certainly a challenging topic. Some special class of Boolean functions with high-order algebraic immunity have been found [5–7]; however, general construction still remains a research topic. Since the algebraic immunity has close relationship with the annihilators of a given Boolean function, construction of annihilators is an approach to study the algebraic immunity. On the other hand, when performing a practical algebraic attack, a concrete annihilator is to be used.

Let $f(x) \in \mathcal{F}_n$. It is obvious that if $g(x)$ is an annihilator of $f(x)$, then $f(x)$ is also an annihilator of $g(x)$.

Let $f(x) \in \mathcal{F}_n$, and denote

$$AN(f) = \{g(x) \in \mathcal{F}_n : g(x)f(x) = 0\} \tag{5.4}$$

as the set of all the annihilators of $f(x)$. Then it is trivial to verify that

Theorem 5.1. *$AN(f)$ with multiplication operation of Boolean functions forms a multiplicative group, and together with the XOR operation, $AN(f)$ forms a ring.*

The conclusion of Theorem 5.1 comes from the simple observation that if $f(x)g_1(x) = 0$ and $f(x)g_2(x) = 0$, then both $f(x)(g_1(x)g_2(x)) = 0$ and $f(x)(g_1(x) \oplus g_2(x)) = 0$ hold.

Theorem 5.2. *Let $f(x) \in \mathcal{F}_n$, and then the minimum degree of the annihilators of $f(x)$ is no larger than $deg(f)$.*

Proof. The conclusion comes from the fact that any Boolean function $f(x)$ has a special annihilator $f(x) \oplus 1$ which has the same algebraic degree as $f(x)$. □

Theorem 5.2 shows that the minimum degree of the annihilators (nonzero ones) of a Boolean function $f(x)$ cannot be larger than the degree of $f(x)$ itself; however, there are cases when it may not be less either. One of the key steps of algebraic attack is to find a low-degree annihilator (of $f(x)$ or of $1 \oplus f(x)$), and how to efficiently find such a low-degree annihilator becomes critical for the algebraic attack to work more efficiently. Here we introduce some constructions of annihilators for an arbitrary Boolean function.

Theorem 5.3. *Let $f(x) \in \mathcal{F}_n$, $g(x)$ be an annihilator of $f(x)$. Then $supp(g)$ must be a subset of $supp(1 \oplus f)$, i.e., $supp(g) \subseteq supp(1 \oplus f)$.*

Proof. For any x with $g(x) = 1$, since $f(x)g(x) = 0$ always hold, we must have $f(x) = 0$, which is equivalent to $1 \oplus f(x) = 1$, i.e., $x \in supp(1 \oplus f)$. This means that $supp(g) \subseteq supp(1 \oplus f)$. □

Some constructions of annihilators of Boolean functions are given by Meier in [16]. Those algorithms are fundamental; hence they are described here.

Algorithm 5.1 (Construction of low-degree annihilators [16]).

Input: A Boolean function $f(x)$ in n variables;
Output: A low-degree annihilator of $f(x)$.
Choose a small integer d, and solve the system of equations composed by $g(x) = 0$ for all $x \in supp(f)$ for the unknown coefficients of $g(x)$, where $g(x)$ is of degree d and can be written as

$$g(x) = a_0 \oplus \bigoplus_{i=1}^{n} a_i x_i \oplus \bigoplus_{1 \le i < j \le n} a_{i,j} x_i x_j \oplus \cdots \oplus \bigoplus_{1 \le i_1 < i_2 < \cdots < i_d \le n} x_{i_1} x_{i_2} \cdots x_{i_d} \quad (5.5)$$

which has $1 + \binom{n}{1} + \binom{n}{2} + \cdots + \binom{n}{d}$ unknown coefficients. If there is a solution for the system of equations (i.e., to find the coefficients of $g(x)$ in (5.5)), such that $g(x) = 0$ hold for all $x \in supp(f)$, then $g(x)$ is an annihilator of $f(x)$, and the degree of $g(x)$ is at most d. If no solution can be found, then increase d by 1, and repeat the above process until a solution can be found.

Algorithm 5.1 is definitely convergent since when d equals $deg(f)$, there must be at least one solution for the equation. However, the computational complexity of the above algorithm can be very high. Below is an improvement of the algorithm:

Algorithm 5.2 (Improvement of Algorithm 5.1 [16]).

Input and output: Same as in Algorithm 5.1.
Step 1: Find from the support of $f(x)$ the vectors whose Hamming weight is 1, since for these vectors v, we must have $g(v) = 0$, this leads to

$$g(e_i) = a_0 \oplus a_i = 0.$$

If $f(x)$ is balanced, then on average there should be half of the coefficients of linear terms of $g(x)$ that can be represented by a_0.
Step 2: For those vectors in $supp(f)$ having Hamming weight 2, $g(x)$ also takes value 0 on these vectors; hence, we have $a_0 \oplus a_i \oplus a_j \oplus a_{ij} = 0$. So a_{ij} can be represented by $a_0 \oplus a_i \oplus a_j$. Following this approach, when examining all the vectors in $supp(f)$ of Hamming weight up to d, we solve the equation of those unknown coefficients of $g(x)$. If there is a solution, then $g(x)$ is an annihilator of $f(x)$ and the degree of $g(x)$ is at most d. If the number of equations about the unknown coefficients are smaller than that of the unknowns, then consider more vectors in $supp(f)$ with larger Hamming weight, which will yield more relations (equations) about the known coefficients, until there are a sufficiently large number of equations to determine the unknown coefficients of Eq. 5.5.

It is seen that Algorithm 5.2 avoids the d to be chosen larger than necessary at the very beginning as in Algorithm 5.1 and ensures the algorithm to output an annihilator of the lowest degree. However, the algebraic degree is one of the important measures of annihilators, and the other properties (e.g., the number of terms in its ANF representation) of the output function of Algorithm 5.2 may not be the best.

Algorithm 5.3 gives another approach, which is also efficient and can produce a large number of annihilators of a given function.

Algorithm 5.3 (Construction of low-degree annihilators).

Input: A Boolean function $f(x)$ in n variables.
Output: A set of low-degree annihilators of $f(x)$.
Step 1: Choose a small integer d, and let F_0 be the set of all the monomials that have degree $\leq d$. Let $supp(f) = \{v_1, v_2, \ldots, v_s\}$.
Step 2: For $i = 1$ to s do{
Let $F_i^0 = \{f \mid f \in F_{i-1}, f(v_i) = 0\}$ and
$F_i^1 = \{f \mid f \in F_{i-1}, f(v_i) = 1\}$
If $|F_i^1| > 1$, then{
set G_i^0 to be the set of even weight linear combinations of functions in F_i^1;
}
else set $G_i^0 = \emptyset$ be the empty set;
set $F_i := F_i^0 \cup G_i^0$
}
Step 3: If F_s is empty, go to Step 1 with d incremented by 1; else output the set F_s.

The basic idea of the algorithm is as follows: let F be initialized as the set of monomials that have degree $\leq d$; the problem is to find all nonzero linear combinations of functions in F, so that the linear combinations take value 0 on all the inputs from $supp(f) = \{v_1, v_2, \cdots, v_s\}$. More precisely, Algorithm 5.3a describes the detailed process:

Algorithm 5.3a

Step 1: Divide the functions in F into two disjoint subsets according to their values at v_1, i.e.,

$$F_1^0 = \{f \in F : f(v_1) = 0\}$$
$$F_1^1 = \{f \in F : f(v_1) = 1\}$$

Step 2: Let even weight linear combinations of the function in F_1^1 form a set G_1^0. Union it with the set F_1^0, and we have a new set $F_1 = F_1^0 \cup G_1^0$.
Step 3: Let $F = F_1$, and divide the functions in F_1 into two disjoint subsets according to their values at v_2, which result in two function sets, i.e.,

$$F_2^0 = \{f \in F : f(x_2) = 0\}$$
$$F_2^1 = \{f \in F : f(x_2) = 1\}$$

Let the even weight linear combinations of functions in F_2^1 form a set G_2^0, and union it with F_2^0; we get F_2.
Step 4: Divide $F = F_2$ into two subsets F_3^0 and F_3^1 according whether they take value 0 or 1 on v_3, and repeat the same process as in Step 1 and Step 2, until we get a set F_s.

The idea of the algorithm is to narrow down the set of functions who take value 0 on one vector in *supp(f)*, on two vectors in *supp(f)*, and so on, until on all the vectors in *supp(f)*, which is the set F_s. However, if $f(x)$ does not exist an annihilator of degree d or less, then obviously F_s will be an empty set; in this case, we need to increase the value of d in order to find a nonzero annihilator of $f(x)$.

In order to demonstrate how the Algorithm 5.3 works, below we give a small example by finding the annihilators of $f(x) = x_1x_2 \oplus x_3$, a Boolean function in three variables, where the algebraic degree of the annihilators is expected to be no more than 2.

It is easy to compute the support of $f(x)$ as $supp(f)$ ={(0,0,1), (0,1,1), (1,0,1), (1,1,0)}. Our task is to find Boolean functions with algebraic degree being no more than 2 who take value 0 on all the elements in *supp(f)*. According to Algorithm 5.3, we first need to list all the monomials whose algebraic degrees are no more than 2. This will result in the following set:

$$F = \{1, x_1, x_2, x_3, x_1x_2, x_1x_3, x_2x_3\}$$

Next, by dividing F into two disjoint subsets according to whether the functions take value 0 or 1 at $v_1 = (0, 0, 1)$, we have

$$F_1^0 = \{x_1, x_2, x_1x_2, x_1x_3, x_2x_3\}$$
$$F_1^1 = \{1, x_3\}$$

Then, when making the union of F_1^0 and the even weight linear combination of functions in F_1^1, we get

$$F_1 = \{x_1, x_2, x_1x_2, x_1x_3, x_2x_3, 1 \oplus x_3\}$$

Without confusion, let $F = F_1 = \{x_1, x_2, x_1x_2, x_1x_3, x_2x_3, 1 \oplus x_3\}$, and then repeat the same process, i.e., divide $F = F_1$ into two disjoint subsets according to whether the functions (elements in F) take value 0 or 1 at $v_2 = (0, 1, 1)$, and we get

$$F_2^0 = \{x_1, x_1x_2, x_1x_3, 1 \oplus x_3\}$$
$$F_2^1 = \{x_2, x_2x_3\}$$

Making the union of F_2^0 and the even weight linear combinations of elements in F_2^1, we get

$$F_2 = \{x_1, x_1x_2, x_1x_3, 1 \oplus x_3, x_2 \oplus x_2x_3\}$$

Again let $F = F_2$, which is then divided into two disjoint subsets according to whether the functions (elements in F) take value 0 or 1 at $v_3 = (1, 0, 1)$, and we get

$$F_3^0 = \{x_1x_2, 1 \oplus x_3, x_2 \oplus x_2x_3\}$$

$$F_3^0 = \{x_1, x_1x_3\}$$

Similar to the above process, we get

$$F_3 = \{x_1x_2, 1 \oplus x_3, x_2 \oplus x_2x_3, x_1 \oplus x_1x_3\}.$$

Divide $F = F_3$ into two disjoint subsets according to whether the functions (elements in F) take value 0 or 1 at $v_4 = (1, 1, 0)$, and we get

$$F_4^0 = \emptyset$$
$$F_4^1 = \{x_1x_2, 1 \oplus x_3, x_2 \oplus x_2x_3, x_1 \oplus x_1x_3\}$$

Finally, we get the union of F_4^0 and the even weight linear combinations of functions in F_4^1, i.e., the set F_4, which has eight nonzero elements, together with the zero function, and it has a basis

$$B = \{1 \oplus x_3 \oplus x_1x_2, 1 \oplus x_2 \oplus x_3 \oplus x_2x_3, 1 \oplus x_1 \oplus x_3 \oplus x_1x_3\}.$$

Therefore, we get eight nontrivial annihilators of $f(x)$ with algebraic degree being no more than 2, and it is noted that their algebraic degrees are exactly 2.

It is interesting to see that in theory Algorithm 5.3 can find all the possible annihilators with degree $\leq d$ of a given function $f(x)$, if they ever exist.

Theorem 5.4. *Algorithm 5.3 can find all the annihilators of any Boolean function $f(x)$ with algebraic degree up to d.*

Proof. Let $g(x)$ be an annihilator of $f(x)$ and $deg(g) \leq d$. Write $g(x)$ in the form of XOR of the monomials, i.e., $g(x) = g_1(x) \oplus g_2(x) \oplus \cdots \oplus g_t(x)$. Then each $g_i(x)$ is a monomial of degree $\leq d$; hence, it is included in the initialization of the set F that starts Algorithm 5.3. When F is divided into two subsets, those $f_j(x)$ that take value 0 on v_1 are in one same set F_1^0, and the rest of $f_j(x)$ that take value 1 on v_1 are in the other set F_1^1, Then a new set F_1 is formed that includes all the elements in F_1^0 and the even weight linear combinations of those in F_1^1. It is noted that F_1 includes all the monomials $g_i(x)$, either in the form of monomials or in the form of a linear combination of them. Note that $g(x)$ must be a linear combination of functions in F_1, since $g(x)$ takes value 0 on v_1. Following Algorithm 5.3, we will get sets F_2, $F_3, \cdots, F_{n-1}$, and similarly it is easy to verify that they all contain $g(x)$ as a linear combination of its elements. Note that Algorithm 5.3 outputs a set, which means that any (nonzero) linear combination of the elements in the output set is an annihilator of the given function $f(x)$. According to the above analysis, the linear combinations will include $g(x)$ as a special linear combination. This concludes Theorem 5.4. □

Define a matrix $P_n^d(S)$ whose column vectors are the values of monomials of degree no more than d on the vectors in S. When $S = GF^n(2)$, we simply write $P_n^d(S)$ as P_n^d. Note that $P_n^d(S)$ can be uniquely determined if vectors in S always follow a specific order. For example, let $S = \{(0,0,1),(0,1,1),(1,0,1),(1,1,0)\}$, where the order of the vectors in S is as how they are written, which can also be treated as the integers in incremental order, where the vectors are the corresponding 2-adic representation. Then $P_3^2(S)$ can be defined as follows:

S			$P_3^2(S)$						
			1	x_1	x_2	x_3	x_1x_2	x_1x_3	x_2x_3
0	0	1	1	0	0	1	0	0	0
0	1	1	1	0	1	1	0	0	1
1	0	1	1	1	0	1	0	1	0
1	1	0	1	1	1	0	1	0	0

It is easy to verify the following conclusion.

Lemma 5.1. *For any $S \subseteq \{0,1\}^n$, a sufficient and necessary condition for the existence of a Boolean function with degree no more than d that takes value 0 on all vectors in S is that the rank of the matrix $P_n^d(S)$ is less than the number of columns of $P_n^d(S)$.*

Proof. Sufficiency: If the rank of $P_n^d(S)$ is less than the number of its columns, it means that the column vectors of $P_n^d(S)$ are linearly dependent, so there is a linear combination that results in a zero column vector. It is trivial to verify that the same linear combination on the monomials will result in a function that takes value 0 on all the rows of S. In fact, the columns of $P_n^d(S)$ can be treated as the truth tables of monomials restricted on set S, and the linear combination on these restricted truth tables is equivalent to the linear combination of these monomials, where the resulted linear combination is a zero function when restricted on S. Since each of the monomials has a degree no more than d, then their linear combination is also a polynomial of degree no more than d.

Necessity: Assume the existence of a polynomial of degree no more than d that takes value 0 on all the vectors in S, and then treat the polynomial as a linear combination of each of the terms (monomials); it corresponds to a linear combination of the columns of $P_n^d(S)$. Since the linear combination is a polynomial that takes value 0 on all the vectors in S, it means that the linear combination of the columns corresponding to the monomials will be a zero vector. This means that the columns of $P_n^d(S)$ must be linearly dependent, and hence the rank of $P_n^d(S)$ must be less than the number of its columns. □

By Lemma 5.1, it is easy to prove the well-known upper bound of algebraic immunity given in [9].

Corollary 5.1. *For any n-variable Boolean function $f(x)$, we have that*

$$AI(f) \leq \left\lceil \frac{n}{2} \right\rceil \tag{5.6}$$

Proof. For any Boolean function $f(x)$ in n variables, one of the supports $supp(f)$ and $supp(1 \oplus f)$ must have no more than 2^{n-1} elements. Without loss of generality, let $|supp(f)| \leq 2^{n-1}$, and denote $S = supp(f)$. The number of monomials (including the constant 1) of degree no more than d is $\sum_{i=0}^{d} \binom{n}{i}$. When $d \geq \frac{n}{2}$, we must have that

$$\sum_{i=0}^{d} \binom{n}{i} > 2^{n-1};$$

in this case, the number of columns in matrix $P_n^d(S)$ must be larger than its rank (since it has no more than 2^{n-1} rows); hence an annihilator of degree no more than d must exist. This means that the algebraic immunity of any Boolean function in n variables is upper bounded by d for all $d \geq \frac{n}{2}$. It is easy to verify that

$$\min\{d : d \geq \frac{n}{2}\} = \left\lceil \frac{n}{2} \right\rceil$$

which is the upper bound of algebraic immunity in general case. □

Lemma 5.1 also leads to a new approach of finding low-degree annihilators of a given Boolean function. Hence, we have the following algorithm:

Algorithm 5.4 (Construction of low-degree annihilator).

Input: A Boolean function $f(x)$ in n variables.
Output: A low-degree annihilator of $f(x)$.
Step 0: If $|supp(f)| \leq 2^{n-1}$, then denote $S = supp(f)$; else write $S = supp(f \oplus 1)$.
Step 1: Choose a small integer d, and write down the matrix $P_n^d(S)$.
Step 2: Check if the rank of $P_n^d(S)$ equals the number of its columns. If so, go to Step 1; else continue;
Step 3: Find a set of columns of $P_n^d(S)$ that are linearly dependent;
Step 4: Find a linear combination of the dependent columns of $P_n^d(S)$, and compose the same linear combination of the monomials corresponding to the chosen linearly dependent columns;
Step 5: Output the linear combination of the monomials.

It is noted that Algorithm 5.4 is easy to implement, and the storage needed in implementation is to hold the matrix $P_n^d(S)$.

Theorem 5.5. *For any $S \subseteq \{0,1\}^n$, if*

$$\sum_{i=0}^{d} \binom{n}{i} > |S|,$$

then there must exist a Boolean function with degree no more than d that takes value 0 on all the vectors in S.

Proof. It is noticed that the number of columns of the matrix $P_n^d(S)$ is larger than that of the rows, so the columns are linearly dependent. By Lemma 5.1, the conclusion follows. □

Theorem 5.5 indicates that, when a Boolean function has very little weight, then it will have low-degree annihilators.

5.5 On the Upper and Lower Bounds of Algebraic Immunity of Boolean Functions

A tight upper bound of the algebraic immunity of Boolean functions in the general case was given in [9, 16] as described in Corollary 5.1, which says that the algebraic immunity of Boolean functions in n variables cannot be larger than $\lceil \frac{n}{2} \rceil$. Then we have

Corollary 5.2. *For any integer d, $0 \leq d \leq n$, denote*

$$\mu(n,d) = \binom{n}{0} + \binom{n}{1} + \cdots + \binom{n}{d}.$$

Let $f(x)$ be a Boolean function in n variables satisfying that $\mu(n,d) \geq |supp(f)|$, and then there exists an annihilator of $f(x)$ with degree lower than or equal to d, i.e.,

$$\min\{deg(g) : g(x) \in AN(f)\} \leq d.$$

Proof. The conclusion comes directly from Theorem 5.5 and the fact that $f(x) \oplus 1$ is an annihilator of $f(x)$. □

With respect to the lower bound of algebraic immunity of Boolean functions, we have

Theorem 5.6. *Let $f(x)$ be a Boolean function in n variables and with algebraic degree being no larger than d. Then we have*

$$2^{n-d} \leq |supp(1 \oplus f)| \leq 2^n - 2^{n-d}. \tag{5.7}$$

and both the lower and the upper bounds are tight.

Proof. For a fixed d, we prove the theorem with an induction on n. If $n = d$, then we must have $1 \leq |supp(1 \oplus f)| \leq 2^n - 1$; hence the conclusion is true. Assume that the conclusion is true for some d, i.e., for all the Boolean functions $f(x)$ with degree no larger than d, the inequality $|supp(1 \oplus f)| \leq 2^n - 2^{n-d}$ holds. Let $f \in \mathcal{F}_{n+1}$ be a Boolean function in $n+1$ variables that has algebraic degree no more than d. Then $f(x)$ must be represented in the following format:

$$f(x_1, \cdots, x_{n+1}) = f'(x_1, \cdots, x_n) \oplus x_{n+1} f''(x_1, \cdots, x_n)$$

where $f', f'' \in \mathcal{F}_n$, $deg(f') \leq d$ and $deg(f'') \leq d - 1$ and either f' or f'' has to be nonzero. Now we consider the following three cases:

(1) If $f' \neq 0, f'' = 0$, then $f(x)$ can be treated as a function in $\mathcal{F}_n$, and by the assumption, we have $supp(1 \oplus f) \leq (2^d - 1)2^{n-d} \leq (2^d - 1)2^{n+1-d}$.
(2) If $f' = 0, f'' \neq 0$, then $supp(1 \oplus f) = \{(x, 0) \cup (x, 1) | x \in supp(1 \oplus f'')\}$, since $deg(f'') \leq d - 1$; hence we have

$$|supp(1 \oplus f)| \leq 2^n + (2^{d-1} - 1)2^{n-(d-1)} = (2^{d-1})2^{n+1-d}.$$

(3) If $f' \neq 0, f'' \neq 0$, then $supp(1 \oplus f)$ can be written as

$$supp(1 \oplus f) = \{(x, 0) \mid x \in supp(1 \oplus f')\} \cup \{(x, 1) \mid x \in supp(1 \oplus f' \oplus f'')\}.$$

If $f' \oplus f'' = 0$, then we get $deg(f') = deg(f'') \leq d - 1$ and

$$\begin{aligned}|supp(1 \oplus f)| &= |supp(1 \oplus f')| + |supp(1 \oplus f' \oplus f'')| \\ &\leq (2^d - 1)2^{n-(d-1)} + 2^n = (2^d - 1)2^{n+1-d}\end{aligned}$$

If $f' \oplus f'' \neq 0$, then we have

$$\begin{aligned}|supp(1 \oplus f)| &= |supp(1 \oplus f')| + |supp(1 \oplus f' \oplus f'')| \\ &\leq (2^d - 1)2^{n-d} + (2^d - 1)2^{n-d} = (2^d - 1)2^{n+1-d}.\end{aligned}$$

In particular, if we choose f' such that $supp(1 \oplus f') = (2^d - 1)2^{n-d}$ and $f'' = 0$, then the upper bound $(2^d - 1)2^{n+1-d}$ can be reached. □

Theorem 5.7. *Let $f(x) \in \mathcal{F}_n$. If $|supp(f)| > 2^n - 2^{n-d}$, then there does not exist an annihilator of $f(x)$ with degree less than or equal to d.*

Proof. Let $g(x)$ be an annihilator of $f(x)$. By $f(x)g(x) = 0$, it is known that $supp(f) \subseteq supp(1 \oplus g)$; hence $|supp(f)| \leq |supp(1 \oplus g)|$. Since $|supp(f)| > 2^n - 2^{n-d}$, we have $|supp(1 \oplus g)| > 2^n - 2^{n-d}$, and by Theorem 5.6, we know that the algebraic degree of $g(x)$ must be larger than d; hence the conclusion of Theorem 5.7 holds. □

5.6 Computing the Annihilators of Boolean Functions

Algebraic attacks have big impact on the security of stream ciphers and block ciphers. Since the key to the success of algebraic attacks is to find low-degree annihilators of a nonlinear Boolean function, many cryptographic researchers have paid much attention on the efficient computation of annihilators of Boolean functions and on the construction of Boolean function that do not have low-degree annihilators. This section is to find some relationships between a Boolean function $f(x)$ and its annihilators and further to design efficient algorithms to find the annihilators of an arbitrarily given Boolean function. The analysis will also provide some advice to the cryptographic designers as what to avoid in order to avoid the use of Boolean functions with low-degree annihilators.

5.6.1 Computing the Annihilators of Boolean Functions: Approach I

It is noticed that the truth table of $f(x) \oplus 1$ is just the complement of that of $f(x)$, so we may call $f(x) \oplus 1$ *the complement function* of $f(x)$ and vice versa.

Definition 5.2. If all the x satisfying $f(x) = 1$ also satisfy that $g(x) = 1$, i.e., $f(x) = 1$ implies that $g(x) = 1$ always holds, then $g(x)$ is called *a cover* of $f(x)$.

If $g(x)$ is an annihilator of $f(x)$, then $f(x)$ is an annihilator of $g(x)$ as well. By Theorem 5.3, it is known that, if $g(x)$ is an annihilator of f, then $g(x) \oplus 1$ must be a cover of $f(x)$.

Since the algebraic degree of $g(x)$ and that of $g(x) \oplus 1$ are the same, if we are given the ANF representation of $g(x) \oplus 1$, then it is easy to get that of $g(x)$. Therefore, in order to find a low-degree annihilator $g(x)$ of $f(x)$, the relationship $supp(g+1) \supseteq supp(f)$ is useful. Our approach is to find a low-degree Boolean function whose support is a subset of $supp(f)$. More precisely, we have the following intuitive but very useful theorem:

Theorem 5.8. *The problem of finding a low-degree annihilator of $f(x)$ is equivalent to finding a low-degree Boolean function whose support is a subset of $supp(f)$.*

By Theorem 5.8, to find an annihilator of $f(x)$ is equivalent to the problems of finding a Boolean function whose support is a subset of $supp(f)$. Let $t = wt(f)$, and then $wt(f \oplus 1) = 2^n - t$. In the support of $f(x) \oplus 1$ that has $2^n - t$ elements, select a nonempty subset as the support of a new Boolean function $g(x)$, and then $g(x)$ is an annihilator of $f(x)$. Hence we get

Theorem 5.9. *Let $f(x)$ be a Boolean function in n variables and $wt(f) = t$. Then the number of annihilators of $f(x)$ equals*

$$\sum_{i=1}^{2^n-t} \binom{2^n - t}{i} = 2^{2^n-t} - 1.$$

Theorem 5.9 gives the total number of annihilators of a given Boolean function, and these annihilators can be constructed using Theorem 5.8. This means that in theory we may construct all the annihilators of an arbitrary Boolean function; however, in practice, only the low-degree annihilators are useful in performing an algebraic attack. Practically, when the algebraic degree of the annihilator (if it is the lowest degree) of a Boolean function is larger than 5, it is difficult to launch an algebraic attack, or the attack is impractical. So with respect to the practical algebraic attacks, only the low-degree annihilators of Boolean functions are of interest. Here we tend to give an efficient and practical algorithm to find low-degree annihilators of Boolean functions.

Theorem 5.10. *Let $f_1(x')$, $f_2(x')$ and $g_1(x')$, $g_2(x')$ be Boolean functions in $n-1$ variables, where $g_1(x')$ is an annihilator of $f_1(x')$ and $g_2(x')$ is an annihilator of $f_2(x')$. Let $f(x) = x_n f_1 \oplus (x_n \oplus 1) f_2$ be a Boolean function in n variables, and then $f(x)$ has an annihilator $g(x) = x_n g_1 \oplus (x_n \oplus 1) g_2$.*

Proof. It is easy to verify that

$$\begin{aligned} g(x)f(x) &= (x_n g_1 \oplus (x_n \oplus 1) g_2)(x_n f_1 \oplus (x_n \oplus 1) f_2) \\ &= x_n f_1 g_1 \oplus (x_n \oplus 1) f_2 g_2 = 0. \end{aligned}$$

This indicates that $g(x)$ is indeed an annihilator of $f(x)$. □

Theorem 5.10 indicates that it is possible to use the decomposition of Boolean functions to find annihilators. If we write $f(x)$ as $f(x) = x_n f_1(x') \oplus (x_n \oplus 1) f_2(x')$, which is called a *cascade representation* of $f(x)$, then the problem of finding annihilators of $f(x)$ can be converted into the problem of finding annihilators of $f_1(x')$ and of $f_2(x')$, respectively.

However, there is a question about whether all the annihilators of $f(x)$ in its cascade representation $f(x) = x_n f_1 \oplus (x_n \oplus 1) f_2$ have an annihilator $g(x)$ which is also in its cascade representation as $g(x) = x_n g_1 \oplus (x_n \oplus 1) g_2$, where g_1 is an annihilator of f_1 and g_2 is an annihilator of f_2. We have the following conclusion:

Theorem 5.11. *Let $f(x) = x_n f_1 \oplus (x_n \oplus 1) f_2$; a Boolean function in n variables in cascade representation has an annihilator $g(x)$. Then there must exist g_1, an annihilator of f_1, and g_2, an annihilator of f_2, such that $g(x)$ can be written as $g(x) = x_n g_1 \oplus (x_n \oplus 1) g_2$*

Proof. Let $g_1 = g(x|x_n = 0)$, $g_2 = g(x|x_n = 1)$, and then it is trivial to verify that the conclusion of Theorem 5.11 holds. □

By Theorems 5.10 and 5.11, the problem of finding annihilators of $f(x)$ can be converted into the ones of finding the annihilators of f_1 and of f_2, respectively. Since the number of variables of f_1 and of f_2 is smaller than that of $f(x)$, hence the problem of finding annihilators for a Boolean function with less number of variables becomes easier on average. Note that our problem is not to find all the annihilators; it is targeted at finding an annihilator with the lowest algebraic degree. Now a question is, if it is possible to find an annihilator with the lowest algebraic degree for both f_1

and f_2, does it mean that we can form an annihilator with the lowest algebraic degree for $f(x)$? First, we give the following conclusion:

Theorem 5.12. *Let $f(x) = x_n f_1 \oplus (x_n \oplus 1) f_2$ be a Boolean function in n variables. Then we have*

$$deg(f) = \begin{cases} deg(f_1) + 1 \text{ if } deg(f_1) > deg(f_2) \\ deg(f_2) + 1 \text{ if } deg(f_1) < deg(f_2) \\ deg(f_1) + c \text{ if } deg(f_1) = deg(f_2) \end{cases} \tag{5.8}$$

where $c \in \{0, 1\}$. Moreover, when f_1 and f_2 have the same term with the highest degree, i.e., when $deg(f_1 \oplus f_2) < deg(f_1) = deg(f_2)$, we have $c = 0$; else $c = 1$.

Proof. Since $f(x) = x_n(f_1 \oplus f_2) \oplus f_2$, when $deg(f_1) \neq deg(f_2)$, it is obvious that $deg(f_1 \oplus f_2) = \max\{deg(f_1), deg(f_2)\}$ must hold. Hence the first two items in Eq. 5.8 hold obviously. With respect to the last item, it is noticed that the degree of $f_1 \oplus f_2$ is less than that of f_2 if and only if f_1 and f_2 have the same term with the highest degree, and in this case, $f(x)$ has the same degree with f_2. Otherwise, we must have $deg(f_1 \oplus f_2) = deg(f_1) = deg(f_2)$, i.e., $deg(f) = deg(f_2) + 1$. Hence the conclusion of Theorem 5.12 is true. □

By Theorem 5.12, we directly have

Corollary 5.3. *Let $f(x) = x_n f_1 \oplus (x_n \oplus 1) f_2$, where f_1 and f_2 are Boolean functions in $n - 1$ variables. Then we have*

$$deg(f) \leq \max\{deg(f_1), deg(f_2)\} + 1.$$

With the above preparation, we are ready to discuss some of the properties of annihilators of Boolean functions in their cascade representation.

Theorem 5.13. *Let $f(x) = x_n f_1 \oplus (x_n \oplus 1) f_2$ be a Boolean function in n variables. Then the minimum degree of annihilators of $f(x)$ is*

$$\min_{g_1 \in AN(f_1),\, g_2 \in AN(f_2)} \{\max\{deg(g_2), deg(g_1 \oplus g_2) + 1\}\} \tag{5.9}$$

Proof. For any $g_1 \in AN(f_1)$ and any $g_2 \in AN(f_2)$, by Theorem 5.11, we know that $g(x) = x_n g_1 \oplus (x_n \oplus 1) g_2$ is an annihilator of $f(x)$. Note that we can write $g(x) = x_n(g_1 \oplus g_2) \oplus g_2$, and it is easy to see that $deg(g) = \max\{deg(g_2), deg(g_1 \oplus g_2) + 1\}$. Take such a function, say $g'(x)$, with the lowest algebraic degree, among all those annihilators in cascade representations, and we get an annihilator of $f(x)$ that has the lowest degree. Further, by Theorem 5.9, these annihilators cover all the possible annihilators of $f(x)$, so $g(x)$ is indeed an annihilator of $f(x)$ with the lowest degree. □

By Theorem 5.13, the correctness of the following algorithm can easily be verified:

Algorithm 5.5 (Construction of annihilator with the lowest degree).

Input: A Boolean function $f(x)$ in n variables.

Output: An annihilator $g(x)$ of $f(x)$ that has the lowest possible degree.

(1) Write $f(x)$ in a cascade representation as $f(x) = x_n(f_1 \oplus f_2) \oplus f_2$.
(2) Find all the possible annihilators $AN(f_1)$ of f_1 and $AN(f_2)$ of f_2, respectively.
(3) For all the possible functions (g_1, g_1), $g_1 \in AN(f_1) and\ g_2 \in AN(f_2)$, compute $d = \max\{deg(g_2), deg(g_1 \oplus g_2) + 1\}$.
(4) Find a function pair (g_1, g_2) corresponding to the smallest possible value of d.
(5) Output $g(x) = x_n g_1 \oplus (x_n \oplus 1)g_2$.

Note that Algorithm 5.5 is meant to find one of the annihilators of $f(x)$ with the lowest degree, which may not be unique. When there are multiple possibilities, it is not sure which annihilator is obtained as the output. In practical applications, an annihilator is only one cryptographic measure, and different annihilators with the same degree may have very different behaviors when performing an algebraic attack. Hence, in practice, more measures need to be taken into consideration, e.g., the number of terms of the final annihilators.

5.6.2 *Computing the Annihilators of Boolean Functions: Approach II*

Lemma 5.2. *Let S be a k-dimensional subspace of $GF^n(2)$ with M being a $2^k \times n$ matrix with row vectors being different elements in S. Then any nonzero column of M must have half of 0's and half of 1's.*

Proof. Let $\alpha_1, \cdots, \alpha_k$ form a basis of S. When $\lambda = (\lambda_1, \lambda_2, \cdots, \lambda_k)$ go through all the possible vectors in $GF^k(2)$, $\lambda \cdot (\alpha_1, \cdots, \alpha_k)^T$ and must meet all the possible row vectors of M, denote by $a_{1j}, a_{2j}, \cdots, a_{kj}$ the j-th coordinate of $\alpha_1, \cdots, \alpha_k$. If the j-th column of M is nonzero, because $\alpha_1, \cdots, \alpha_k$ is a basis of S, it means that $(a_{1j}, a_{2j}, \cdots, a_{kj}) \neq \mathbf{0}$. Hence the linear combination

$$\lambda_1 a_{1j} \oplus \lambda_2 a_{2j} \oplus \cdots \oplus \lambda_k a_{kj} = 0$$

has 2^{k-1} solutions in $GF^k(2)$, i.e., $a_{1j}, a_{2j}, \cdots, a_{kj}$ has 2^{k-1} linear combinations that result in 0, while other 2^{k-1} linear combinations will result in value 1. □

Definition 5.3. Let V be a d-dimensional subspace of $GF^n(2)$ and c be a nonzero vector in $GF^n(2)$. Then the set $\{c \oplus v, v \in V\}$ is called a *d-dimensional affine subspace* of $GF^n(2)$.

The concept of affine subspace is the same as that of coset as defined in Definition 2.5; however, the former emphasizes the dimension while the latter one does not. Note that an $(n-1)$-dimensional subspace (linear or affine) is also called a hyperplane.

Theorem 5.14 ([17]). *If the support of a Boolean function $f(x)$ in n variables is a k-dimensional subspace or a k-dimensional affine subspace of $GF^n(2)$, then the algebraic degree of $f(x)$ is $n-k$.*

Proof. Let $\{(a_{i_1}, a_{i_2}, \ldots, a_{i_n} : 1 \leq i \leq 2^n - 1\}$ be a k-dimensional subspace of $GF^n(2)$. Let $S = \{(a_{i1} \oplus c_1, \cdots, a_{in} \oplus c_n)\}$, where $c = (c_1, c_2, \cdots, c_n)$ is a constant vector in $GF^n(2)$, and let $g(x)$ be a Boolean function with S being its support. Then $g(x)$ can be written in minterm representation as

$$g(x_1, x_2, \cdots, x_n) = \bigoplus_{(a_{i1} \oplus c_1, \cdots, a_{in} \oplus c_n) \in S} \prod_{l=1}^{n} (x_l \oplus a_{il} \oplus c_l \oplus 1), \tag{5.10}$$

where $c = (c_1, c_2, \cdots, c_n)$. For the convenience of writing, we write $y_l = x_l \oplus c_l \oplus 1$ in Eq. 5.10. Then we have

$$\begin{aligned} h(y_1, y_2, \cdots, y_n) &= \bigoplus_{(a_{i1} \oplus c_1, \cdots, a_{in} \oplus c_n) \in S} \prod_{l=1}^{n} (y_l \oplus a_{il}) \\ &= \bigoplus_{l=1}^{n} (\bigoplus_{i=1}^{2^k} a_{ij_{l+1}} a_{ij_{l+2}} \cdots a_{ij_n}) y_{j_1} y_{j_2} \cdots y_{j_l}, \end{aligned} \tag{5.11}$$

where $1 \leq l \leq n$, and $j_{l+1}, j_{l+2} \cdots j_n$ is a permutation of $1, 2, \cdots, n$, i.e.,

$$\{j_{l+1}, j_{l+2} \cdots j_n\} = \{1, 2, \cdots, n\} \backslash \{j_1, \cdots, j_l\}.$$

It is easy to see that $deg(h) = deg(g)$. Now it leaves to prove that the algebraic degree of $h(x)$ is exactly $n-k$. Forming a matrix M with the elements in S as row vectors, we get

$$M = \begin{pmatrix} a_{11} & a_{12} & \cdots a_{1k} & \cdots a_{1n} \\ a_{21} & a_{22} & \cdots a_{2k} & \cdots a_{2n} \\ a_{31} & a_{32} & \cdots a_{3k} & \cdots a_{3n} \\ \cdots & \cdots & \cdots\cdots & \cdots\cdots \\ a_{2^k 1} & a_{2^k 2} & \cdots a_{2^k k} & \cdots a_{2^k n} \end{pmatrix}.$$

Then the coefficients of $y_{j_1} y_{j_2} \cdots y_{j_l}$, $(l \leq n)$ are equal to the sum of the products of same row elements in the $j_{l+1}, j_{l+2} \cdots j_n$-th columns of M. Below we prove that, in the ANF of $h(x)$, all the terms with degree larger than $n-k$ have a zero coefficient.

Case (1): Considering the term with the highest degree in $h(x)$, because the matrix M has 2^k rows, the coefficient of $y_1 y_2 \cdots, y_n$ therefore equals 0.

Case (2): Considering the terms with degree $n-1$ in $h(x)$, the coefficient of any such a term $y_{j_1} y_{j_2} \cdots y_{j_{n-1}}$ can be written as $\bigoplus_{i=1}^{2^k} a_{ij_n}$, which is the XOR of all the elements in the j_n-th column of M. If all the elements in the j_n-th column are all zero, then the coefficient is obviously 0. Otherwise, by Lemma 5.2, we know that the j_n-th column of M has 2^{n-1} of 1's, and when $n > 1$, their XOR will also result in 0. This means that all the coefficients of the terms of $h(x)$ with degree $n-1$ are 0.

Case (3): Considering the terms with degree of $n-2$ in $h(x)$, then the coefficient of such a term $y_{j_1} y_{j_2} \cdots y_{j_{n-2}}$ is

$$\bigoplus_{i=1}^{2^k} a_{ij_{n-1}} a_{ij_n},$$

and it is easy to see that the above XOR equals the inner product of the j_{n-1}-th column and the j_n-th column of the matrix M.
If these two vectors (columns in M) are linearly dependent, then there are two cases in which either one of the vectors is a zero vector, in which case the above equals zero, or the two vectors are equal, in which case similar to the discussion in case (2), and we have

$$\sum_{i=1}^{2^k} a_{ij_{n-1}} a_{ij_n} = \sum_{i=1}^{2^k} a_{ij_{n-1}} = 2^{k-1} \equiv 0 \pmod 2, \tag{5.12}$$

where $k > 1$. If these two vectors are independent, let $Prob(a_{ij} = 1)$ represent the probability that the (i, j)-th element in M is 1; then by Lemma 5.2, we have

$$Prob(a_{ij_{n-1}} = 1) = Prob(a_{ij_n} = 1) = \frac{1}{2}$$

and further we have

$$\begin{aligned}\sum_{i=1}^{2^k} a_{ij_{n-1}} a_{ij_n} &= P\{a_{ij_{n-1}} a_{ij_n} = 1\} \times 2^k \\ &= 2^{k-2} \equiv 0 \pmod 2,\end{aligned} \tag{5.13}$$

where $k > 2$.

Case (4): Considering the terms with degree of $n-3$ in $h(x)$, in this case, we further consider the following three cases to discuss the coefficient of $y_{j_4} \cdots y_{j_n}$ in Eq. 5.11:

1. The j_1, j_2, j_3-th columns of M are linearly independent;
2. One of the columns is the XOR of the other two columns;
3. Two columns are equal.

For the first case, similar to Eq. 5.13, we have, for $k > 3$,

$$\sum_{i=1}^{2^k} a_{ij_{n-2}} a_{ij_{n-1}} a_{ij_n} = P\{a_{ij_{n-2}} a_{ij_{n-1}} a_{ij_n} = 1\} \times 2^k$$
$$= 2^{k-3} \equiv 0 \pmod 2.$$

For the second case, we have

$$\sum_{i=1}^{2^k} a_{ij_{n-2}} a_{ij_{n-1}} a_{ij_n} = \sum_{i=1}^{2^k} a_{ij_{n-2}} a_{ij_{n-1}} (a_{ij_{n-2}} + a_{ij_{n-1}})$$
$$\equiv 0 \pmod 2. \tag{5.14}$$

And for the third case, it can be converted into a case either in Eq. 5.12 or Eq. 5.13.

Case (5): Considering the terms with degree of l ($l > n - k$) in $h(x)$, and in this case, we use induction to prove that the coefficient of the term $y_{j_1} y_{j_2} \cdots y_{j_l}$ is zero. If the rest $n - l$ columns of M are linearly independent, then we have

$$\sum_{i=1}^{2^k} a_{ij_{l+1}} \cdots a_{ij_n} = 2^{k-(n-l)} \equiv 0 \pmod 2,$$

where $n - l < k$. If these $n - l$ columns are linearly dependent, let $\alpha_{j1}, \cdots, \alpha_{jt}$, $t < n - l$ be a basis of these column vectors. Then we have

$$\sum_{i=1}^{2^k} a_{ij_{l+1}} \cdots a_{ij_n}$$
$$= \sum_{i=1}^{2^k} a_{ij'_1} \cdots a_{ij'_t} \lambda^{j_{t+1}} \cdot (a_{ij'_1}, \cdots, a_{ij'_t}) \cdots$$
$$\cdot \lambda^{j_{n-l}} \cdot (a_{ij'_1}, \cdots, a_{ij'_t}). \tag{5.15}$$

here $\lambda^{j_{t+1}}, \cdots, \lambda^{j_l} \in GF^k(2)$, $\alpha_{j_1} = (a_{1j'_1}, \cdots, a_{2^k j'_1})^T, \cdots, \alpha_{j_t} = (a_{1j'_t}, \cdots, a_{2^k j'_t})^T$. In the representation of $\{\lambda^{j_{t+1}}, \cdots, \lambda^{j_{n-l}}\}$, there is a λ^{jz} that has even Hamming weight; hence similar to Eq. 5.14, for any $1 \leq i \leq 2^k$, we have

$$a_{ij'_1} \cdots a_{ij'_t} \lambda^{j_{t+1}} \cdot (a_{ij'_1}, \cdots, a_{ij'_t}) \cdots \lambda^{j_{n-l}} \cdot (a_{ij'_1}, \cdots, a_{ij'_t}) \equiv 0 \pmod 2.$$

If all these $\lambda^{j_{t+1}}, \cdots, \lambda^{j_{n-l}}$ have odd Hamming weight, then we get

$$a_{ij'_1} \cdots a_{ij'_t} \lambda^{j_{t+1}} \cdot (a_{ij'_1}, \cdots, a_{ij'_t}) \cdots \lambda^{j_{n-l}} \cdot (a_{ij'_1}, \cdots, a_{ij'_t}) = a_{ij'_1} \cdots a_{ij'_t};$$

hence, for $t < k$, we have

$$\sum_{l=1}^{2^k} a_{ij_{l+1}} \cdots a_{ij_n} = \sum_{l=1}^{2^k} a_{ij'_1} \cdots a_{ij'_t} = 2^{k-t} \equiv 0 \pmod 2.$$

Combining the cases (1)–(5), we have proved that all the terms in $h(x)$ with degree larger than $n-k$ have a zero coefficient.

Now we prove that, if the $j_{n-k+1}, j_{n-k+2}, \cdots, j_n$-th columns in M are linearly independent, then in the ANF representation of $h(x)$, the coefficient of the term $x_{j_1}x_{j_2}\cdots x_{j_{n-k}}$ is 1. In fact, in the sub-matrix composed by the $j_{n-k+1}, j_{n-k+2}, \cdots, j_n$-th columns of M, there is only one row with all 1's; hence

$$\bigoplus_{i=1}^{2^k} a_{ij_{n-k+1}} a_{ij_{n-k+2}} \cdots a_{j_n} = 1.$$

Therefore, the algebraic degree of $h(x)$ is indeed $n-k$, which also equals the degree of $g(x)$, and hence the conclusion of the theorem follows. □

Definition 5.4. Let A be a subset of $GF^n(2)$, and then function

$$f_A(x) = \begin{cases} 1 & \text{if } x \in A \\ 0 & \text{if } x \in GF^n(2) - A \end{cases}$$

is called the *characteristic function* of the set A.

Theorem 5.15 ([17]). *If there exist some disjoint d-dimensional subspaces or affine subspaces of $GF^n(2)$, denoted by A_k, such that the union of these subspaces $\cup A_k$ is a cover of the support of Boolean function $f(x)$ in n variables. Then there must exist an annihilator of $f(x)$ with degree $n-d$. A special case is to let the union of the above subspaces be the support of $g(x)$; in this case, $g(x)$ is an annihilator of $f(x)$ and is of degree $n-d$. More specifically, we can write*

$$g(x) = 1 \oplus \bigoplus_{\tau \in \cup A_k} \prod_{i=1}^{n} (x_i \oplus \tau_i \oplus 1), \tag{5.16}$$

where $\tau = (\tau_1, \tau_2, \cdots, \tau_n)$.

Proof. It is easy to verify that the characteristic function of any A_k is

$$\bigoplus_{\tau \in A_k} \prod_{i=1}^{n} (x_i \oplus \tau_i \oplus 1)$$

where $\tau = (\tau_1, \tau_2, \cdots, \tau_n)$, and by Theorem 5.14, their algebraic degrees are all equal to $n-d$; hence their XOR is

$$\bigoplus_{\tau \in \cup A_k} \prod_{i=1}^{n} (x_i \oplus \tau_i \oplus 1)$$

which is also a Boolean function in n variables with algebraic degree no larger than $n - d$; hence the conclusion of the theorem follows. □

Example 5.1. Let

$$f(x) = x_1 \oplus x_3 \oplus x_5 \oplus x_1x_3 \oplus x_1x_4 \oplus x_1x_5 \oplus x_2x_4 \oplus x_2x_5 \oplus x_1x_2x_3 \oplus x_2x_3x_4x_5,$$
$$g(x) = 1 \oplus x_1 \oplus x_2 \oplus x_3 \oplus x_5 \oplus x_1x_2 \oplus x_1x_3 \oplus x_1x_5 \oplus x_2x_3 \oplus x_2x_5.$$

Use the vectors in the support of $f(x)$ as row vectors to form a matrix as

$$M = \begin{bmatrix} 0\,0\,0\,0\,1 \\ 0\,0\,0\,1\,1 \\ 0\,0\,1\,0\,0 \\ 0\,0\,1\,1\,0 \\ 1\,1\,0\,0\,0 \\ 1\,1\,0\,1\,0 \\ 1\,1\,1\,0\,1 \\ 0\,1\,0\,1\,0 \\ 0\,1\,0\,1\,1 \\ 0\,1\,1\,0\,0 \\ 0\,1\,1\,0\,1 \\ 0\,1\,1\,1\,1 \\ 1\,0\,0\,0\,0 \\ 1\,0\,0\,0\,1 \\ 1\,0\,1\,0\,0 \\ 1\,0\,1\,0\,1 \end{bmatrix}$$

The vectors in the support of $f(x)$ can be expanded into a union of the following affine subspaces of dimension 3:

$$\{(00001), (00011), (00100), (00110), (11000), (11010), (11101), (11111)\},$$
$$\{(10000), (10001), (10010), (10011)(10100), (10101), (10110), (10111)\}$$

and

$$\{(01000), (01001), (01010), (01011)(01100), (01101), (01110), (01111)\}.$$

Therefore, $f(x)$ has an annihilator of degree 2. Furthermore, since this is the only 3-dimensional subspace of $GF^5(2)$ composed of the union of some 3-dimensional affine subspaces and 3-dimensional subspaces constituting a cover of the support of $f(x)$, hence it is the only annihilator of degree 2 for $f(x)$.

Note that in Theorem 5.15, $\cup A_k$ are disjoint d-dimensional subspaces or affine subspaces; the number of elements of these sets is 2^d or integral multiple of 2^d. Hence the annihilators of $f(x)$ constructed by Theorem 5.15 all have a Hamming weight of a multiple of 2^d.

By Theorem 5.15, we can give the following algorithm to compute the annihilators of Boolean functions:

Algorithm 5.6 (Construction of annihilator with the lowest degree).

Input: A Boolean function $f(x)$ in n variables.
Output: An annihilator $g(x)$ of $f(x)$ that has the lowest possible degree.
(1) Represent $f(x)$ by its support, and write $E = supp(f)$;
(2) From the union of some d-dimensional subspaces and affine subspaces of $GF^n(2)$, find a nontrivial subset that covers $supp(f)$; If $\alpha \oplus \beta_1 \in supp(f)$, $\alpha \oplus \beta_2 \in supp(f)$, then construct $\alpha \oplus \beta_1 \oplus \beta_2$ and add it into E, regardless whether $\alpha \oplus \beta_1 \oplus \beta_2$ belongs to $supp(f)$;
(3) Let E be the support of $g(x)$ and output $g(x) \oplus 1$.

In fact, Theorem 5.15 can be generalized as follows:

Theorem 5.16. *Let A_1, A_2 be two subspaces or affine subspaces of $GF^n(2)$ with dimension no less than d. Write $A = (A_1 \cup A_2) \setminus (A_1 \cap A_2)$. If $supp(f) \subseteq A$, then $f(x)$ must exist an annihilator with degree no more than $n-d$. More specifically, the Boolean function with $\bar{A}$ as support is an annihilator of $f(x)$ and has degree $n-d$, i.e.,*

$$g(x) = 1 \oplus \bigoplus_{\tau \in (A_1 \cup A_2) \setminus (A_1 \cap A_2)} \prod_{i=1}^{n} (x_i \oplus \tau_i \oplus 1) \tag{5.17}$$

where $\tau = (\tau_1, \tau_2, \cdots, \tau_n)$.

Proof. Write $\tau = (\tau_1, \tau_2, \cdots, \tau_n)$; then the characteristic functions of A_1 and A_2 are

$$g_{A_1}(x) = \bigoplus_{\tau \in A_1} \prod_{i=1}^{n} (x_i \oplus \tau_i \oplus 1)$$

and

$$g_{A_2}(x) = \bigoplus_{\tau \in A_2} \prod_{i=1}^{n} (x_i \oplus \tau_i \oplus 1),$$

respectively. By Theorem 5.14, we know that $g_{A_1}(x)$ and $g_{A_2}(x)$ are both Boolean functions with degree no more than $n-k$; hence their XOR is

$$g_{A_1}(x) \oplus g_{A_2}(x) = \bigoplus_{\tau \in A_1} \prod_{i=1}^{n} (x_i \oplus \tau_i \oplus 1) \oplus \bigoplus_{\tau \in A_2} \prod_{i=1}^{n} (x_i \oplus \tau_i \oplus 1)$$

$$= \bigoplus_{\tau \in (A_1 \cup A_2) \setminus (A_1 \cap A_2)} \prod_{i=1}^{n} (x_i \oplus \tau_i \oplus 1) \qquad (5.18)$$

which has a degree no more than $n-d$. The correctness of Eq. 5.18 is due to that the elements in $A_1 \cap A_2$ appear both in the characteristic functions of A_1 and that of A_2, which yield their XOR to be zero. Let $g(x) = 1 \oplus g_{A_1}(x) \oplus g_{A_2}(x)$; then the algebraic degree of $g(x)$ is no more than $n-d$, and the support of $g(x)$ is $\bar{A}$. Since $supp(f) \subseteq A$, it is trivial to verify that $g(x)$ is an annihilator of $f(x)$. □

Example 5.2. Let

$$A_1 = \{(1100), (1101), (1110), (1111)\},$$
$$A_2 = \{(1111), (0011), (1011), (0111)\}$$

be two affine subspaces of $GF^4(2)$ with dimension $d = 2$. Then

$$A = (A_1 \cup A_2) \setminus (A_1 \cap A_2) = \{(1100), (1101), (1110), (0011), (1011), (0111)\}.$$

Let $f(x) = x_1x_2 \oplus x_1x_2x_3x_4$, then $supp(f) \subset A$. Since the characteristic function of A is $x_1x_2 \oplus x_3x_4$ which has degree $n-d = 2$, by Theorem 5.16, $g(x) = 1 \oplus x_1x_2 \oplus x_3x_4$ is an annihilator of $f(x)$.

Example 5.3. Define three subsets in $GF^6(2)$ as

$$A_1 = \left\{ \begin{array}{l} (1\,1\,0\,0\,0\,0) \\ (1\,1\,0\,0\,0\,1) \\ (1\,1\,0\,0\,1\,0) \\ (1\,1\,0\,0\,1\,1) * \\ (1\,1\,0\,1\,0\,0) \\ (1\,1\,0\,1\,0\,1) \\ (1\,1\,0\,1\,1\,0) \\ (1\,1\,0\,1\,1\,1) * \end{array} \right\}$$

$$A_2 = \left\{ \begin{array}{l} (1\ 0\ 1\ 1\ 0\ 0) \\ (1\ 0\ 1\ 1\ 0\ 1) \\ (1\ 0\ 1\ 1\ 1\ 0) \\ (1\ 0\ 1\ 1\ 1\ 1)\ * \\ (1\ 1\ 1\ 1\ 0\ 0) \\ (1\ 1\ 1\ 1\ 0\ 1) \\ (1\ 1\ 1\ 1\ 1\ 0) \\ (1\ 1\ 1\ 1\ 1\ 1)\ * \\ (0\ 1\ 1\ 1\ 0\ 0) \\ (0\ 1\ 1\ 1\ 0\ 1) \\ (0\ 1\ 1\ 1\ 1\ 0) \\ (0\ 1\ 1\ 1\ 1\ 1)\ * \\ (0\ 0\ 1\ 1\ 0\ 0) \\ (0\ 0\ 1\ 1\ 0\ 1) \\ (0\ 0\ 1\ 1\ 1\ 0) \\ (0\ 0\ 1\ 1\ 1\ 1)\ * \end{array} \right\}$$

$$A_3 = \left\{ \begin{array}{l} (0\ 0\ 0\ 0\ 1\ 1) \\ (0\ 0\ 0\ 1\ 1\ 1) \\ (0\ 0\ 1\ 0\ 1\ 1) \\ (0\ 0\ 1\ 1\ 1\ 1)\ * \\ (0\ 1\ 0\ 0\ 1\ 1) \\ (0\ 1\ 0\ 1\ 1\ 1) \\ (0\ 1\ 1\ 0\ 1\ 1) \\ (0\ 1\ 1\ 1\ 1\ 1)\ * \\ (1\ 0\ 0\ 0\ 1\ 1) \\ (1\ 0\ 0\ 1\ 1\ 1) \\ (1\ 0\ 1\ 0\ 1\ 1) \\ (1\ 0\ 1\ 1\ 1\ 1)\ * \\ (1\ 1\ 1\ 0\ 1\ 1) \\ (1\ 1\ 0\ 0\ 1\ 1)\ * \\ (1\ 1\ 0\ 1\ 1\ 1)\ * \\ (1\ 1\ 1\ 1\ 1\ 1)\ * \end{array} \right\}$$

Then the characteristic functions of A_1, A_2, A_3 are $x_1x_2(x_3 \oplus 1)$, x_3x_4, and x_5x_6, respectively. Let $\bigcup_{i=1}^{3} A_i$ be the set of elements belonging to $A_i, i = 1, 2, 3$, where the elements appear in two sets are double counted. Then the characteristic function of $\bigcup_{i=1}^{3} A_i$ is $x_1x_2x_3 \oplus x_1x_2 \oplus x_3x_4 \oplus x_5x_6$, which is a bent function in six variables.

Since the elements marked with asterisk ($*$) in the matrix appear for even number of times, so all the rest vectors without the asterisk form the support of function $x_1x_2x_3 \oplus x_1x_2 \oplus x_3x_4 \oplus x_5x_6$.

From Example 5.3 above, we can see that, although the vectors without an asterisk ($*$) in each of the sets do not form a d-dimensional subspace and neither a d-dimensional affine subspace, however, by adding some vectors into each of the sets, such that among the whole matrix, each new vector is added for even number of times, then the characteristic function of the whole matrix forms a Boolean function with algebraic degree no more than d.

References

1. Armknecht, F.: Improving fast algebraic attacks. In: Fast Software Encryption 2004. LNCS 3017, pp. 65–82. Springer, Berlin/New York (2004)
2. Armknecht, F., Krause, M.: Algebraic attacks on combiners with memory. In: Advances in Cryptology, Proceedings of Crypto'03. LNCS 2729, pp. 162–175. Springer, Berlin (2003)
3. Armknecht, F., et al.: Efficient computation of algebraic immunity of algebraic and fast algebraic attacks. In: Advances in Cryptology, Proceedings of Eurocrypt'2006. LNCS 4004, pp. 147–164. Springer, Berlin/New York (2006)
4. Berlekamp, E.: Algebraic Coding Theory. McGraw-Hill, New York (1968)
5. Braeken, A., Preneel, B.: On the algebraic immunity of symmetric Boolean functions. In: Proceedings of Indocrypt 2005. LNCS 3797, pp. 35–48. Springer, Berlin/New York (2005)
6. Carlet, C., Feng, K.: An infinite class of balanced functions with optimal algebraic immunity, good immunity to fast algebraic attacks and good nonlinearity. In: Advances in Cryptology, Proceedings of Asiacrypt 2008. LNCS 5350, pp. 425–440. Springer, Berlin (2008)
7. Carlet, C., Dalai, D.K., Gupta, G.C., Maitra, S.: Algebraic immunity for cryptographically significant Boolean functions: analysis and construction. IEEE Trans. Inf. Theory **IT-52**(7), 3105–3121 (2006)
8. Courtois, N.: Fast algebraic attacks on stream ciphers with linear feedback. In: Advances in Cryptology, Proceedings of Crypto'03. LNCS 2729, pp. 176–194. Springer, Berlin/New York (2003)
9. Courtois, N., Meier, W.: Algebraic attacks on stream ciphers with linear feedback. In: Advances in Cryptology, Proceedings of Eurocrypt'03. LNCS 2656, pp. 345–359. Springer, Berlin (2003)
10. Courtois, N., Pieprzyk, J.: Cryptanalysis of block ciphers with overdefined systems of equations. In: Advances in Cryptology, Proceedings of Asiacrypt 2002. LNCS 2501, pp. 267–287. Springer, Berlin/New York (2002)
11. Dalai, D.K., Gupta, K.C., Maitra, S.: Results on algebraic immunity for cryptographically significant Boolean functions. In: Proceedings of Indocrypt 2004. LNCS 3348, pp. 92–106. Springer, Berlin/New York (2004)
12. Dalai, D.K., Gupta, K.C., Maitra, S.: Cryptographically significant Boolean functions: construction and analysis in terms of algebraic immunity. In: Fast Software Encryption 2005. LNCS 3557, pp. 98–111. Springer, Berlin/New York (2005)
13. Lee, D.H., Kim, J., Hong, J., Han, J.W., Moon, D.: Algebraic attacks on summation generators. In: Fast Software Encryption 2004. LNCS 3017, pp. 34–48. Springer, Berlin/New York (2004)
14. Lobanov, M.S.: Exact relation between nonlinearity and algebraic immunity. Discret. Math. Appl. **16**(5), 453–460 (2006)
15. Massey, J.L.: Shift-register synthesis and BCH decoding. IEEE Trans. Inf. Theory **IT-15**(1), 122–127 (1969)
16. Meier, W., Pasalic, E., Carlet, C.: Algebraic attacks and decomposition of Boolean functions. In: Advances in Cryptology, Proceedings of Eurocrypt'04. LNCS 3027, pp. 474–491. Springer, Berlin/New York (2004)
17. Zhang, W., Wu, C.K., Yu, J.: On the annihilators of cryptographic Boolean functions. Acta Electron. Sin. **34**(1), 51–54 (2006) (in Chinese)
18. Zhang, W., Wu, C.K., Liu, X.: Construction and enumeration of Boolean functions with maximum algebraic immunity. Sci. China (Ser. F) **52**(1), 32–40 (2009)

Chapter 6
The Symmetric Property of Boolean Functions

Symmetric property is a special property of Boolean functions, which has attracted much study on it. This chapter presents fast Walsh transforms of symmetric Boolean functions, correlation immunity of symmetric functions, construction of symmetric resilient Boolean functions, and some cryptographic properties of majority functions being a special class of symmetric Boolean functions. The study on the ε-correlation immunity of majority functions shows that majority functions have good asymptotical behavior of ε-correlation immunity, i.e., although they are not correlation immune, they have, however, asymptotical correlation immunity.

6.1 Basic Properties of Symmetric Boolean Functions

Symmetric Boolean functions are a class of functions with the special property that they are indistinguishable to different inputs with the same Hamming weight. In other words, when a permutation applies on the input variables, the outputs of the function always remain the same as the original inputs. This property has been regarded as being favorable, as it meets, in one aspect, Shannon's "confusion" requirement [15] when some keys act as part of the input bits, and the key bits play the same importance as the message bits. However, another argument about the symmetry property of Boolean functions is something to be avoided in the design of ciphers. Regardless whether the symmetry property of Boolean functions is favorable or on the contrary, this special property has attracted much interests in theoretical research [14, 21, 24]. Some work on cryptographic properties of symmetric Boolean functions can be found in public literatures, for example, [1, 2, 8–10, 22].

C.-K. Wu, D. Feng, *Boolean Functions and Their Applications in Cryptography*,
Advances in Computer Science and Technology, DOI 10.1007/978-3-662-48865-2_6

Definition 6.1. Let $f(x) \in \mathcal{F}_n$. Then $f(x)$ is called a *symmetric Boolean function* if for any permutation π on $\{1, 2, \ldots, n\}$, we have

$$f(x_{\pi(1)}, x_{\pi(2)}, \ldots, x_{\pi(n)}) = f(x_1, x_2, \ldots, x_n).$$

For a symmetric function f, it is known that there is an integral function I_f : $\{0, 1, \ldots, n\} \longrightarrow \{0, 1\}$ such that

$$f(x) = I_f(k) \text{ if and only if } wt(x) = k,$$

where $wt(x)$ is the Hamming weight of x. Let $\varphi_i(x)$, $i = 0, 1, \ldots, n$, be the homogeneous symmetric function which is composed of all the terms of degree i, i.e.,

$$\begin{aligned}
\varphi_0(x) &= 1, \\
\varphi_1(x) &= x_1 \oplus x_2 \oplus \cdots \oplus x_n, \\
\varphi_2(x) &= x_1x_2 \oplus x_1x_3 \oplus \cdots \oplus x_1x_n \oplus x_2x_3 \oplus \cdots \oplus x_{n-1}x_n, \\
&\cdots, \\
\varphi_n(x) &= x_1x_2 \cdots x_n.
\end{aligned}$$

Let $\lambda_j(x)$, $j = 0, 1, \ldots, n$, be the symmetric function satisfying that $\lambda_j(x) = 1$ if and only if $wt(x) = j$. Then we have

Lemma 6.1. *The set $\mathcal{S}_n$ of all the symmetric Boolean functions in $\mathcal{F}_n$ forms an $(n+1)$-dimensional vector space over $GF(2)$ with $\{\varphi_i(x)\}_{i=0}^n$ and $\{\lambda_i(x)\}_{i=0}^n$ as two different bases.*

Proof. First of all, it is trivial to verify that both $\{\varphi_i(x)\}_{i=0}^n$ and $\{\lambda_i(x)\}_{i=0}^n$ form a class of independent Boolean functions, i.e., no function from the set can be represented by a linear combination of the rest. Now we show that any symmetric Boolean function can be represented by a linear combination of $\{\varphi_i(x)\}_{i=0}^n$ and $\{\lambda_i(x)\}_{i=0}^n$. Let $f(x) \in \mathcal{S}_n$ be an arbitrary Boolean function. If $f(x)$ has a term of degree k in its algebraic normal form, then due to the symmetry, it must have all the terms of degree k in its algebraic normal form representation, which means that $f(x)$ can be written as $f(x) = \varphi_k(x) \oplus g(x)$, where $g(x)$ is also symmetric and does not have any term of degree k in its algebraic normal form. When we consider all the possible values of k ($k \in \{0, 1, 2, \ldots, n\}$), a subset of $\{\varphi_i(x)\}_{i=0}^n$ is selected where the XOR of this subset is $f(x)$. This means that any symmetric Boolean function $f(x)$ can be represented by a linear combination of $\{\varphi_i(x)\}_{i=0}^n$, and hence $\{\varphi_i(x)\}_{i=0}^n$ forms a basis of S_n.

Similarly we can prove that $\{\lambda_i(x)\}_{i=0}^n$ forms a basis of S_n. □

Though trivial, it is worth to point out the following:

Lemma 6.2. $\lambda_i(x)\lambda_j(x) \equiv 0$ *if and only if* $i \neq j$.

Lemma 6.2 shows that $\{\lambda_i(x)\}_{i=0}^n$ is actually an orthogonal basis. By using this orthogonal basis to represent other symmetric functions, the study of their other cryptographic properties can become more convenient.

Furthermore, since the product of two symmetric functions is also a symmetric function, $\mathcal{S}_n$ is virtually a ring with the multiplication operation. Let $f(x) \in \mathcal{S}_n$ be symmetric; it is easy to verify that $f(x)$ can be written as

$$f(x) = \bigoplus_{k=0}^{n} I_f(k)\lambda_k(x). \tag{6.1}$$

For a vector x with $wt(x) = k$, there are $\binom{k}{i}$ subsets of coordinates with i coordinates having value 1. So for this specific x, the value of $\varphi_i(x)$ is $\varphi_i(x) = \binom{k}{i} \bmod 2$, i.e., $I_{\varphi_i}(k) = \binom{k}{i} \bmod 2$. Taken into Eq. 6.1, we have in general case

$$\varphi_i(x) = \bigoplus_{k=i}^{n} (\binom{k}{i} \bmod 2)\lambda_k(x). \tag{6.2}$$

Denote the binary representation of integer a be $(a_{n-1}, a_{n-2}, \ldots, a_0)$, i.e., $a = \sum_{i=0}^{n-1} a_i 2^i$, where $a_i \in GF(2)$. For two integers of binary length no more than n, define *a to be a cover of b* if $a_i \geq b_i$ holds for all $i = 0, 1, \ldots, n$ and is denoted as $a \succeq b$ or equivalently as $b \preceq a$. If there exists at least one i such that $a_i > b_i$, then a is called an *absolute cover* of b and is denoted as $a \succ b$, or equivalently as $b \prec a$. By the well-known Lucas theorem (named after the French mathematician François Édouard Anatole Lucas, 1842–1891):

$$\binom{k}{i} \bmod 2 = 1 \text{ if and only if } a \succeq b$$

Eq. 6.2 can be written as

$$\varphi_i(x) = \bigoplus_{k \succeq i} \lambda_k(x). \tag{6.3}$$

Equation 6.2 or equivalently Eq. 6.3 is the representation of $\varphi_i(x)$ in terms of $\lambda_k(x)$.

Summing up Eq. 6.3 for all $i \succeq t$, we get

$$\bigoplus_{i \succeq t} \varphi_i(x) = \bigoplus_{i \succeq t} \bigoplus_{k \succeq i} \lambda_k(x). \tag{6.4}$$

On the right-hand side of Eq. 6.4, $\lambda_k(x)$ appears for even number of times for every $k \succ t$, and hence it vanishes after the XOR operation. This is due to the following lemma.

Lemma 6.3. *Let k be an integer of binary length n. Then there are even number of k' such that $k' \succeq k$ except $k = 2^n - 1$.*

Proof. Trivial and obvious. □

By Lemma 6.3, the right-hand side of Eq. 6.4 only leaves $\lambda_t(x)$, which gives the expression of $\lambda_t(x)$ in terms of $\varphi_i(x)$'s. To be consistent with the indexes used in Eq. 6.3, we can write

$$\lambda_i(x) = \bigoplus_{k \succeq i} \varphi_k(x). \tag{6.5}$$

Equation 6.5 is the representation of $\lambda_i(x)$ in terms of $\varphi_k(x)$. So Eqs. 6.4 and 6.5 will enable us to convert the representations of a symmetric Boolean function in terms of $\varphi_k(x)$ to $\lambda_t(x)$ and vice versa.

6.2 Computing the Walsh Transform of Symmetric Boolean Functions

The Walsh transform has been widely used in studying cryptographic properties of Boolean functions [11, 23]. Efficient algorithms for performing Walsh transforms are very useful. Noticing the particular properties of symmetric Boolean functions, it is possible to present efficient algorithms to perform Walsh transforms on symmetric Boolean functions.

For a Boolean function $f(x) \in \mathcal{F}_n$, it is noticed that

$$S_f(w) = \sum_{x \in supp(f)} (-1)^{\langle w, x \rangle} = wt(f) - 2 \sum_{\substack{x \in supp(f) \\ \langle w, x \rangle = 1}} 1. \tag{6.6}$$

So the Walsh transform $S_f(w)$ of $f(x)$ can be calculated by the following steps:

1. Let X_f be the characteristic matrix of $f(x)$, i.e., X_f is a $wt(f) \times n$ matrix with its row vectors being the vectors in $supp(f)$.
2. For every $w_i = 0$, delete the i-th column of X_f to get a target matrix A.
3. The number of even weight rows of A minus the number of odd weight rows of A yields the value of $S_f(w)$.

6.2.1 Walsh Transforms on Symmetric Boolean Functions

By Lemmas 6.1 and 6.2 and Theorem 1.1, we have

$$S_f(w) = \sum_{k=0}^{n} I_f(k) S_{\lambda_k}(w). \tag{6.7}$$

Denote by $w^{(i)}$ the vector with i consecutive ones followed by zeros, and let

$$A_{ik} = \sum_{x \in supp(\lambda_k)} \langle w^{(i)}, x \rangle, \quad i, k \in \{0, 1, \ldots, n\}. \tag{6.8}$$

It is known that the Hamming weight of symmetric function $\lambda_k(x)$ is

$$wt(\lambda_k) = \binom{n}{k}.$$

By Eq. 6.6 we have

$$S_{\lambda_k}(w^{(i)}) = \binom{n}{k} - 2A_{ik}. \tag{6.9}$$

Substituting Eq. 6.7 by Eq. 6.9, we have

$$S_f(w^{(i)}) = \sum_{k=0}^{n} I_f(k) \binom{n}{k} - 2 \sum_{k=0}^{n} I_f(k) A_{ik}. \tag{6.10}$$

In particular, for the functions $\sigma_j(x)$, since

$$\sigma_j(x) = \bigoplus_{k=j}^{n} (\binom{k}{j} \bmod 2) \lambda_k(x) = \bigoplus_{k \succeq j} \lambda_k(x), \tag{6.11}$$

we have

$$\begin{aligned} S_{\sigma_j}(w^{(i)}) &= \sum_{k=j}^{n} \left[\left(\binom{k}{j} \bmod 2 \right) \left(\binom{n}{k} - 2A_{ik} \right) \right] \\ &= \sum_{k \succeq j} \binom{n}{k} - 2A_{ik}. \end{aligned} \tag{6.12}$$

It is noticed that for a symmetric function $f(x)$, its Walsh transform $S_f(w)$ is also a symmetric function. So when the Walsh transforms on the vectors $w^{(0)}, w^{(1)}, \ldots, w^{(n)}$ have been calculated out, the whole Walsh spectrum of this symmetric function f becomes clear. It is seen from Eq. 6.10 that the computational complexity for the computation of Walsh transforms of symmetric functions depends largely on that for the computation of the value of A_{ik}. So more attention should be paid to the computation of A_{ik}. Write $\binom{n}{k} = 0$ whenever $k > n$ or $k < 0$. Then we have

Lemma 6.4.

$$A_{ik} = \sum_{1 \leq j \leq n,\ j \text{ is odd}} \binom{i}{j} \binom{n-i}{k-j}. \tag{6.13}$$

Proof. Recall that $supp(\lambda_k)$ is the set of x with $wt(x) = k$. The value of A_{ik} is actually the number of those vectors in $supp(\lambda_k)$ which have an odd number of ones in the first i positions. This is the expression of Eq. 6.13. □

It is interesting to notice that the value of A_{ik} is just the coefficient of x^k in the expansion of the multiplication of $(1+x)^{n-i}$ and all terms with odd degree of $(1+x)^i$. It is easy to show that

Lemma 6.5. *All terms with odd degree of* $(1 + x)^m$ *form a polynomial*

$$\frac{1}{2}(1 + x)^m - \frac{1}{2}(1 - x)^m.$$

So, by Lemma 6.5, the value of A_{ik} is the coefficient of x^k in the expansion of

$$A_i(x) = \left(\frac{1}{2}(1 + x)^i - \frac{1}{2}(1 - x)^i\right)(1 + x)^{n-i}. \tag{6.14}$$

From Eq. 6.14 we obtain $A_{ik} = \frac{1}{2}(\binom{n}{k} - \sum_{j=0}^{i}(-1)^j\binom{i}{j}\binom{n-i}{k-j})$. By substituting this expression in Eq. 6.10, we have

$$S_f(w^{(i)}) = \sum_{k=0}^{n} I_f(k) \sum_{j=0}^{i} (-1)^j \binom{i}{j} \binom{n-i}{k-j}. \tag{6.15}$$

Now we can sketch the algorithm for computing Walsh transforms of symmetric Boolean functions. Given a symmetric Boolean function $f(x) \in \mathcal{F}_n$, then its Walsh transform at a single vector w and the whole Walsh spectrums can be calculated by the following algorithms:

Algorithm 6.1 (Walsh transform at a single vector).

(1) Compute $I_f(k)$ for $k = 0, 1, \ldots, n$.
(2) Let $i = wt(w)$.
(3) Use Eq. 6.15 to compute $S_f(w^{(i)})$.

Algorithm 6.2 (whole Walsh spectrums).

(1) Compute $I_f(k)$ for $k = 0, 1, \ldots, n$.
(2) Use Eq. 6.15 to compute $S_f(w^{(i)})$ for $i = 0, 1, \ldots, n$.
(3) Then $S_f(w) = S_f(w^{(i)})$ if and only if $wt(w) = i$.

In the algorithms above, we need first to compute the value of $I_f(k)$ for $k = 0, 1, \ldots, n$. If a Boolean function is given by its truth table representation, it is easy to get the value of $I_f(k)$ from the truth table of $f(x)$. If the function is given by a polynomial form, instead of computing $f(x)$ for some x with $wt(x) = k$ to get $I_f(k)$, we use functions $\{\sigma_j(x)\}_{j=0}^{n}$. From the polynomial representation, it is very easy to get the linear combination

$$f(x) = \bigoplus_{j=0}^{n} c_j \sigma_j(x).$$

By Eq. 6.11 we have

$$f(x) = \bigoplus_{j=0}^{n} \bigoplus_{k=j}^{n} c_j \left(\binom{k}{j} \bmod 2 \right) \lambda_k(x) = \bigoplus_{j=0}^{n} \bigoplus_{k \succeq j} \lambda_k(x). \tag{6.16}$$

From the expansion of Eq. 6.16, we can easily get the value of $I_f(k)$ for $k = 0, 1, \ldots, n$.

It should be noted that in general it is not much easier to determine the value of A_{ik} from the expansion of Eq. 6.14 than from Eq. 6.13. However, Eq. 6.14 can be used to compute certain particular values. For example, when $i = \frac{n}{2}$ (n even), we have $A_i(x) = \frac{1}{2}(1+x)^n - \frac{1}{2}(1-x^2)^{\frac{n}{2}}$. So

$$A_{ik} = \begin{cases} \frac{1}{2}\binom{n}{k}, & k \text{ odd}, \\ \frac{1}{2}\binom{n}{k} - \frac{(-1)^{\frac{k}{2}}}{2}\binom{n/2}{k/2}, & k \text{ even}. \end{cases}$$

Similarly for $i = \frac{n-1}{2}$ (n odd), we have

$$A_{ik} = \frac{1}{2}\binom{n}{k} - \frac{1}{2}(-1)^{\lfloor \frac{k}{2} \rfloor} \binom{i}{\lfloor \frac{k}{2} \rfloor},$$

and for $i = \frac{n+1}{2}$ (n odd), we have

$$A_{ik} = \frac{1}{2}\binom{n}{k} - \frac{1}{2}(-1)^{k+\lfloor \frac{k}{2} \rfloor} \binom{i-1}{\lfloor \frac{k}{2} \rfloor},$$

where $\lfloor a \rfloor$ means the largest integer which is less than or equal to a. In general our suggestion is to use Eq. 6.15 to compute the Walsh transforms of symmetric Boolean functions.

6.2.2 Computational Complexity

At the first sight, Algorithm 6.2 is just $n + 1$ times repeat of Algorithm 6.1. It turns out that its computational complexity is not simply $n + 1$ times the complexity of Algorithm 6.1. In order to assess the computational complexity of the two algorithms, we treat both the multiplication and the division of two integers as a computing unit. Then it is obvious that the complexity for computing $\binom{n}{t}$ is at most $2t$ units. So the complexity for computing $\binom{i}{j}\binom{n-i}{k-j}$ is at most $2j+2(k-j) = 2k$ units.

The computational complexity of Algorithm 6.1 varies for different symmetric functions and different vectors $w^{(i)}$. Note that the value of $I_f(k)$ can be determined by comparing a certain coordinate of the truth table of $f(x)$ when $f(x)$ is given by truth table representation and by expanding Eq. 6.16 when $f(x)$ is given by polynomial representation. The worst case for the complexity of computing all values of $I_f(k)$, from Eq. 6.16, is at most $\frac{n^2}{2}$ times that of computing $\binom{k}{j}$ mod 2. Write both k and j as binary integers: $k = k_0k_1 \ldots k_{t-1}$ and $j = j_0j_1 \ldots j_{t-1}$, where $t = \lfloor \log_2 k \rfloor \leq \lfloor \log_2 n \rfloor$. Then by the well-known Lucas theorem, we have $\binom{k}{j}(\text{mod}2) = 1$ if $k_i \geq j_i$ for every $0 \leq i \leq t-1$ and $\binom{k}{j}(\text{mod}2) = 0$ otherwise. By comparing the value of k_i and j_i for every $i = 0, 1, \ldots, t-1$, we then have the value of $\binom{k}{j}$ mod 2 or equivalently of $c_j(\binom{k}{j}$ mod 2). So the computational complexity for computing $c_j(\binom{k}{j}$ mod 2) is at most t times comparison of integers from $\{0, 1\}$. Hence the complexity for computing the whole $I_f(k)$ is upper bounded by $\frac{1}{2}n^2\lfloor \log_2 n \rfloor$. The lower bound for the complexity of Eq. 6.15 is when $I_f(k) \equiv 1$. Since the computational complexity for computing $\sum_{j=0}^{i}(-1)^j\binom{i}{j}\binom{n-i}{k-j}$ from Eq. 6.15 is upper bounded by

$$\sum_{j=0}^{i}(2k) = 2k(i+1) \leq 2k(n+1),$$

the computational complexity for computing Eq. 6.15 is then upper bounded by

$$\sum_{k=0}^{n} 2k(n+1) = n(n+1)^2.$$

Sum them together we have that the complexity of Algorithm 6.1, in the worst case, is $\frac{n^2\lfloor \log_2 n \rfloor}{2}$ times binary integer comparison and $n(n+1)^2$ units of computation.

Now we consider the computational complexity of Algorithm 6.2. In the first step, the complexity is the same as that of Algorithm 6.1. The second step also depends on the particular symmetric function. We are most interested in the expected value or, in other words, the weighted average value of computational complexity. Assume all symmetric functions are chosen at random. Since for each k with $I_f(k) = 1$ there are $\binom{n}{k}$ vectors x on which $f(x)$ has value 1 and the Hamming weight of such an x is k, we then have that the weighted average complexity for the computation of a symmetric Boolean function is upper bounded by

$$\frac{\sum_{k=0}^{n} \left(k\binom{n}{k} \cdot \sum_{i=0}^{n} 2k(i+1)\right)}{2^n} = \frac{n(n+1)^2(n+2)}{4}. \tag{6.17}$$

Taking into account the complexity of step one, we have the weighted average complexity of Algorithm 6.2. By dividing by $n+1$ we obtain the average weighted complexity for computing the Walsh transform of a symmetric function on a single vector.

In order to specify this complexity more precisely, we define *bit addition* as the complexity unit when addition is performed between two integers from $\{0, 1\}$. Then binary integer comparison can be treated as being equivalent to bit addition. It is known that the complexity of the multiplication of two integers is no larger than n, or equivalently the complexity unit defined above is upper bounded by $n \cdot \log_2 n$ bit additions. When Eq. 1.12 is used to compute the Walsh transform, since each $\langle w, x\rangle$ needs n bit additions and the summation in Eq. 1.12 should be performed through all $x \in GF^n(2)$, the complexity would be $n \cdot 2^n$ bit additions. This does not take into account the complexity for computing the value of $f(x)$ when it is given by polynomial representation. If Eq. 6.6 rather than Eq. 1.12 is used, then the weighted average Hamming weight of a symmetric Boolean function is

$$\sum_{k=0}^{n} k\binom{n}{k} = 2^{n-1}.$$

The weighted average complexity for the computation of $\sum_{x \in supp(f)} \langle w, x\rangle$ then is $n \cdot 2^{n-1}$ bit additions. In addition we need to determine the value of $wt(f)$ for a particular symmetric Boolean function. Note that

$$wt(f) = \sum_{k=0}^{n} k\binom{n}{k} I_f(k),$$

and the complexity for the computation of $\binom{n}{k}$ is $2k$ units. Hence, the weighted average complexity for the computation of $wt(f)$ is

$$\frac{1}{2}\sum_{k=0}^{n}(2k+1) = \frac{(n+1)^2}{2}$$

units or roughly $\frac{1}{2}n(n+1)^2\log_2 n$ bit additions. It can be seen that the computational complexity of Eq. 6.6 is exponential on n, the number of variables of the Boolean function.

Now we show that the computational complexities of both Algorithms 6.1 and 6.2 are polynomial on n. Note that the complexity assessment of computing $I_f(k)$ is already counted by bit addition. So the complexity of Algorithms 6.1 and 6.2 is upper bounded by

$$\frac{1}{2}n^2\lfloor\log_2 n\rfloor + n^2(n+1)^2\log_2 n$$

and

$$\frac{n^2\lfloor\log_2 n\rfloor}{2} + \frac{n^2(n+1)^2(n+2)\log_2 n}{4}$$

bit additions, respectively. Summing up the discussions above, we know that the complexities of Walsh transforms of symmetric functions on one vector are as follows:

- Exhaustive search by Eq. 6.6:

$$C_S = n\cdot 2^{n-1} + \frac{1}{2}n(n+1)^2\log_2 n$$

  bit additions.
- Worst case (upper bound) by Algorithm 6.1:

$$C_U = \frac{1}{2}n^2\lfloor\log_2 n\rfloor + n^2(n+1)^2\log_2 n$$

  bit additions.
- Weighted average complexity from Algorithm 6.2:

$$C_E = \frac{n^2\lfloor\log_2 n\rfloor}{2(n+1)} + \frac{n^2(n+1)(n+2)\log_2 n}{4}$$

  bit additions.

Comparison of complexities are shown in Table 6.1.
From Table 6.1 it is seen that the method introduced here is not efficient for small n. But when n is large, the average complexity is much lower than the values when either Eq. 1.12 or Eq. 6.6 is used in computing the Walsh transforms of symmetric Boolean functions.

Table 6.1 Complexities of Walsh transforms of symmetric functions at a single vector

n	C_S	C_U	C_E	C_E/C_S
2	17	38	13	0.7647
4	132	816	244	1.8485
8	1996	15648	4331	2.1698
16	533536	296448	78366	0.1460
32	2.15×10^9	5578240	1436338	0.0007
64	5.91×10^{20}	103845888	26357949	4.5×10^{-14}
128	2.18×10^{40}	1.19×10^9	480829884	2.2×10^{-32}

6.3 Correlation Immunity of Symmetric Functions

Correlation immunity of Boolean functions with its cryptographic significance is first studied in [16] and has attracted a wide attention. A Boolean function $f(x) \in \mathcal{F}_n$ is called *correlation immune* (CI) of order m if for every m indices $1 \leq i_1 < i_2 < \cdots < i_m \leq n$ and for every $(a_1, a_2, \ldots, a_m) \in GF^m(2)$ we have

$$Prob(f(x) = 1|(x_{i_1}, \ldots, x_{i_m}) = (a_1, \ldots, a_m)) = Prob(f(x) = 1).$$

By the probabilistic identity,

$$\begin{aligned} &Prob(f(x) = 1|(x_{i_1}, \ldots, x_{i_m}) = (a_1, \ldots, a_m)) \cdot Prob((x_{i_1}, \ldots, x_{i_m}) = (a_1, \ldots, a_m)) \\ &= Prob((x_{i_1}, \ldots, x_{i_m}) = (a_1, \ldots, a_m)|f(x) = 1) \cdot Prob(f(x) = 1); \end{aligned}$$

the above implies that if $f(x)$ is m-th order correlation immune, then the following equation must be true provided that $f(x) \not\equiv 0$.

$$\begin{aligned} &Prob((x_{i_1}, \ldots, x_{i_m}) = (a_1, \ldots, a_m)|f(x) = 1) \\ &= Prob((x_{i_1}, \ldots, x_{i_m}) = (a_1, \ldots, a_m)) = \frac{1}{2^m}. \end{aligned} \tag{6.18}$$

This means that if we put all the values of $x \in supp(f) = \{x \in supp(f)\}$ to form an $n \times wt(f)$ matrix with each column of the matrix being the binary representation of value of that x, then for an arbitrary sub-matrix composed of any m rows, the columns of the sub-matrix have equal number of every possible binary string of length m.

In the following, $f(x)$ is assumed to be a symmetric function of $\mathcal{F}_n$ unless specified otherwise. In this case Eq. 6.18 can simply be written as

$$Prob((x_1, \ldots, x_m) = (a_1, \ldots, a_m)|f(x) = 1) = \frac{1}{2^m}.$$

Denote by $\alpha = (a_1, \ldots, a_m)$. Then the number of x's with $wt(x) = k$ and $(x_1, \ldots, x_m) = (a_1, \ldots, a_m)$ is $\binom{n-m}{k-wt(\alpha)}$. The range of possible Hamming weight of α is from 0 to m when α takes all possible strings. Define $\binom{n}{t} = 0$ for the cases when $t < 0$ and when $t > n$. By Eq. 6.18 we know that a necessary condition for $f(x)$ to be correlation immune of order m is

$$\sum_{t=0}^{n} \binom{n-m}{t} I_f(t) = \sum_{t=0}^{n} \binom{n-m}{t-1} I_f(t) = \cdots = \sum_{t=0}^{n} \binom{n-m}{t-m} I_f(t). \tag{6.19}$$

Note that an m-th order correlation immune function is also i-th correlation immune for every $i \leq m$. So another stronger necessary condition for $f(x)$ to be correlation immune of order m is

$$\sum_{t=0}^{n} \binom{n-i}{t} I_f(t) = \sum_{t=0}^{n} \binom{n-i}{t-1} I_f(t) = \cdots = \sum_{t=0}^{n} \binom{n-i}{t-i} I_f(t), \ i = 1, 2, \ldots, m. \tag{6.20}$$

By the binomial coefficient identity

$$\binom{n}{k} = \binom{n-1}{k} + \binom{n-1}{k-1},$$

we know that Eq. 6.20 is equivalent to

$$\sum_{t=0}^{n-i} \binom{n-i}{t} I_f(t) = \sum_{t=1}^{n-i+1} \binom{n-i}{t-1} I_f(t), \ i = 1, 2, \ldots, m. \tag{6.21}$$

Now we will show that Eq. 6.21 is also sufficient for $f(x)$ to be correlation immune of order m.

By Lemma 2.4, if Eq. 6.20 or equivalently Eq. 6.21 holds, it means that *supp(f)* contains an equal number of vectors of which the corresponding segment vectors covering positions 1, 2, ..., j are of Hamming weight 0, 1, ..., j. So the number of even weight and odd weight of such vectors are equal. By Lemma 2.4 we know that the coordinates 1, 2, ..., j of *supp(f)* cover $GF^j(2)$ or its multiple copies. Since $f(x)$ is symmetric, such a property of *supp(f)* also includes the case when the positions are $i_1, i_2, \ldots, i_j$ instead. By Eq. 6.18 we know that $f(x)$ is indeed an m-th order correlation immune function. This proves the following conclusion.

Theorem 6.1. *Let $f(x) \in S_n$ be a symmetric function corresponding to an integer function $I_f(t)$ from $\{0, 1, \ldots, n\}$ to $\{0, 1\}$, i.e., $f(x) = I_f(t)$ if and only if $wt(x) = t$. Then $f(x)$ is m-th order correlation immune if and only if Eq. 6.21 holds.*

In the later part of this section, we will mainly consider symmetric functions with correlation immunity of order 1 or 2. By Theorem 6.1, a symmetric function $f(x) \in \mathcal{F}_n$ is 1-st order correlation immune if and only if the following equation holds,

$$\sum_{t=0}^{n-1} \binom{n-1}{t} I_f(t) = \sum_{t=1}^{n} \binom{n-1}{t-1} I_f(t). \tag{6.22}$$

It is known that $I_f(t) \equiv 0$ and $I_f(t) \equiv 1$ are two trivial roots of the above equation. Moreover, the alteration of 0 and 1 is also a root of Eq. 6.22, corresponding to $\varphi_1(x)$ or $\varphi_1(x) \oplus 1$.

Write $\Gamma = I_f^{-1} = \{t : I_f(t) = 1\}$. Then Eq. 6.22 can be written as

$$\sum_{t\in\Gamma} \binom{n-1}{t} = \sum_{t\in\Gamma} \binom{n-1}{t-1}. \tag{6.23}$$

Now the problem of constructing symmetric correlation immune functions is converted to the construction of such sets Γ so that equality of Eq. 6.23 holds. It seems that explicit solutions of Eq. 6.23 are very hard to derive. We will derive some sporadic solutions of this equation.

6.3.1 When n Is Odd

By noticing $\binom{n-1}{\frac{n-1}{2}+1} = \binom{n-1}{\frac{n-1}{2}-1}$, it is known that

$$\Gamma_1 = \{\frac{n-1}{2} + 1, \frac{n-1}{2}\}$$

is a solution of Eq. 6.23. In general, it is easy to check that

$$\Gamma_k = \left\{\frac{n-1}{2} + k, \frac{n-1}{2} + k - 1, \frac{n-1}{2} + k - 2, \ldots, \frac{n-1}{2} - k + 1\right\}$$

is a solution of Eq. 6.23 for every $k = 1, 2, \ldots, \frac{n-1}{2}$. It is also noticed that $\Gamma_1 \subset \Gamma_2 \subset \cdots \subset \Gamma_{\frac{n-1}{2}}$. We have the following result.

Lemma 6.6. *Let $f_1, f_2 \in \mathcal{F}_n$ with supp(f_1) $\cap$ supp(f_2) $= \phi$. Then the fact that any two functions from $f_1, f_2, f_1 \oplus f_2$ are m-th order correlation immune implies that the third one is also such a function.*

By Lemma 6.6 we can calculate the number of constructed functions as follows: Firstly Γ_1 is a solution of Eq. 6.23. By employing Γ_2, we get two more solutions: Γ_2 and $\Gamma_2 - \Gamma_1$; by employing Γ_3, we get four more solutions: Γ_3, $\Gamma_3 - \Gamma_1$, $\Gamma_3 - \Gamma_2$, and $\Gamma_3 - (\Gamma_2 - \Gamma_1)$. By the same method, it can be shown that with each forthcoming Γ_i, we can get 2^{i-1} more solutions. It can easily check that all of those solutions obtained by those Γ_i's together with their subtractions and additions are different. So the number of different solutions of Eq. 6.23 that we have obtained is

$$\Delta_1 = \sum_{k=1}^{\frac{n-1}{2}} 2^{k-1} = 2^{\frac{n-1}{2}} - 1.$$

Note that each of those functions is a delegate of a couple, itself and its complement, and they still do not include $\varphi_1(x)$ and $\varphi_1(x) \oplus 1$. So, the number of symmetric 1-st order correlation immune functions (including 0 and 1) for odd integer n is lower bounded by

$$B_1 = 2\Delta_1 + 4 = 2^{\frac{n+1}{2}} + 2.$$

6.3.2 *When n Is Even*

Similar to the case when n is odd, it can be verified that $\Gamma_k = \{\frac{n}{2} + k - 1, \frac{n}{2} + k - 2, \ldots, \frac{n}{2} - k + 1\}$ is a solution of Eq. 6.23 for every $k = 1, 2, \ldots, \frac{n}{2}$. It is also noticed that $\Gamma_1 \subset \Gamma_2 \subset \cdots \subset \Gamma_{\frac{n}{2}}$. By Lemma 6.6 and the similar idea as in the last section, we have that the number of different solutions of Eq. 6.23 is

$$\Delta_2 = \sum_{k=1}^{n/2} 2^{k-1} = 2^{\frac{n}{2}} - 1.$$

Note that different from the case when n is odd, the affine symmetric functions $\varphi_1(x)$ and $\varphi_1(x) \oplus 1$ are covered by the subtraction/addition of Γ_i's, i.e., they are $\Gamma_{\frac{n}{2}} - \Gamma_{\frac{n}{2}-1} + \Gamma_{\frac{n}{2}-2} - \cdots + (-1)^{n/2+1}\Gamma_1$ and its complement. Therefore, the number of symmetric 1-st order correlation immune functions (including 0 and 1) in this case is lower bounded by

$$B_2 = 2\Delta_2 + 2 = 2^{\frac{n}{2}+1}.$$

Table 6.2 Comparison of lower bounds with the exact numbers of symmetric 1-st order correlation immune functions

n even	B_2	Total	n odd	B_1	Total
2	4	4	3	6	6
4	8	8	5	10	10
6	16	20	7	18	26
8	32	48	9	34	42
10	64	64	11	66	66
12	128	144	13	130	178
14	256	452	15	258	428
16	512	576	17	514	514
18	1024	1072	19	1026	1442
20	2048	2864	21	2050	2534
22	4096	4608	23	4098	6402
24	8192	12448	25	8194	9350
26	16384	16648	27	16386	16522
28	32768	32768	29	32770	36866
30	65536	82496	31	65538	77186
32	131072	132352	33	131074	148170
34	262144	393216			

Comparison of the lower bounds as described above and the actual number of symmetric 1-st order correlation immune functions, as found by exhaustive computing search, is shown in Table 6.2. It is noticed that in certain cases, the lower bound just meets the total number of symmetric 1-st order correlation immune functions (e.g., when n =2, 3, 4, 5, 10, 11, 17, 28). It seems that the method can be used to construct most of the symmetric 1-st order correlation immune functions. The efficiency of the method could be assessed if an upper bound on the number of symmetric correlation immune functions is found. This problem needs further research.

6.3.3 *Higher-Order Correlation Immunity*

For symmetric functions with correlation immunity of order larger than one, besides the exclusive-or of all the variables and its negation, there indeed exist such nonlinear functions. The investigation of this problem seems much more complicated. We have not yet found an efficient construction method. In the following table, we list some results which were found by exhaustive computing search. Numbers in the column initiated by "n" mean the number of variables, those in the column initiated by "*CI order*" mean the order of correlation immunity, and those in the column initiated by "*Tally*" mean the total number of such functions with the corresponding correlation immunity. The exhaustive search was conducted for all n up to 34 (see Table 6.3). We have omitted from the table the cases where the

Table 6.3 High-order correlation immunity of symmetric functions

n	CI order	Tally	n	CI order	Tally	n	CI order	Tally
7	2	4	8	2	4	8	3	2
9	2	4	9	3	2	10	3	2
13	2	4	14	2	8	14	3	2
15	2	18	15	3	4	16	3	8
19	2	8	20	2	8	20	3	4
21	2	8	21	3	4	22	3	4
23	2	4	24	2	4	24	3	2
25	2	16	25	3	2	26	2	8
26	3	6	27	2	4	27	3	4
28	3	2	31	2	12	32	2	8
32	3	6	33	2	8	33	3	4
34	2	4	34	3	4			

tally is zero. It is seen that the highest correlation immune order of those functions available by our computing search is three, and in general only very few symmetric functions are correlation immune of order higher than one (Table 6.3).

6.4 On Symmetric Resilient Functions

When correlation immune Boolean functions are balanced, they are also called resilient functions. Symmetric resilient functions have some particular interesting properties [7]. In 1985, Chor et al. conjectured in [3] that the only 1-resilient symmetric Boolean functions are the exclusive-or of all n variables and its negation. This conjecture was disproved by Gopalakrishnan, Hoffman, and Stinson in [8] where a class of infinite counterexamples were found. The following were also proposed in [8] as two open problems:

- Find the smallest constant t such that the Statement, *the only t-resilient symmetric Boolean functions are the exclusive-or of all n variables,* is true, or disprove this statement.
- Other than the exclusive-or of all n variables, its negation, and those corresponding to the infinite class presented in [8], are there any symmetric functions which are 1-resilient or 2-resilient?

In addition to the infinite class of nonlinear symmetric resilient functions introduced in [8], we will introduce two other infinite classes of nonlinear symmetric resilient functions, and it is interesting to find that one of the introduced class of resilient functions are all 1-resilient, while the other class of functions are all 2-resilient. We will also point out that the infinite class of resilient functions presented in [8] are all 1-resilient, and none is 2-resilient, and hence the second resiliency statement of theorem 3.1 in [8] is proved to be incorrect. What makes it more interesting is that there is an example introduced in [8] that is indeed 2-resilient; however, that example does not belong to the general class of resilient functions as constructed in [8].

In this section we will study the construction of balanced correlation immune symmetric Boolean functions.

6.4.1 Constructions of Symmetric Resilient Boolean Functions

Let $f(x) \in \mathcal{F}_n$. Then by Definition 4.1, $f(x)$ is t-resilient, if and only if for every $(a_1, \ldots, a_t) \in GF^t(2)$ and for every $c \in GF(2)$, we have

$$Prob(f(x) = c|(x_{i_1}, \ldots, x_{i_t}) = (a_1, \ldots, a_t)) = \frac{1}{2}.$$

Recall that a t-resilient Boolean function is also a balanced Boolean function with correlation immunity of order t. By the following probability identity

$$\begin{aligned} &Prob((x_{i_1}, \ldots, x_{i_t}) = (a_1, \ldots, a_t)) \cdot Prob(f(x) = c|(x_{i_1}, \ldots, x_{i_t}) = (a_1, \ldots, a_t)) \\ &= Prob(f(x) = c) \cdot Prob((x_{i_1}, \ldots, x_{i_t}) = (a_1, \ldots, a_t)|f(x) = (y_1, \ldots, y_m)), \end{aligned}$$

we have

Lemma 6.7. *Let $f(x) \in \mathcal{F}_n$. Then $f(x)$ is t-resilient if and only if $f(x)$ is balanced, and for every t-subset $\{i_1, \ldots, i_t\} \subset \{1, 2, \ldots, n\}$ and for any $c \in GF(2)$, we have*

$$Prob((x_{i_1}, \ldots, x_{i_t}) = (a_1, \ldots, a_t)|x \in supp(f)) = \frac{1}{2^t}.$$

The following result was proved in [8] and is included here.

Lemma 6.8 ([8]). *There exists a symmetric 1-resilient function in $\mathcal{F}_n$ if and only if the following equations have a solution:*

$$\begin{cases} \sum_{i=0}^{n} \binom{n}{i} \times I_f(i) = 2^{n-1} \\ \sum_{i=0}^{n-1} \binom{n-1}{i} \times I_f(i) = 2^{n-2} \end{cases} \tag{6.24}$$

where $I_f = (I_f(i))$ is a binary string of length $n+1$ which is to be determined.

It is also noticed that the exclusive-or of all n variables, i.e., the function $\varphi_1(x)$, satisfies that $\varphi_1(x) = 1$ if and only if $wt(x)$ is odd. So Eq. 6.24 is guaranteed to have a solution $I_{\varphi_1} = (0, 1, 0, 1, \ldots)$ which is an alternating 0-1 vector, and the complement of $\varphi_1(x)$ corresponds to another solution of Eq. 6.24, that is $(1, 0, 1, 0, \ldots)$ which is also an alternating of 0-1 vector.

It is easy to show that the roots of Eq. 6.24 always appear in couple. In general we have

Lemma 6.9. *If $I_f = (I_f(i))$ is a solution of Eq. 6.24, then*

$$I_g = (I_g(i)) = (I_f(i) + 1) \mod 2$$

is another solution.

Proof. The conclusion follows from the identity $\sum_{i=0}^{n} \binom{n}{i} = 2^n$. □

6.4.2 Searching for More Solutions

In [8], besides the affine symmetric resilient functions, an infinite class of nonlinear symmetric resilient functions were found by a way which can be described as follows: Set

$$I_f(i) = \begin{cases} I_{\varphi_1}(i) + 1 \mod 2 & \text{if } i \in \{k, k+1, n-k, n-k+1\} \\ I_{\varphi_1}(i) & \text{otherwise} \end{cases}$$

where k is to be determined. By solving Eq. 6.24 with this restriction, an infinite class of symmetric resilient functions were found which have parameters $n = r^2 - 2$, $k = \frac{1}{2}(r-2)(r-1)$, where $r > 2$ is an even integer. The smallest example in this class corresponds to the vector $I_f = (0,1,0,1,0,0,1,1,0,0,1,1,0,1,0)$. We will search for other classes of counterexamples for n being even and odd, respectively, and the following methods come from [22].

6.4.2.1 Type-A: When n Is Even

Let

$$I_f(i) = \begin{cases} I_{\varphi_1}(i) + 1 \mod 2 & \text{if } i \in \{k, k-1, n-k, n-k-1\} \\ I_{\varphi_1}(i) & \text{otherwise} \end{cases}$$

where k is to be determined. Then we have

$$\sum_{i=0}^{n} \binom{n}{i} \times I_f(i) = 2^{n-1} \pm \left[\binom{n}{k} - \binom{n}{k-1} + \binom{n}{n-k} - \binom{n}{n-k-1} \right]$$

and

$$\sum_{i=0}^{n-1} \binom{n-1}{i} \times I_f(i) = 2^{n-2} \pm \left[\binom{n-1}{k} - \binom{n-1}{k-1} + \binom{n-1}{n-k} - \binom{n-1}{n-k-1} \right].$$

In order for $I_f = (I_f(i))$ to be a solution of Eq. 6.24, the following two equations must hold:

$$\binom{n}{k} - \binom{n}{k-1} + \binom{n}{n-k} - \binom{n}{n-k-1} = 0, \qquad (6.25)$$

$$\binom{n-1}{k} - \binom{n-1}{k-1} + \binom{n-1}{n-k} - \binom{n-1}{n-k-1} = 0. \qquad (6.26)$$

Notice that Eq. 6.26 is an identity, so only Eq. 6.25 needs to be solved. Dividing by $\binom{n}{k}$ throughout Eq. 6.25, we have

$$1 - \frac{k}{n-k+1} + 1 - \frac{n-k}{k+1} = 0.$$

By solving this equation, we have two solutions in the form

$$k = \frac{1}{2}(n \pm \sqrt{n+2}). \qquad (6.27)$$

The roots of Eq. 6.27 are k and $n-k$. So we have from Eq. 6.27 only one desired set of $\{k, k-1, n-k, n-k-1\}$, and hence we have a solution for Eq. 6.24.

Equation 6.27 can be written in another form as

$$\begin{cases} n = 4r^2 - 2 \\ k = 2r^2 - r - 1 \end{cases}$$

where $r \geq 2$ is an arbitrary integer. Set $r = 2$, we have $n = 14$ and $k = 5$, which yields the smallest solution of Eq. 6.24 in this case as $I_f = (0, 1, 0, 1, 1, 0, 0, 1, 1, 0, 0, 1, 0, 1, 0)$.

6.4.2.2 Type-B: When n Is Odd

Let

$$I_f(i) = \begin{cases} I_{\varphi_1}(i) + 1 \bmod 2 \text{ if } i \in \{k, k+1, n-k, n-k-1\}, \\ I_{\varphi_1}(i) \qquad\qquad\qquad \text{otherwise.} \end{cases}$$

Then we have

$$\sum_{i=0}^{n}\binom{n}{i}\times I_f(i)=2^{n-1}\pm\left[\binom{n}{k+1}-\binom{n}{k}+\binom{n}{n-k}-\binom{n}{n-k-1}\right]$$

and

$$\sum_{i=0}^{n-1}\binom{n-1}{i}\times I_f(i)=2^{n-2}\pm\left[\binom{n-1}{k+1}-\binom{n-1}{k}+\binom{n-1}{n-k}-\binom{n-1}{n-k-1}\right].$$

In order for $I_f=(I_f(i))$ to be a solution of Eq. 6.24, the following equations must hold:

$$\binom{n}{k+1}-\binom{n}{k}+\binom{n}{n-k}-\binom{n}{n-k-1}=0, \tag{6.28}$$

$$\binom{n-1}{k+1}-\binom{n-1}{k}+\binom{n-1}{n-k}-\binom{n-1}{n-k-1}=0. \tag{6.29}$$

Noticing that Eq. 6.28 is an identity, only Eq. 6.29 needs to be solved. Similar to the procedure above, we have a solution of Eq. 6.29 in the form

$$\begin{cases} n=4r^2-1 \\ k=2r^2-r-1 \end{cases} \tag{6.30}$$

where $r\geq 2$ is an arbitrary integer. If set $r=2$, we have $n=15,\ k=5$, which yields the smallest solution of this case as I_f=(0,1,0,1,0,0,1,1,0,0,1,1,0,1,0,1).

Note that this class of counterexamples can be seen as being derived from the construction in [8] where n is replaced with $n-1$ and hence becomes an odd number. This observation has been noticed but not yet made explicit in [20].

6.4.3 The Exact Resiliency of Constructed Resilient Functions

For the above constructed resilient functions, it is interesting to know whether they have higher-order resiliency. Let $f(x)$ be a symmetric t-resilient function corresponding to a vector $I_f=(I_f(i))$. Then we have

$$\sum_{i=0}^{n-m}\binom{n-m}{i}(I_f(i)-I_{\varphi_1}(i))=\sum_{i=0}^{n-m}\binom{n-m}{i-1}(I_f(i)-I_{\varphi_1}(i))$$

$$=\cdots=\sum_{i=0}^{n-m}\binom{n-m}{i-t}(I_f(i)-I_{\varphi_1}(i)),\quad m=0,1,\ldots,t. \tag{6.31}$$

By the binomial coefficient identity

$$\binom{n}{i} = \binom{n-1}{i} + \binom{n-1}{i-1},$$

we know that Eq. 6.31 is equivalent to

$$\sum_{i=0}^{n-k} \binom{n-k}{i} (I_f(i) - I_{\varphi_1}(i)) = 0, \ k = 0, 1, \ldots, t. \tag{6.32}$$

It should be noticed that Eq. 6.32 describes exactly the 0-resiliency of $f(x)$ (or equivalently $f(x)$ is balanced) when $m = 0$ and the 1-resiliency of $f(x)$ when $m = 1$. We write this conclusion in the following theorem.

Theorem 6.2. *Let $f \in \mathcal{F}_n$ be a symmetric function corresponding to a vector $I_f = (I_f(i))$. Then $f(x)$ is a t-resilient function if and only if Eq. 6.32 holds.*

For the abovementioned constructed resilient functions, we consider whether their exact resiliency is higher than one. It is already known that they are 1-resilient, which means that Eq. 6.32 holds for $m = 0$ and $m = 1$. We only need to consider the case when $m > 1$. This will be treated in the following three cases.

6.4.3.1 On the Construction of [8]

In the construction of [8], n is even and $I_f(i) \neq I_{\varphi_1}(i)$ if and only if $i \in \{k, k+1, n-k, n-k+1\}$. So when $m = 2$, Eq. 6.32 becomes

$$\binom{n-2}{k+1} - \binom{n-2}{k} + \binom{n-2}{n-k+1} - \binom{n-2}{n-k} = 0. \tag{6.33}$$

It is easy to notice that this can be simplified as

$$(n-2k-3)(n-k)((n-k)^2-1) + k(k-1)(k-2)(k+1) - k(k^2-1)(n-k+1) = 0.$$

Replacing n and k by $r^2 - 2$ and $\frac{1}{2}(r-2)(r+1)$, respectively, we have

$$(r-3)(r^2+r-2)(r^4+2r^3-3r^2-4r) = (r^2-r-2)(r^2-r-4)(r^2-r)(r+3).$$

By solving this equation, we have $r = 0, 1, -1$. This is a contradiction to the statement that $r > 2$ be even. So we can conclude that:

Theorem 6.3. *The functions constructed in [8] are all 1-resilient functions, and none is 2-resilient.*

6.4.3.2 On the Construction of *Type-A* Resilient Functions

Type-A resilient functions are in even number of variables, and $I_f(i) \neq I_{\varphi_1}(i)$ if and only if $i \in \{k, k-1, n-k, n-k+1\}$. So, if $I_f(i)$ corresponds to a 2-resilient function, according to Theorem 6.2, the following equation must hold:

$$\binom{n-2}{k} - \binom{n-2}{k-1} + \binom{n-2}{n-k} - \binom{n-2}{n-k-1} = 0. \tag{6.34}$$

By solving Eq. 6.34, we have

$$k = \frac{1}{2}(n \pm \sqrt{n}).$$

Combining this with Eq. 6.27, we know that no solution exists, i.e., no function in this class is 2- resilient.

6.4.3.3 On the Construction of *Type-B* Resilient Functions

Type-B resilient functions are in odd number of variables, and $I_f(i) \neq I_{\varphi_1}(i)$ if and only if $i \in \{k, k+1, n-k, n-k-1\}$. So if I_f corresponds to a 2-resilient function, then by Theorem 6.2, the following equation must hold:

$$\binom{n-2}{k+1} - \binom{n-2}{k} + \binom{n-2}{n-k} - \binom{n-2}{n-k-1} = 0. \tag{6.35}$$

Surprisingly, Eq. 6.30 makes Eq. 6.35 an identity. Since the functions we are considering are symmetric, properties which apply to (x_1, x_2) are also valid for every pair (x_i, x_j). Hence by Lemma 6.7 we know that the functions constructed in Sect. 6.4.2.2 are all 2-resilient functions.

However, when $m = 3$, it is easy to verify that Eq. 6.32 cannot be hold simultaneously with the cases for $m = 0, 1, 2$. This means that all functions in this class are exactly 2-resilient.

6.5 Basic Properties of Majority Functions

The development of cryptographic algorithms have experienced different attacks. As a result of the attacks, different measure about the resistance against the corresponding attacks are proposed. When correlation attack [17–19] was treated as a threat, the concept of correlation immunity was proposed in [16] as a measure about the resistance that a nonlinear combination function has against the correlation attack. In [4] a new attack known as the algebraic attack is proved to be very effective to many stream ciphers as well as to some block ciphers. As a measure of the resistance of a nonlinear function against the algebraic attack, another measure known as algebraic immunity is proposed. As studied in Chap. 5, the idea of algebraic attack is to find a low-degree annihilator of the targeting combining function. By doing so, the process of algebraic attack is to solve a system of nonlinear equations. When the algebraic degree of the annihilator is low, the computational complexity to solve such a system of nonlinear equations is also low. So the effectiveness of algebraic attack depends on whether one can find such an annihilator with low algebraic degree. On the other hand, when the combining function is of high algebraic immunity, the algebraic degree of any of its annihilators cannot be very low. Hence, a significant job for the designers is to find combining functions with highest possible algebraic immunity. It has been proved [12] that the order of the algebraic immunity of a Boolean function in n variables cannot exceed $\lceil \frac{n}{2} \rceil$. If a Boolean function has algebraic immunity of order $\lceil \frac{n}{2} \rceil$, then this function is said to have the highest algebraic immunity.

In 2004, Dalai [5] studied the majority functions in odd number of variables to be a class of Boolean functions with highest algebraic immunity, and it was further proved in [9] that the majority functions and their complements are the only symmetric Boolean functions in odd number of variables with maximum algebraic immunity. While algebraic immunity is an important cryptographic measure, very often the best performance with one cryptographic measure will sacrifice the performance with other cryptographic measures. Here we study the correlation immunity of the generalized majority functions, which include the majority functions in odd number of variables and newly defined such functions in even number of variables.

Definition 6.2. Let n be an odd number. The following defined Boolean function $f(x)$ in n variables is called a *majority function*:

$$f(x) = \begin{cases} 0, & \text{if } wt(x) \leq \frac{n-1}{2}; \\ 1, & \text{if } wt(x) \geq \frac{n+1}{2}. \end{cases} \tag{6.36}$$

The natural meaning of the above-defined majority function is that when the majority of the n-bit input has value 1, then the function outputs 1 which means a TRUE value, and when the majority of the input has value 0, the function outputs 0 which means FALSE.

The definition of majority function is very natural for the case when the input has odd number of coordinates, i.e., the number of inputs n of the function is an odd number, and in this case, it is a symmetric Boolean function. When n is even, there is no natural way of defining majority functions, as there are cases where the input has equal number of 0 values and 1 values. For this case, we generalize the concept of majority function as follows:

Definition 6.3. Let n be an even number. Define $S = \{x \in GF^n(2) : wt(x) = \frac{n}{2}\}$ and let $A \subseteq S$ be a subset of S. Then

$$f_A(x) = \begin{cases} 0, & \text{if } wt(x) < \frac{n}{2} \text{ or } x \in A; \\ 1, & \text{if } wt(x) > \frac{n}{2} \text{ or } x \in (S \setminus A) \end{cases} \tag{6.37}$$

is called a *set-majority function.*

Definition 6.3 generalizes the concept of majority function in odd number of variables to the case in even number of variables, and hence without confusion, the set-majority function may simply be called the majority function and is simply denoted as $f(x)$ (without the subindex "A"). The definition of Eq. 6.37 can be treated as universal (i.e., it applies to odd and even number of variables) since when n is odd, the set S is an empty set and so is A.

There are two extreme cases of the set-majority functions, that is, when $A = S$ and when $A = \phi$ which is an empty set. When $A = S$, Eq. 6.37 becomes

$$f_1(x) = \begin{cases} 0, & \text{if } wt(x) \leq \frac{n}{2}; \\ 1, & \text{if } wt(x) > \frac{n}{2}, \end{cases} \tag{6.38}$$

which is called a *strict majority function.* When $A = \phi$, Eq. 6.37 becomes

$$f_0(x) = \begin{cases} 0, & \text{if } wt(x) < \frac{n}{2}; \\ 1, & \text{if } wt(x) \geq \frac{n}{2}, \end{cases} \tag{6.39}$$

which is called a *loose majority function.* The meaning of the above two extreme cases can be interpreted as follows: The strict majority function has value 1 only when there are absolutely more 1 values than 0 values in the input; otherwise, it has value 0, including the case when the input has equal number of 0's and 1's. The loose majority function has value 1 as long as the number of 1 value inputs is no less than that of 0 value inputs, including the case when their numbers are equal, and it takes 0 only when there are absolutely less 1's than 0's in the input. In general case, the set-majority function has value 1 when there are absolutely more 1's than 0's in the input, and it has value 0 when there are absolutely less 1's than 0's in the input, and in the case when the input has equal number of 0's and 1's, it has to check if the input is from set A or $S \setminus A$. For the former case, the function has value 0 and otherwise it has value 1.

For any given even number n, the strict majority function and the loose majority function are uniquely determined, just as the case of majority function defined for odd n. However, in general, the set-majority function is not uniquely determined yet, as it depends on the set A. Note that when n is even, $f_A(x)$ is symmetric if and only if $A = \phi$ or $A = S$. So in general, $f_A(x)$ is not a symmetric Boolean function. However, our study will be on the general case, where all the induced conclusions will apply to the cases when $A = \phi$ and $A = S$ as well.

Theorem 6.4. *When n is odd, the majority functions in n variables are all balanced; when n is even, the (set) majority functions in n variables are balanced if and only if $|A| = \frac{|S|}{2}$, where $|A|$ is the cardinality of set A.*

Proof. When n is odd, by the Definition 6.2, the Hamming weight of the majority function $f(x)$ is $wt(f) = \binom{n}{\frac{n+1}{2}} + \binom{n}{\frac{n+1}{2}+1} + \cdots + \binom{n}{n}$. Note that

$$\begin{aligned}
2^n &= \binom{n}{0} + \binom{n}{1} + \cdots + \binom{n}{\frac{n-1}{2}} \\
&\quad + \binom{n}{\frac{n+1}{2}} + \binom{n}{\frac{n+1}{2}+1} + \cdots + \binom{n}{n} \\
&= \binom{n}{n} + \binom{n}{n-1} + \cdots + \binom{n}{\frac{n+1}{2}} \\
&\quad + \binom{n}{\frac{n+1}{2}} + \binom{n}{\frac{n+1}{2}+1} + \cdots + \binom{n}{n} \\
&= 2wt(f)
\end{aligned}$$

Hence, we have $wt(f) = 2^{n-1}$ which means that $f(x)$ is balanced.

When n is even, the Hamming weight of majority function $f_A(x)$ is $wt(f_A) = \binom{n}{\frac{n}{2}+1} + \binom{n}{\frac{n}{2}+2} + \cdots + \binom{n}{n} + |S \setminus A|$. For convenience of writing, let $\Delta = \binom{n}{\frac{n}{2}+1} + \binom{n}{\frac{n}{2}+2} + \cdots + \binom{n}{n}$ and $A' = S \setminus A$. Then Δ can also be expressed as:

$$\Delta = \binom{n}{0} + \binom{n}{1} + \cdots + \binom{n}{\frac{n}{2}-1}.$$

Hence, we have

$$\begin{aligned}
2^n &= \binom{n}{0} + \binom{n}{1} + \cdots + \binom{n}{\frac{n}{2}-1} \\
&\quad + \binom{n}{\frac{n}{2}} + \binom{n}{\frac{n}{2}+1} + \cdots + \binom{n}{n} \\
&= \Delta + \binom{n}{\frac{n}{2}} + \Delta
\end{aligned}$$

Note that $|A| + |A'| = |S| = \binom{n}{\frac{n}{2}}$, from the above we have

$$2^n = 2\Delta + |A| + |A'|.$$

So, $f_A(x)$ is balanced $\Longleftrightarrow wt(f_A) = \Delta + |A'| = 2^{n-1} \Longleftrightarrow \Delta + |A| = 2^{n-1} \Longleftrightarrow |A| = |A'| \Longleftrightarrow |A| = \frac{|S|}{2}$. □

Table 6.4 The number of balanced majority functions in even number of variables

n	C(n)
2	2
4	20
6	184756
8	1121862778166628454 32
10	3.63×10^{74}
12	3.72×10^{276}
14	1.85×10^{1031}
16	1.26×10^{3872}
18	4.33×10^{14633}
20	2.32×10^{55614}

It is seen from the above theorem that there is a strict restriction on the size of A when the majority function is required to be balanced. What is the number of such balanced functions for a given even n? Since A is any subset of S that has half of the elements in S, there can be $\binom{|S|}{|S|/2}$ choices of A. To distinguish this special case with the general case, we call this case as balanced majority functions, because this class of functions are all balanced.

Denote by $C(n)$ the number of balanced majority functions. Then $C(n) = \binom{0}{0} = 1$ for any odd value n. When n is even, it is easy to prove that

$$C(n) = \binom{\binom{n}{n/2}}{\binom{n}{n/2}/2}.$$

From Table 6.4 it can be seen that the size of $C(n)$ increases very fast with the increase of n.

For the general case, by Stirling formula, $n! \approx \sqrt{2\pi} n^{n+\frac{1}{2}} e^{-n+\frac{1}{12n}}$, we can get an approximation:

$$\binom{n}{n/2} \approx \frac{2^{n+1}}{\sqrt{2\pi n e^{\frac{1}{4n}}}} \approx \frac{2^{n+1}}{\sqrt{2\pi n}}$$

and hence

$$C(n) = \binom{\binom{n}{n/2}}{\binom{n}{n/2}/2} \approx 2^{\frac{2^{n+1}}{\sqrt{2n\pi}} - \frac{n}{2} + \frac{1}{4}} n^{\frac{1}{4}} \pi^{-\frac{1}{4}} e^{\frac{1}{8n}}$$

which increases super exponentially with the increase of n. In this sense, the generalized majority functions in even number of variables are more applicable in practice for their large number of supplies.

6.6 The Walsh Spectrum of Majority Functions

Walsh transform has been a very useful tool in analyzing cryptographic properties of Boolean functions. Here we use Walsh transform to study the correlation immunity of majority functions.

Since the definition of majority functions differs much for the cases when n is odd and when n is even, our discussion will treat each of the cases, respectively. Note that, when we write the XOR of two vectors such as $x \oplus s$, it means the bitwise XOR of vectors x and s. Let $\mathbf{1}$ be the all-one vector in $GF^n(2)$, then we use $x \oplus \mathbf{1}$ to denote the complement of x, i.e., all the coordinates of x is taken the complement by XORing with 1.

6.6.1 When n Is Odd

First we notice the following property of this class of functions:

Theorem 6.5. *When n is odd, a Boolean function $f(x)$ defined in Definition 6.2 satisfies:*

$$f(x \oplus \mathbf{1}) = f(x) \oplus 1.$$

Proof. By Definition 6.2, $f(x) = 0 \Longleftrightarrow wt(x) \leq (n-1)/2 \Longleftrightarrow wt(x \oplus \mathbf{1}) \geq (n+1)/2 \Longleftrightarrow f(x \oplus \mathbf{1}) = 1$. Similarly, $f(x) = 1 \Longleftrightarrow f(x \oplus \mathbf{1}) = 0$. □

Theorem 6.6. *Let $f(x)$ be a majority function in n variables, and then the Walsh transform of $f(x)$ satisfies:*

$$S_{(f)}(w) = \begin{cases} 0, & \text{if } wt(w) \text{ is even;} \\ 2 \sum_{wt(x) \leq \frac{n-1}{2}} (-1)^{\langle w, x \rangle}, & \text{if } wt(w) \text{ is odd.} \end{cases} \tag{6.40}$$

Proof. Since $f(x)$ is a majority function in odd number of variables, we have

$$\begin{aligned} S_{(f)}(w) &= \sum_{x \in GF^n(2)} (-1)^{f(x) + \langle w, x \rangle} \\ &= \sum_{wt(x) \leq \frac{n-1}{2}} (-1)^{\langle w, x \rangle} - \sum_{wt(x) \geq \frac{n+1}{2}} (-1)^{\langle w, x \rangle} \\ &= \sum_{wt(x) \leq \frac{n-1}{2}} (-1)^{\langle w, x \rangle} - \sum_{wt(x) \leq \frac{n-1}{2}} (-1)^{\langle w, (\mathbf{1} \oplus x) \rangle} \end{aligned}$$

$$= \sum_{wt(x)\le\frac{n-1}{2}} (-1)^{\langle w,x\rangle} - \sum_{wt(x)\le\frac{n-1}{2}} (-1)^{\langle w,\mathbf{1}\rangle+\langle w,x\rangle}$$

$$= \sum_{wt(x)\le\frac{n-1}{2}} (-1)^{\langle w,x\rangle} - \sum_{wt(x)\le\frac{n-1}{2}} (-1)^{wt(w)+\langle w,x\rangle}$$

$$= \begin{cases} 0, & \text{if } wt(w) \text{ is even;} \\ 2\sum_{wt(x)\le\frac{n-1}{2}} (-1)^{\langle w,x\rangle}, & \text{if } wt(w) \text{ is odd.} \end{cases}$$

□

6.6.2 When n Is Even

From Definition 6.3 it is known that the majority function in even number of variables is not uniquely determined; it depends on the set A. Denote by $A' = S \setminus A = \{x : x \in S \text{ and } x \notin A\}$ to be the complement set of A with respect to S, and define

$$A_1 = \{x : x \in GF^n(2) \text{ and } wt(x) < \frac{n}{2}\}$$

$$A_2 = \{x : x \in GF^n(2) \text{ and } wt(x) > \frac{n}{2}\}$$

$$A_3 = \{x : x \in A \setminus (A \cap \bar{A})\}$$

$$A_4 = \{x : x \in A' \setminus (A' \cap \bar{A}')\}$$

$$A_5 = \{x : x \in A \cap \bar{A}\}$$

$$A_6 = \{x : x \in A' \cap \bar{A}'\}$$

where $\bar{A} = \{x \oplus \mathbf{1} : x \in A\}$. Then we have

Theorem 6.7. *The above defined sets satisfy the following:*

1. $|A_1| = |A_2|, |A_3| = |A_4|$, *where* $|A|$ *means the cardinality of set* A, *i.e., the number of elements in* A. *Furthermore, if* $|A| = \frac{|S|}{2}$, *then we also have* $|A_5| = |A_6|$.
2. $f(x)|_{A_1} = 0, f(x)|_{A_2} = 1, f(x)|_{A_3} = 0, f(x)|_{A_4} = 1, f(x)|_{A_5} = 0, f(x)|_{A_6} = 1$, *where* $f(x)|_A$ *represents the constraint function of* $f(x)$ *whose variable* x *can only take values from* A.
3. *Define a map* $\phi(x) = x \oplus \mathbf{1}$. *It maps every coordinate of* x *to its complement, and for a set* $B \subset GF^n(2)$, *we denote* $\phi(B) = \{y = \phi(x) : x \in B\}$. *Then we have* $\phi^2(x) = x$, $\phi^2(B) = B$, *and* $\phi(A_1) = A_2$, $\phi(A_3) = A_4$, $\phi(A_5) = A_5$, $\phi(A_6) = A_6$.

Proof.

1. By the definition of S, it is known that for any $x \in S$, we have $wt(x) = \frac{n}{2}$, and hence $wt(x \oplus \mathbf{1}) = \frac{n}{2}$, or $(x \oplus \mathbf{1}) \in S$. This means that the elements in S appear in pairs that are complement to each other, i.e., for any $x \in S$, there must exist $y \in S$ such that $x \oplus y = \mathbf{1}$.
   By the definitions of A_i ($i = 1 \sim 6$) above, we have that A has $|A_5|$ elements whose complement is also in A and $|A_3|$ elements whose complement is not in A (and therefore must be in A'). So we have $|A| = |A_3| + |A_5|$. Similarly we have $|A'| = |A_4| + |A_6|$. Since the complement of every element in A_3 must be in A_4 and vice versa, we have $|A_3| = |A_4|$. Furthermore, if $|A| = \frac{|S|}{2}$, then $|A| = |A'|$ and hence $|A_5| = |A_6|$.
2. From Definition 6.3 we know that $f(x)|_{A_1} = 0, f(x)|_{A_2} = 1$, and $f(x)|_A = 0$, and hence $f(x)|_{A'} = 1$. The conclusion comes from the fact that $A_3 \subset A$, $A_5 \subset A$, $A_4 \subset A'$, and $A_6 \subset A'$.
3. It is obvious that $\phi^2(x) = x$ and $\phi^2(B) = B$. From the definitions of A_1 to A_6, it is trivial to verify that $\phi(x)$ is a one-to-one mapping from A_1 to A_2, from A_3 to A_4, from A_5 onto itself, and from A_6 onto itself, and hence the conclusion follows.

□

Based on Theorem 6.7, we can formulate the Walsh transform of the majority functions in even number of variables. First we give:

Lemma 6.10. *Let $V \subset GF^n(2)$ be a self-complement set, i.e., $\bar{V} = \{x \oplus \mathbf{1} : x \in V\} = V$, then for any odd Hamming weight vector $w \in GF^n(2)$, we have* $\sum_{x \in V} (-1)^{\langle w,x \rangle} = 0$.

Proof. Denote $\delta = \sum_{x \in V} (-1)^{\langle w,x \rangle}$, then

$$
\begin{aligned}
\delta &= \sum_{x \in V} (-1)^{\langle w, (x \oplus \mathbf{1}) \rangle} \\
&= \sum_{x \in V} (-1)^{\langle w,x \rangle + wt(w)} \\
&= (-1)^{wt(w)} \sum_{x \in V} (-1)^{\langle w,x \rangle} \quad \text{(since } wt(w) \text{ is odd)} \\
&= -\delta.
\end{aligned}
$$

Hence, $\delta = 0$. □

Theorem 6.8. *Let $f_A(x)$ be a majority function in n variables. Then the Walsh transform of $f_A(x)$ is:*

$$
S_{(f_A)}(w) = \begin{cases} \sum_{x \in A_5} (-1)^{\langle w,x \rangle} - \sum_{x \in A_6} (-1)^{\langle w,x \rangle}, & \textit{if } wt(w) \textit{ is even;} \\ 2 \sum_{x \in A_3 \text{ or } wt(x) < \frac{n}{2}} (-1)^{\langle w,x \rangle}, & \textit{if } wt(w) \textit{ is odd.} \end{cases} \tag{6.41}
$$

Proof. By the definition of $f_A(x)$ with respect to the sets A and S, and note that $S = A \cup A' = A_3 \cup A_5 \cup A_4 \cup A_6$ and $GF^n(2) = S \cup A_1 \cup A_2 = \bigcup_{i=1}^{6} A_i$, by Lemma 6.10 and Theorem 6.7, we have

$$
\begin{aligned}
S_{(f_A)}(w) &= \sum_{x \in GF^n(2)} (-1)^{f_A(x)+\langle w,x\rangle} \\
&= \sum_{x\in A_1}(-1)^{\langle w,x\rangle} - \sum_{x\in A_2}(-1)^{\langle w,x\rangle} + \sum_{x\in A_3}(-1)^{\langle w,x\rangle} \\
&\quad - \sum_{x\in A_4}(-1)^{\langle w,x\rangle} + \sum_{x\in A_5}(-1)^{\langle w,x\rangle} - \sum_{x\in A_6}(-1)^{\langle w,x\rangle} \\
&= \sum_{x\in A_1}(-1)^{\langle w,x\rangle} - \sum_{x\in A_1}(-1)^{\langle w,(x\oplus \mathbf{1})\rangle} + \sum_{x\in A_3}(-1)^{\langle w,x\rangle} \\
&\quad - \sum_{x\in A_3}(-1)^{\langle w,(x\oplus \mathbf{1})\rangle} + \sum_{x\in A_5}(-1)^{\langle w,x\rangle} - \sum_{x\in A_6}(-1)^{\langle w,x\rangle} \\
&= \sum_{x\in A_1}(-1)^{\langle w,x\rangle} - (-1)^{wt(w)}\sum_{x\in A_1}(-1)^{\langle w,x\rangle} \\
&\quad + \sum_{x\in A_3}(-1)^{\langle w,x\rangle} - (-1)^{wt(w)}\sum_{x\in A_3}(-1)^{\langle w,x\rangle} \\
&\quad + \sum_{x\in A_5}(-1)^{\langle w,x\rangle} - \sum_{x\in A_6}(-1)^{\langle w,x\rangle} \\
&= \begin{cases} \sum\limits_{x\in A_5}(-1)^{\langle w,x\rangle} - \sum\limits_{x\in A_6}(-1)^{\langle w,x\rangle}, & \text{if } wt(w) \text{ is even;} \\ 2\sum\limits_{x\in A_1\cup A_3}(-1)^{\langle w,x\rangle}, & \text{if } wt(w) \text{ is odd} \end{cases} \\
&= \begin{cases} \sum\limits_{x\in A_5}(-1)^{\langle w,x\rangle} - \sum\limits_{x\in A_6}(-1)^{\langle w,x\rangle}, & \text{if } wt(w) \text{ is even;} \\ 2\sum\limits_{x\in A_3 or wt(x)<\frac{n}{2}}(-1)^{\langle w,x\rangle}, & \text{if } wt(w) \text{ is odd} \end{cases}
\end{aligned}
$$

□

6.7 The Correlation Immunity of Majority Functions

A common method to study the correlation immunity of Boolean functions is to use Walsh transforms, this is due to Xiao-Massey theorem (see Lemma 4.13) which gives a clear representation of correlation immunity in terms of Walsh spectrum.

Note that in Lemma 4.13, any type of Walsh spectrum can be used, as it only considers the Walsh values on vectors with nonzero Hamming weight, and for any nonzero w, we always have that $S_f(w) = 0$ if and only if $S_{(f)}(w) = 0$. In order to check if the majority functions are correlation immune of any order at all, we first look at whether they are correlation immune of order 1; for this purpose, we only need to verify their Walsh spectrum on a vector w with Hamming weight 1. Without loss of generality, let this vector be e_i whose i-th coordinate is 1 and 0 elsewhere.

Regarding the correlation immunity of majority functions, we have the following conclusion.

Theorem 6.9. *None of the majority functions defined in Definitions 6.2 and 6.3 is correlation immune.*

Proof. When n is odd, by Theorem 6.6 we have

$$S_{(f)}(e_i) = 2 \sum_{wt(x) \le \frac{n-1}{2}} (-1)^{\langle e_i, x \rangle} = 2 \sum_{wt(x) \le \frac{n-1}{2}} (-1)^{x_i}$$

Among all the n-dimensional vectors x with $wt(x) \le \frac{n-1}{2}$, the number of such vectors that also satisfy that the i-th coordinate is 1 (and the other $n-1$ coordinates can have $0 \sim \frac{n-3}{2}$ of 1's) is

$$\binom{n-1}{0} + \binom{n-1}{1} + \binom{n-1}{2} + \cdots + \binom{n-1}{\frac{n-3}{2}},$$

and the number of such vectors whose i-th coordinate is 0 (and the other $n-1$ coordinates can have $1 \sim \frac{n-1}{2}$ of 1's) is

$$\binom{n-1}{1} + \binom{n-1}{2} + \binom{n-1}{3} + \cdots + \binom{n-1}{\frac{n-1}{2}}.$$

Therefore,

$$\begin{aligned} S_{(f)}(e_i) &= 2\left[\left(\binom{n-1}{1} + \binom{n-1}{2} + \binom{n-1}{3} + \cdots + \binom{n-1}{\frac{n-1}{2}}\right)\right. \\ &\quad \left. - \left(\binom{n-1}{0} + \binom{n-1}{1} + \binom{n-1}{2} + \cdots + \binom{n-1}{\frac{n-3}{2}}\right)\right] \\ &= 2\left[\binom{n-1}{\frac{n-1}{2}} - \binom{n-1}{0}\right] \\ &= 2\left[\binom{n-1}{\frac{n-1}{2}} - 1\right] \end{aligned}$$

Obviously the above is not zero for $n > 1$, which means that the majority function $f(x)$ in odd number of variables is not correlation immune (note that no Boolean function in one variable is correlation immune).

When n is even, by Theorem 6.8 we have

$$S_{(f_A)}(e_i) = 2 \sum_{x \in A_1 \cup A_3} (-1)^{\langle e_i, x \rangle} = 2 \left(\sum_{wt(x) < \frac{n}{2}} (-1)^{x_i} + \sum_{x \in A_3} (-1)^{x_i} \right)$$

Similar to the case when n is odd, we have

$$\sum_{wt(x) < \frac{n}{2}} (-1)^{x_i} = \binom{n-1}{\frac{n}{2}-1} - 1,$$

therefore,

$$S_{(f_A)}(e_i) = 2 \left(\binom{n-1}{\frac{n}{2}-1} - 1 + \sum_{x \in A_3} (-1)^{x_i} \right).$$

We show that the above is not always zero, i.e., if the above is zero for some i, then there must exist j such that $S_{(f_A)}(w_j) \neq 0$. Denote by $A_3^{i1} = \{x \in A_3 : x_i = 1\}$ and $A_3^{i0} = \{x \in A_3 : x_i = 0\}$, then $A_3 = A_3^{i0} \cup A_3^{i1}$.

Assume for some i, $S_{(f_A)}(e_i) = 0$, and then $\sum_{x \in A_3} (-1)^{x_i} = 1 - \binom{n-1}{\frac{n}{2}-1}$. This means that $|A_3^{i1}| - |A_3^{i0}| = \binom{n-1}{\frac{n}{2}-1} - 1$. Note that when the i-th coordinate is fixed to be 1, the number of such vectors in S is $\binom{n-1}{\frac{n}{2}-1}$ (the other $n-1$ coordinates has $\frac{n}{2}-1$ of 1's). Since $|A_3^{i1}|$ cannot be larger than $\binom{n-1}{\frac{n}{2}-1}$, then there are only two possible cases: (1) $|A_3^{i1}| = \binom{n-1}{\frac{n}{2}-1}$ and $|A_3^{i0}| = 1$ or (2) $|A_3^{i1}| = \binom{n-1}{\frac{n}{2}-1} - 1$. We show that in both of the cases, there must exist a j such that $S_{(f_A)}(w_j) \neq 0$ and hence induces the conclusion of the theorem.

If case (1) is true, then the other $n-1$ coordinates (except i) of the vectors in A_3 have all the possible vectors of Hamming weight $\frac{n}{2}-1$. So for any $j \neq i$, there are $\binom{n-2}{\frac{n}{2}-1}$ elements in A_3^{i1} whose j-th coordinate is 0 (let the other $n-2$ coordinates take $\frac{n}{2}-1$ of 1's), and there are $\binom{n-2}{\frac{n}{2}-2}$ elements in A_3^{i1} whose j-th coordinate is 1 (let the other $n-2$ coordinates take $\frac{n}{2}-2$ of 1's). Since the j-th coordinate of the element in A_3^{i0} may be 0 or 1, we have

$$\sum_{x \in A_3} (-1)^{x_j} = \binom{n-2}{\frac{n}{2}-1} - \binom{n-2}{\frac{n}{2}-2} - c = \frac{(n-2)!}{\frac{n}{2}!(\frac{n}{2}-1)!} - c,$$

where $c \in \{0, 1\}$. It is easy to verify that the above expression is larger than or equal to 0 when $n > 2$, and hence $S_{(f_A)}(w_j) > 0$.

If case (2) is true, then similarly other $n-1$ coordinates (except i) have all but one of the possible vectors of Hamming weight $\frac{n}{2}-1$. So for any $j \neq i$, there are $\binom{n-2}{\frac{n}{2}-1} - c_1$ elements in A_3^{i1} whose j-th coordinate is 0 (let the other $n-2$ coordinates take $\frac{n}{2}-1$ of 1's, taking away one such vector), and there are $\binom{n-2}{\frac{n}{2}-2} - c_2$ elements in A_3^{i1} whose j-th coordinate is 1 (let the other $n-2$ coordinates take $\frac{n}{2}-2$ of 1's, taking away one such vector). Hence, we have

$$\sum_{x \in A_3}(-1)^{x_j} = \binom{n-2}{\frac{n}{2}-1} - c_1 - \left(\binom{n-2}{\frac{n}{2}-2} - c_2\right) = \frac{(n-2)!}{\frac{n}{2}!(\frac{n}{2}-1)!} + c_2 - c_1,$$

where $c_1, c_2 \in \{0, 1\}$. As in case (1), when $n > 2$, it always results in $S_{(f_A)}(w_j) > 0$. When $n = 2$, all the possible majority functions in 2 variables are $f_1(x) = x_1$, $f_2(x) = x_2$, $f_3(x) = x_1x_2$, and $f_4(x) = x_1 \oplus x_2 \oplus x_1x_2$. It is easy to verify that none of these functions is correlation immune, and hence the conclusion of the theorem is true. □

6.8 The ε-Correlation Immunity of Majority Functions

It is known that the majority functions defined in Definition 6.2 (and those defined in Definition 6.3 as well) have good algebraic Immunity; their correlation immunity, however, as shown by Theorem 6.9, is not so good. Note that the correlation immunity is a cryptographic measure about the resistance against correlation attack; there can be cases where although a combining function is not correlation immune, the correlation attack, however, still consumes large amount of computation due to the function being "near" to be correlation immune, i.e., when the ε-correlation immunity is near to 1. Now we compute the ε-correlation immunity of the majority functions using Eq. 4.47 or Eq. 4.48.

6.8.1 *When n Is Odd*

By Definition 6.2 it is known that $f(x) = 1$ if and only if $wt(x) \geq \frac{n+1}{2}$. Similar to the discussion above, among the vectors with Hamming weight being larger than or equal to $\frac{n+1}{2}$, the number of such vectors where the i-th coordinate is 0 (and the rest $n-1$ coordinates can have $\frac{n+1}{2} \sim n-1$ of 1's) is $\binom{n-1}{n-1} + \binom{n-1}{n-2} + \cdots + \binom{n-1}{\frac{n+1}{2}}$, and the number of such vectors where the i-th coordinate is 1 (and the rest $n-1$ coordinates can have $\frac{n-1}{2} \sim n-1$ of 1's) is $\binom{n-1}{n-1} + \binom{n-1}{n-2} + \cdots + \binom{n-1}{\frac{n-1}{2}}$. Therefore,

$$
\begin{aligned}
S_{(f)}(e_i) &= -2 \sum_{x \in supp(f)} (-1)^{x_i} \\
&= 2\left(\binom{n-1}{n-1} + \binom{n-1}{n-2} + \cdots + \binom{n-1}{\frac{n-1}{2}}\right) \\
&\quad -2\left(\binom{n-1}{n-1} + \binom{n-1}{n-2} + \cdots + \binom{n-1}{\frac{n+1}{2}}\right) \\
&= 2\binom{n-1}{\frac{n-1}{2}}
\end{aligned}
$$

Note that here the value of $S_{(f)}(e_i)$ is independent of i; hence, by Eq. 4.48 we have

$$
\begin{aligned}
CI_{\varepsilon}(f) &= 1 - \frac{1}{2wt(f)} \max_i |S_{(f)}(e_i)| \\
&= 1 - \frac{1}{wt(f)} \cdot \binom{n-1}{\frac{n-1}{2}} \\
&= 1 - \frac{1}{wt(f)} \binom{n-1}{\frac{n-1}{2}}
\end{aligned} \tag{6.42}
$$

By Definition 6.2 we know that when n is odd, the majority functions are balanced; hence, $wt(f) = 2^{n-1}$, and hence, the above becomes

$$
CI_{\varepsilon}(f) = 1 - \frac{1}{2^{n-1}} \binom{n-1}{\frac{n-1}{2}}.
$$

By Stirling formula, $n! \approx \sqrt{2\pi} n^{n+\frac{1}{2}} e^{-n+\frac{1}{12n}}$, we further have

$$
\begin{aligned}
\binom{n}{\frac{n}{2}} &= \frac{n!}{(\frac{n}{2})^2} \\
&\approx \frac{\sqrt{2\pi} n^{n+\frac{1}{2}} e^{-n+\frac{1}{12n}}}{2\pi (\frac{n}{2})^{n+1} e^{-n+\frac{1}{3n}}} \\
&= \frac{2^{n+1} e^{-\frac{1}{4n}}}{\sqrt{2\pi n}}
\end{aligned}
$$

Then,

$$\binom{n-1}{\frac{n-1}{2}} \approx \frac{2^n}{\sqrt{2\pi(n-1)}} e^{-\frac{1}{4(n-1)}}$$

So,

$$1 - \frac{1}{2^{n-1}} \binom{n-1}{\frac{n-1}{2}} \approx 1 - \frac{2}{\sqrt{2\pi(n-1)}} e^{-\frac{1}{4(n-1)}} \approx 1 - \frac{2}{\sqrt{2\pi(n-1)}}.$$

Summarize the discussion above, we have

Theorem 6.10. *When n is odd, the ε-correlation immunity of the majority functions is*

$$CI_\varepsilon(f) = 1 - \frac{1}{2^{n-1}} \binom{n-1}{\frac{n-1}{2}} \approx 1 - \frac{2}{\sqrt{2\pi(n-1)}}. \tag{6.43}$$

6.8.2 When n Is Even

By Theorem 6.8 we have

$$S_{(f_A)}(e_i) = 2 \sum_{x \in A_3 or wt(x) < \frac{n}{2}} (-1)^{\langle e_i, x \rangle} = 2 \left(\sum_{x \in A_3} (-1)^{x_i} + \sum_{wt(x) < \frac{n}{2}} (-1)^{x_i} \right).$$

It is easy to verify that, among the n-dimensional vectors of Hamming weight less than $\frac{n}{2}$, the number of such vectors where the i-th coordinate is 0 (and the rest $n-1$ coordinates can have $0 \sim (\frac{n}{2} - 1)$ of 1's) is $\binom{n-1}{0} + \binom{n-1}{1} + \cdots + \binom{n-1}{\frac{n}{2}-1}$, and the number of such vectors where the i-th coordinate is 1 (and the rest $n-1$ coordinates can have $0 \sim \frac{n}{2} - 2$ of 1's) is $\binom{n-1}{0} + \binom{n-1}{1} + \cdots + \binom{n-1}{\frac{n}{2}-2}$. Therefore

$$\begin{aligned}
\sum_{wt(x) < \frac{n}{2}} (-1)^{x_i} &= \left(\binom{n-1}{0} + \binom{n-1}{1} + \cdots + \binom{n-1}{\frac{n}{2}-1} \right) \\
&\quad - \left(\binom{n-1}{0} + \binom{n-1}{1} + \cdots + \binom{n-1}{\frac{n}{2}-2} \right) \\
&= \binom{n-1}{\frac{n}{2}-1}.
\end{aligned}$$

Note from the definition that $|S| = \binom{n}{\frac{n}{2}}$ and A_3 cannot have more than half of the elements in S (otherwise A_3 would have at least a pair of complement vectors which contradicts with the definition), i.e. $|A_3| \le |S| = \binom{n}{\frac{n}{2}}/2$. Since

$$-|A_3| \le \sum_{x \in A_3} (-1)^{\langle e_i, x\rangle} \le |A_3|,$$

so we get a lower bound of $S_{(f_A)}(e_i)$:

$$\begin{aligned} S_{(f_A)}(e_i) &= 2\left(-|A_3| - \binom{n-1}{\frac{n}{2}-1}\right) \\ &\ge 2\left(-\binom{n}{\frac{n}{2}}/2 - \binom{n-1}{\frac{n}{2}-1}\right) \\ &= -2\binom{n-1}{\frac{n}{2}} = -\binom{n}{\frac{n}{2}} \end{aligned}$$

and an upper bound

$$\begin{aligned} S_{(f_A)}(e_i) &= 2\left(|A_3| - \binom{n-1}{\frac{n}{2}-1}\right) \\ &\le 2\left(\binom{n}{\frac{n}{2}}/2 - \binom{n-1}{\frac{n}{2}}\right) \\ &= 0. \end{aligned}$$

By Definition 6.3, we have

$$\begin{aligned} wt(f_A) &= \binom{n}{\frac{n}{2}+1} + \binom{n}{\frac{n}{2}+2} + \cdots + \binom{n}{n} + |A_3| \\ &\ge 2^{n-1} - \binom{n}{\frac{n}{2}}/2. \end{aligned}$$

Therefore, by Eq. 4.48 we have

$$\begin{aligned} CI_\varepsilon(f_A) &= 1 - \frac{1}{2wt(f_A)} \max_i |S_{(f_A)}(e_i)| \\ &\ge 1 - \frac{1}{2^n - \binom{n}{\frac{n}{2}}} \cdot \binom{n}{\frac{n}{2}} \end{aligned}$$

$$\approx 1 - \frac{2}{\sqrt{2\pi n} - 2}.$$

Summarizing the discussion above, we have

Theorem 6.11. *When n is even, then the ε-correlation immunity of majority functions in n variables $f_A(x)$ satisfies*

$$CI_\varepsilon(f_A) \geq 1 - \frac{1}{2^n - \binom{n}{\frac{n}{2}}}\binom{n}{\frac{n}{2}} \approx 1 - \frac{2}{\sqrt{2\pi n} - 2}. \tag{6.44}$$

Noticing that when n is odd, by Theorem 6.10 we have

$$\lim_{n\to\infty} CI_\varepsilon(f) = 1,$$

and when n is even, by Theorem 6.11 we have

$$\lim_{n\to\infty} CI_\varepsilon(f_A) = 1;$$

this yields the following conclusion.

Theorem 6.12. *The ε-correlation immunity of the majority functions defined in Definitions 6.2 and 6.3 approaches to 1 with the increase of the number of variables.*

Theorem 6.12 means that the majority functions are almost correlation immune, and the approximation becomes more precise with the increase of n. This property is called *asymptotical correlation immunity*.

6.9 Remarks

It has been shown that for symmetric Boolean functions, the method for computing their Walsh transforms described in this chapter is much faster than traditional methods for general Boolean functions. This is not surprising because symmetric Boolean functions are just a special class of Boolean functions, and the number of those functions is equivalent to the number of affine ones. The computational complexity is shown to be a polynomial of n for the computation of Walsh transform (and the whole spectrums as well) of any symmetric Boolean function in $\mathcal{F}_n$, while it is exponential in n for the general cases. In [11] it is shown that the nonlinearity of a Boolean function depends only on the maximum absolute value of its Walsh transforms, and in [6] it is shown that from a maximum absolute value of Walsh transforms of a Boolean function, a best affine approximation function can easily be found. The best affine approximation attack is a potential attack to some stream ciphers and block ciphers [6, 13]. In this fashion the results suggest that symmetric Boolean functions should be avoided in those relevant generators.

With respect to the correlation immunity of symmetric Boolean functions, we have studied the constructions of such functions, and some interesting results are obtained. As has been mentioned above, constructions of symmetric functions with higher-order correlation immunity need to be studied further. As for symmetric correlation immune functions with the property of being balanced, besides the exclusive-or of all n variables, its complement, and those presented in [8], two more infinite classes of such functions have been presented which are correlation immune of order one for n being even and correlation immune of order two for n being odd, respectively, which answers the second open problem proposed in [8]. It is also shown that the two classes of such functions (one class is from [8], and the other class is introduced above) are only 1-resilient; hence, the conclusion of theorem 3.1 in [8] is adjusted. The functions constructed in this chapter for n being odd, however, are exactly 2-resilient. It is also noticed that in [20], one of such examples has been found, but the method was not clear enough to induce the whole infinite class.

Apart from the general symmetric Boolean functions, another special class of symmetric Boolean functions have attracted more attention; they are majority functions. It is shown that majority functions do not have correlation immunity. However, by using the concept of ε-correlation immunity, it is shown that although none of the majority functions is correlation immune in the traditional sense, the ε-correlation immunity of the majority functions in both odd and even number of variables will approach 1 with the increase of the number of variables. This means that when the number of variables n is large enough, although the majority functions are not immune against correlation attack due to their zero correlation immunity, the cost of the correlation attack, however, would be very high due to their ε-correlation immunity being approaching 1.

References

1. Braeken, A., Preneel, B.: On the algebraic immunity of symmetric Boolean functions. In: Proceedings of Indocrypt 2005. LNCS 3797, pp. 35–48. Springer, Berlin/Heidelberg (2005)
2. Canteaut, A., Videau, M.: Symmetric Boolean functions. IEEE Trans. Inf. Theory **IT-51**(8), 2791–2811 (2005)
3. Chor, B., Goldreich, O., Hastad, J., Friedman, J., Rudich, S., Smolensky, R.: The bit extraction problem or t-resilient functions. In: Proceedings of 26th IEEE Symposium on Foundations of Computer Science, Portland, pp. 396–407 (1985)
4. Courtois, N., Meier, W.: Algebraic attacks on stream ciphers with linear feedback. In: Advances in Cryptology – Proceedings of Eurocrypt'03. LNCS 2656, pp. 345–359. Springer, Berlin/New York (2003)
5. Dalai, D.K., Maitra, S., Sarkar, S.: Basic theory in construction of Boolean functions with maximum possible annihilator immunity. Des. Codes Cryptogr. **40**(1), 41–58 (2006)
6. Ding, C., Shan, W., Xiao, G.: The Stability Theory of Stream Ciphers. LNCS 561, Springer, Berlin/New York (1991)

7. Friedman, J.: On the bit extraction problem. In: Proceedings 33rd IEEE Symposium on Foundations of Computer Science, Pittsburgh, pp. 314–319 (1992)
8. Gopalakrishnan, K., Hoffman, D.G., Stinson, D.R.: A note on a conjecture concerning symmetric resilient functions. Inf. Process. Lett. **47**, 139–143 (1993)
9. Li, N., Qi, W.-F.: Symmetric Boolean functions depending on an odd number of variables with maximum algebraic immunity. IEEE Trans. Infor. Theory **IT-52**(5), 2271–2273 (2006)
10. Maitra, S., Sarkar, P.: Maximum nonlinearity of symmetric boolean functions on odd number of variables. IEEE Trans. Inf. Theory **IT-48**(9), 2626–2630 (2002)
11. Meier, W., Staffelbach, O.: Nonlinearity criteria for cryptographic functions. In: Advances in Cryptology – Proceedings of Eurocrypt'89. LNCS 434, pp. 549–562. Springer, Berlin/Heidelberg (1990)
12. Meier, W., Pasalic, E., Carlet, C.: Algebraic attacks and decomposition of Boolean functions. In: Advances in Cryptology – Proceedings of Eurocrypt'04. LNCS 3027, pp. 474–491. Springer, Berlin/Heidelberg (2004)
13. Nyberg, K.: Linear approximation of block ciphers, In: Advances in Cryptology – Proceedings of Eurocrypt'94. LNCS 950, pp. 439–444. Springer, Berlin/Heidelberg (1995)
14. Qu, C., Seberry, J., Pieprzyk, J.P.: On the symmetric property of homogeneous Boolean functions. In: Proceedings of Australian Conference on Information Security and Privacy (ACISP'99). LNCS 1587, pp. 26–35. Springer, Berlin/Heidelberg (1999)
15. Shannon, C.E.: Communication theory of secrecy systems. Bell Syst. Tech. J. **28**, 59–88 (1949)
16. Siegenthaler, T.: Correlation-immunity of nonlinear combining functions for cryptographic applications. IEEE Trans. Inf. Theory **IT-30**(5), 776–780 (1984)
17. Siegenthaler, T.: Decrypting a class of stream ciphers using ciphertext only. IEEE Trans. Comput. **C-34**(1), 81–85 (1985)
18. Siegenthaler, T.: Cryptanalysts' representation of nonlinearly filtered m-sequences. In: Advances in Cryptology – Proceedings of Eurocrypt'85. LNCS 219, pp. 103–110. Springer, Heidelberg (1986)
19. Siegenthaler, T.: Design of combiners to prevent divide and conquer attacks. In: Advances in Cryptology, Proceedings of Crypto'85. LNCS 218, pp. 237–279. Springer, Berlin/Heidelberg/New York (1986)
20. Stinson, D.R., Massey, J.L.: An infinite class of counterexamples to a conjecture concerning nonlinear resilient functions. J. Cryptol. **8**, 167–173 (1995)
21. Stockmeyer, L.J.: On the combinational complexity of certain symmetric Boolean functions. Math. Syst. Theory **10**, 323–336 (1977)
22. Wu, C.K., Dawson, E.: Correlation immunity and resiliency of symmetric Boolean functions. Theor. Comput. Sci. **312**, 321–335 (2004)
23. Xiao, G.Z., Massey, J.L.: A spectral characterization of Correlation-immune combining functions. IEEE Trans. Inf. Theory **IT-34**(3), 569–571 (1988)
24. Zhao, Y., Li, H.: On bent functions with some symmetric properties. Discret. Appl. Math. **154**, 2537–2543 (2006)

Chapter 7
Boolean Function Representation of S-Boxes and Boolean Permutations

S-boxes are often the core nonlinear component in many encryption algorithms. By using vector Boolean functions to represent S-boxes, cryptographic properties as well as constructions can be made possible. This chapter studies the S-boxes by the view of vector Boolean functions, with focus being on Boolean permutations, which are a special class of vector Boolean functions. Properties and constructions of Boolean permutations are studied; computation of inverses of Boolean functions is also studied. The concept of one-way trapdoor Boolean permutation is proposed. Construction of Boolean permutations using function composition is studied which enables the construction of one-way trapdoor Boolean permutations.

7.1 Vectorial Boolean Function Representation of S-Boxes

In the design of cryptographic algorithms, particularly the algorithm of block ciphers, S-boxes play an essential role in ensuring the security of the algorithms. For example, both the DES [4] and AES [3] symmetric key encryption algorithms (block ciphers) use S-boxes as their nonlinear components of transformations. It is known that many stream cipher algorithms also use S-boxes as important nonlinear components, and practical good S-boxes need to possess some cryptographic properties [1, 7]. Given the importance of S-boxes, their study has induced many research publications (see, e.g., [2, 5, 6, 9, 10, 12, 14–17, 19, 21, 23, 26]). Since an S-box can be represented by a vectorial Boolean function, cryptographic properties can be represented by the corresponding properties of vectorial Boolean functions, although this is not always the best representation in terms of complexity both in presentation and implementation. This chapter studies the vectorial Boolean function representation of S-boxes, particularly the construction of Boolean permutations as a special class of S-boxes, and some basic properties of Boolean permutations.

C.-K. Wu, D. Feng, *Boolean Functions and Their Applications in Cryptography*, Advances in Computer Science and Technology, DOI 10.1007/978-3-662-48865-2_7

A cryptographic S-box, or simply called S-box, is such a function that takes as input a string of length n and outputs a string of length m. This means that an S-box is a mapping $F(x)$ from $GF^n(2)$ to $GF^m(2)$, and for this sake, an S-box is also called an *(n, m)-Boolean function*. In the following description, to be more precise, we will use the notions of (n, m)-Boolean functions and S-boxes interchangeably.

Note that an (n, m)-Boolean function can always be represented as a collection of m Boolean functions from $\mathcal{F}_n$, and we write

$$F(x) = [f_1(x), f_2(x), \cdots, f_m(x)],$$

where each $f_i \in \mathcal{F}_n$, $i = 1, 2, \ldots, m$, is a Boolean function in n variables, and it is called a *coordinate function* of $F(x)$. It appears that the study of such (n, m)-Boolean functions can be converted into the study of individual coordinate Boolean functions; however, as a whole, (n, m)-Boolean function may have many properties that cannot be reflected from any of its individual coordinate Boolean functions.

Considering the output of an (n, m)-Boolean function, any such an output is a vector in $GF^m(2)$. Then there is a probability for each of the vectors to be the output of the function when the input variable x goes through all the possible values in $GF^n(2)$. If $n < m$, then the input space is smaller than the output space, i.e., the (n, m)-Boolean function $F(x)$ maps $GF^n(2)$ into a subset of $GF^m(2)$; in this case, the output of $F(x)$ does not have the same chances to be any value in $GF^m(2)$. This kind of S-boxes is called *expansion S-boxes*.

If $n > m$, then the input space is larger than the output space; this means that a subset of the input may result in all possible outputs in $GF^m(2)$. It is noted that even in such case, the output of $F(x)$ may also have more chances to be some of the vectors in $GF^m(2)$ and less chances to be some other vectors. This kind of S-boxes are called *compression S-boxes*. In order to study the probabilistic behavior, let the input x of $F(x)$ be a random variable, which is a collection of n independent binary variables that take values either 0 or 1 with equal probability. Then the output of $F(x)$ can be regarded as a collection of m random variables, but they are in general not independent.

If $n = m$, then this special case can either be called a compression S-box or an expansion S-box wherever convenient.

7.2 Boolean Function Representation of S-Boxes

From cryptographic point of view, a secure S-box is expected to have the property that any subset of its output gives no information about other bits of the output. This means that an (n, m)-Boolean function that represents an S-box has the property that the m output functions, $f_1(x), f_2(x), \ldots, f_m(x)$, are statistically independent of each other. Theorem 2.39 says that this happens only when $n \geq m$. We will ignore the trivial case when any of the $f_i(x)$ is a constant which is not what cryptographically desired.

Definition 7.1. Let an S-box be represented by an (n, m)-Boolean function $F(x)$.

If $n >= m$, then $F(x)$ (and hence the S-box) is called *unbiased*, if for any $(a_1, a_2, \cdots, a_m) \in GF^m(2)$, the following equality always holds:

$$Prob\,(F(x) = (a_1, a_2, \cdots, a_{n-m})) = \frac{1}{2^m}.$$

If $n < m$, then $F(x)$ (and hence the S-box) is called *unbiased*, if any n coordinate functions of $F(x)$ form an unbiased (n, n)-Boolean function.

Theorem 7.1. *Let $F(x) = [f_1(x), f_2(x), \cdots, f_m(x)]$ be an (n, m)-Boolean function representation of an S-box, where $n \geq m$. Then $F(x)$ is unbiased if and only if:*

(1) *$f_1(x), f_2(x), \cdots, f_m(x)$ form a statistically independent Boolean function family.*
(2) *$f_1(x), f_2(x), \cdots, f_m(x)$ are all balanced.*

Proof. Necessity: Since $[f_1(x), f_2(x), \cdots, f_m(x)]$ is unbiased, by Definition 7.1, for any $(a_1, a_2, \ldots, a_m) \in GF^m(2)$, $Prob((f_1(x), f_2(x), \cdots, f_m(x)) = (a_1, a_2, \ldots, a_m)) = \frac{1}{2^m}$ is a fixed constant.

Let x go through all the vectors in $GF^n(2)$; then for each $(a_1, a_2, \ldots, a_m) \in GF^m(2)$, there are 2^{n-m} values of x such that $(f_1(x), f_2(x), \cdots, f_m(x)) = (a_1, a_2, \ldots, a_m)$ holds. This means that when x goes through all the values in $GF^n(2)$, $(f_1(x), f_2(x), \cdots, f_m(x))$ also goes through every vector in $GF^m(2)$ for exactly 2^{n-m} times. It is known that when $(a_1, a_2, \ldots, a_m)$ goes through all the vectors in $GF^m(2)$, each of its coordinates a_i has equal chances to be 0 or 1. This is the same when $(a_1, a_2, \ldots, a_m)$ goes through all the vectors in $GF^m(2)$ for 2^{n-m} times. This proves that every $f_i(x)$ is balanced, and it is easy to verify that in this case, we have

$$\begin{aligned} &Prob((f_1(x), f_2(x), \cdots, f_m(x)) = (a_1, a_2, \ldots, a_m)) \\ &= \prod_{i=1}^{m} Prob(f_i(x) = a_i) = 2^{-m} \end{aligned}$$

holds; by Definition 2.11, $f_1(x), f_2(x), \cdots, f_m(x)$ form a statistically independent Boolean function family.

Sufficiency: For any $(a_1, a_2, \ldots, a_m) \in GF^m(2)$, by Definition 2.11, we have

$$\begin{aligned} Prob((f_1(x), f_2(x), \cdots, f_m(x)) &= (a_1, a_2, \ldots, a_m)) \\ &= \prod_{i=1}^{m} Prob(f_i(x) = a_i) = 2^{-m}. \end{aligned}$$

This means that $(f_1(x), f_2(x), \cdots, f_m(x))$ has equal chances to take any values in $GF^m(2)$. Because

$$\sum_{(a_1,a_2,\ldots,a_m)\in GF^m(2)} Prob((f_1(x),f_2(x),\cdots,f_m(x)) = (a_1,a_2,\ldots,a_m)) = 1,$$

so for each $(a_1, a_2, \ldots, a_m) \in GF^m(2)$, we have that

$$Prob((f_1(x),f_2(x),\cdots,f_m(x)) = (a_1,a_2,\ldots,a_m)) = 2^{-m}.$$

By Definition 7.1, $[f_1(x),f_2(x),\cdots,f_m(x)]$ is unbiased. □

7.2.1 *On the Properties of* (n, n)*-Boolean Permutations*

A special case of (n, m)-Boolean functions is when $m = n$, which means that the output domain of the function is equal to the input domain. This is actually a special case covered by the case when $n \geq m$, which means that all the conclusions for the case of $n \geq m$ are true for this case. Because it is a special case, we hereby give it a special consideration, and here we will only consider the unbiased (n, n)-Boolean functions.

Let $F(x)$ be an (n, m)-Boolean function. In the case of $m = n$ and the function is unbiased, different inputs will yield different outputs. By treating each input-output as the binary representation of an integer within $S = \{0, 1, \ldots, 2^n - 1\}$, the above function F performs a permutation on S. We refer to such a permutation on S in Boolean function representation as an (n, n)-*Boolean permutation*. For simplicity, we also call it a Boolean permutation in n variables. Since any Boolean permutation can be represented as a collection of Boolean functions in n variables, we can write it as

$$F(x) = [f_1(x),f_2(x),\cdots,f_n(x)]. \tag{7.1}$$

Note that not every collection of Boolean functions forms a Boolean permutation; they must satisfy certain conditions. The following is a necessary and sufficient condition for a collection of Boolean functions to be a Boolean permutation.

Theorem 7.2. *Let* $F(x) = [f_1(x),f_2(x),\cdots,f_n(x)]$ *be an* (n, n)*-Boolean function, where* $f_i(x) \in \mathcal{F}_n$, $i = 1, 2, \ldots, n$. *Then* $F(x)$ *is a Boolean permutation if and only if any nonzero linear combination (i.e., the X-or) of* $f_1(x),f_2(x),\cdots,f_n(x)$ *is a balanced Boolean function, i.e., for any nonzero vector* $c = (c_1, c_2, \cdots, c_n) \in \{0, 1\}^n$, *we have*

$$wt\left(\bigoplus_{i=1}^{n} c_i f_i\right) = 2^{n-1} \tag{7.2}$$

Before we present a proof of Theorem 7.2, the following two lemmas are needed.

Lemma 7.1. *Let $f(x) \in \mathcal{F}_n$ be a Boolean function in n variables. Denote $f^0(x) = 1 \oplus f(x)$, $f^1(x) = f(x)$. Then for any $a \in \{0,1\}$, we have that $f^a(x) = a$ holds if and only if $f(x) = 1$ holds. Similarly we have that $f^a(x) = 1$ holds if and only if $f(x) = a$ holds.*

Proof. The correctness of the lemma can be verified by trivially checking the cases when $a = 0$ and $a = 1$. □

Lemma 7.2. *Let $f_i \in \mathcal{F}_n$, $i = 1, 2, \ldots, n$. Then $f_1(x), f_2(x), \cdots, f_n(x)$ satisfy Eq. 7.2, if and only if for any $a \in \{0,1\}$ and for any $i \in \{1, 2, \cdots, n\}$, functions*

$$f_1(x), \cdots, f_{i-1}(x), f_i^a(x), f_{i+1}(x), \cdots, f_n(x)$$

also satisfy Eq. 7.2.

Proof. By Lemma 7.1, the sufficiency and necessity of Lemma 7.2 are symmetric. So we only need to prove the necessity. When $a = 1$, the conclusion is trivially true. Let $a = 0$; then for any i with $1 \leq i \leq n$, we have

$$wt\left(\bigoplus_{k,k\neq i} c_k f_k \oplus c_i f_i^0\right) = wt\left(\bigoplus_{k=1}^{n} c_k f_k \oplus c_i\right)$$

$$= \begin{cases} wt\left(\bigoplus_{k=1}^{n} c_k f_k\right) & \text{if } c_i = 0 \\ 2^n - wt\left(\bigoplus_{k=1}^{n} c_k f_k\right) & \text{if } c_i = 1 \end{cases}$$

$$= 2^{n-1}$$

So the conclusion of the lemma follows. □

Proof of Theorem 7.2:

Necessity: Treat each output of $(f_1(x), f_2(x), \cdots, f_n(x))$ as the binary representation of an integer in $S = \{0, 1, \cdots, 2^n - 1\}$; then the output of $f_i(x)$ is the ith coordinate of the binary representation of this integer. When x goes from 0 to $2^n - 1$, because $F(x)$ is a Boolean permutation which is an unbiased (n, n)-Boolean function, by Definition 7.1, the output of $F(x)$ also goes through every element in S exactly once. So the truth table of $F(x) = [f_1(x), f_2(x), \cdots, f_n(x)]$ is a permutation of the truth table of $x = [x_1, x_2, \cdots, x_n]$. Therefore, the truth table of $f'(x) = \bigoplus_{i=1}^{n} c_i f_i(x)$, a nonzero linear combination of $f_1(x), f_2(x), \cdots, f_n(x)$, is a permutation of the truth table of $\bigoplus_{i=1}^{n} c_i x_i$, the same nonzero linear combination of $x_1, x_2, \ldots, x_n$, which is obviously a balanced Boolean function. This means that the necessity of Theorem 7.2 holds.

Sufficiency: By Eq. 7.2 and by choosing the coefficient vector to be the special case whose Hamming weight is 1, we have

$$wt(f_i) = 2^{n-1},\ i = 1, 2, \ldots, n$$

Since $wt(f_i \oplus f_j) = wt(f_i) + wt(f_j) - 2wt(f_if_j)$, we have

$$wt(f_if_j) = 2^{n-2},\ i \neq j.$$

Assume that $wt(f_{i_1}f_{i_2}\cdots f_{i_t}) = 2^{n-t}$ holds for $t = 1, 2, \cdots, k$, where $1 \leq i_1 < i_2 < \cdots < i_t \leq n$, since

$$\begin{aligned} & wt(f_1 \oplus f_2 \oplus \cdots \oplus f_{k+1}) \\ &= \sum_{i=1}^{k+1} wt(f_i) - 2 \sum_{1\leq i<j\leq n} wt(f_if_j) \\ &+ \cdots + (-1)^k 2^k wt(f_1f_2\cdots f_{k+1}) \end{aligned}$$

which is equivalent to

$$\begin{aligned} 2^{n-1} &= (k+1)2^{n-1} - \binom{k+1}{2} 2^{n-1} + \cdots \\ &+ (-1)^{k-1}\binom{k+1}{k} 2^{n-1} + (-1)^k wt(f_1f_2\cdots f_{k+1}) \end{aligned}$$

we hence have

$$wt(f_1f_2\cdots f_{k+1}) = 2^{n-(k+1)}.$$

It is noted that the order of the functions $f_1(x), f_2(x), \cdots, f_n(x)$ that satisfy Eq. 7.2 does not matter; hence, the above means that for any $k+1$ coordinate function of $F(x)$, we have

$$wt(f_{i_1}f_{i_2}\cdots f_{i_{k+1}}) = 2^{n-(k+1)}$$

holds. According to the principle of induction, we have for the case of $k = n-1$, the following also holds

$$wt(f_1f_2\cdots f_n) = 2^{n-n} = 1.$$

This means that there exists only one x satisfying that $f_1(x)f_2(x)\cdots f_n(x) = 1$. For any $(a_1, a_2, \cdots, a_n) \in \{0,1\}^n$, by using Lemma 7.2 repeatedly, we know that $f_1^{a_1}(x), f_2^{a_2}(x), \cdots, f_n^{a_n}(x)$ also satisfy Eq. 7.2. This means that there exists only one x such that $f_1^{a_1}(x)f_2^{a_2}(x)\cdots f_n^{a_n}(x) = 1$ holds, i.e., $f_i^{a_i}(x) = 1$ holds. By Lemma 7.1, we have $f_i(x) = a_i$. This shows that the output of $F(x) = [f_1(x), f_2(x), \cdots, f_n(x)]$ has exactly one chance to be any value in S when x goes through all the possible values in S, and hence $F(x)$ is a permutation on S, i.e., $F(x)$ is a Boolean permutation in n variables.

In light of the above, the conclusion of Theorem 7.2 is true. □

Theorem 7.2 is actually the fundamental XOR lemma for the case of (n,n)-Boolean permutations. Let $z_i = f_i(x)$, $i = 1,2,\ldots,n$, be a system of (nonlinear) equations. Since F is a Boolean permutation, there must be a unique solution to the equation system, say, $x_i = g_i(z)$, where $z = (z_1,\ldots,z_n)$, and $[g_1(z),g_2(z),\ldots,g_n(z)]$ is called the inverse Boolean permutation of $F(x)$ and is denoted as $F^{-1}(z)$.

In the following discussion, we will use the notation $P(x)$ to represent a Boolean permutation and $P^{-1}(x)$ the inverse Boolean permutation of $P(x)$.

7.3 Properties of Boolean Permutations

Apart from Theorem 7.2, some fundamental properties of Boolean permutations are listed below. These properties will be helpful to understand and manipulate the use of Boolean permutations.

Theorem 7.3. *Let $P = [f_1,f_2,\cdots,f_n]$ be a Boolean permutation and σ_n be a permutation on the set {0, 1, . . ., n}. Then*

$$\sigma_n(P) = [f_{\sigma_n(1)},f_{\sigma_n(2)},\cdots,f_{\sigma_n(n)}] \tag{7.3}$$

is also a Boolean permutation.

Theorem 7.3 states that a permutation on the index of a Boolean permutation yields another Boolean permutation. A generalization of this result leads to the following theorem.

Theorem 7.4. *Let $P = [f_1,f_2,\cdots,f_n]$ be a Boolean permutation, $D = (d_{ij})$ an $n \times n$ binary matrix, and $C = (c_1,c_2,\ldots,c_n) \in GF^n(2)$. Then*

$$PD \oplus C = \left[\bigoplus_{i=1}^{n} d_{i1}f_i \oplus c_1, \bigoplus_{i=1}^{n} d_{i2}f_i \oplus c_2, \cdots, \bigoplus_{i=1}^{n} d_{in}f_i \oplus c_n\right] \tag{7.4}$$

is a Boolean permutation if and only if D is nonsingular.

Proof. It is easy to verify that $P = [f_1,f_2,\cdots,f_n]$ is a Boolean permutation if and only if for any vector $\alpha = (a_1,a_2,\ldots,a_n)$, $P \oplus \alpha = [f_1 \oplus a_1,f_2 \oplus a_2,\cdots,f_n \oplus a_n]$ is also a Boolean permutation. So we only need to prove the case when $C = 0$.

Necessity: Suppose that D is a singular matrix. Then there must exist a nonzero vector $B = (b_1,b_2,\ldots,b_n)$ such that $DB^T = 0$; hence,

$$[f_1,f_2,\cdots,f_n]DB^T = \sum_{j=1}^{n} b_j \sum_{i=1}^{n} d_{i,j}f_i = 0.$$

This indicates that the nonzero linear combination of the coordinates of $[f_1,f_2,\cdots,f_n]D$ with coefficient vector B is zero rather than a balanced Boolean function. By Theorem 7.2, we know that $[f_1,f_2,\cdots,f_n]D$ is not a Boolean permutation.

Sufficiency: Suppose D is nonsingular. Then for any nonzero vector $B \in GF^n(2)$, $DB^T \neq 0$. Therefore,

$$[f_1,f_2,\cdots,f_n]DB^T = \sum_{i=1}^{n} f_i \sum_{j=1}^{n} d_{i,j}b_j$$

is a nonzero linear combination (with the coordinates of DB^T as coefficients) of f_i. Since P is a Boolean permutation, by Theorem 7.2, we have

$$wt\left(\sum_{i=1}^{n} f_i \sum_{j=1}^{n} d_{i,j}b_j\right) = 2^{n-1}.$$

Given the arbitrariness of B and using Theorem 7.2 again, we know that $[f_1,f_2,\cdots,f_n]D$ is a Boolean permutation. □

Theorem 7.5. *Let $P = [f_1,f_2,\cdots,f_n]$ be a Boolean permutation, $D = (d_{ij})$ be an $n \times n$ binary matrix, and $C = (c_1,c_2,\ldots,c_n) \in GF^n(2)$. Then*

$$P(xD \oplus C) = [f_1(xD \oplus C), f_2(xD \oplus C), \ldots, f_n(xD \oplus C)] \tag{7.5}$$

is a Boolean permutation if and only if D is nonsingular.

Proof. Denote $y = (y_1,y_2,\ldots,y_n) = (x_1,x_2,\ldots,x_n)D \oplus C$. Then it is easy to see that $y_1,y_2,\ldots,y_n$ are n independent variables if and only if D is nonsingular. Since $P = [f_1,f_2,\cdots,f_n]$ is a Boolean permutation, $[f_1(y),f_2(y),\cdots,f_n(y)]$ is also a Boolean permutation if and only if $y_1,y_2,\ldots,y_n$ are n independent variables. □

Theorems 7.4 and 7.5 show that linear transformations on the coordinate functions or variables of a Boolean permutation will yield a new Boolean permutation. Now, we consider the composition of Boolean permutations.

Theorem 7.6. *Let $P = [f_1,f_2,\cdots,f_n]$ and $Q = [g_1,g_2,\cdots,g_n]$ be two Boolean permutations. Then their composition*

$$P(Q) = [f_1(g_1,g_2,\cdots,g_n), f_2(g_1,g_2,\cdots,g_n), \cdots, f_n(g_1,g_2,\cdots,g_n)] \tag{7.6}$$

is a new Boolean permutation.

Proof. This result comes from the fact that an (n,n)-Boolean function is a Boolean permutation if and only if it is a one-to-one mapping from its inputs to its outputs. □

Now, we introduce a new operation, concatenation of Boolean permutations. Concatenation of two functions $F_1(x)$ and $F_2(x)$ involves independent variables. For example, the concatenation of $F_1(x) = [x_1, x_1 \oplus x_2]$ and $F_2(x) = [x_1 \oplus x_2x_3, x_2, x_2 \oplus x_3]$ forms a new function $F(x) = [F_1; F_2] = [x_1, x_1 \oplus x_2, x_3 \oplus x_4x_5, x_4, x_4 \oplus x_5]$.

Theorem 7.7. *Let $P_1 = [f_1, \cdots, f_{n_1}]$ and $P_2 = [g_1, \cdots, g_{n_2}]$ be two Boolean permutations in n_1 and n_2 variables, respectively. Then their concatenation $P = [P_1, P_2]$ forms a Boolean permutation in $n = n_1 + n_2$ variables.*

As a direct corollary of Theorems 7.7 and 7.6, we have the following:

Corollary 7.1. *Let $P = [f_1, \cdots, f_n]$ be a Boolean permutation in n variables and $R_i = [g_{i,1}, \ldots, g_{i,n_i}]$ a Boolean permutation in n_i variables for $i = 1, 2, \ldots, k$, where $n_1 + n_2 + \cdots + n_k = n$. Then*

$$\begin{aligned} Q = [&g_{1,1}(f_1, \ldots, f_{n_1}), \cdots, g_{1,n_1}(f_1, \ldots, f_{n_1}), \\ &g_{2,1}(f_{n_1+1}, \ldots, f_{n_1+n_2}), \cdots, g_{2,n_2}(f_{n_1+1}, \ldots, f_{n_1+n_2}), \\ &\cdots, g_{k,1}(f_{n_1+n_2+\ldots+n_{k-1}+1}, \ldots, f_n), \cdots, \\ &g_{k,n_k}(f_{n_1+n_2+\ldots+n_{k-1}+1}, \ldots, f_n)] \end{aligned} \quad (7.7)$$

is a Boolean permutation in n variables.

The above conclusions are just a few simple operations on Boolean permutations. Complex operations can be achieved by combining these operations.

Theorem 7.8. *Let $P = [f_1, \cdots, f_n]$ be a Boolean permutation in n variables. If there exists a subset $A = \{f_{i_1}, f_{i_2}, \cdots, f_{i_t}\}$ of the coordinate functions of P, such that for any x_i, either all functions in A are independent of x_i or all functions in $\{f_1, f_2, \cdots, f_n\} - A$ are independent of x_i; then $[f_{i_1}, f_{i_2}, \cdots, f_{i_t}]$ forms a degenerate Boolean permutation (i.e., when those variables x_k that all f_{i_j} are independent of are ignored), and $\{f_1, f_2, \cdots, f_n\} - \{f_{i_1}, f_{i_2}, \cdots, f_{i_t}\}$ forms another degenerate Boolean permutation.*

Theorem 7.8 is just to treat Theorem 7.7 from a different angle. Actually when the coordinate functions of the Boolean permutation described in Theorem 7.7 perform a permutation, then the result is a permutation having the properties as stated in Theorem 7.8.

7.4 Inverses of Boolean Permutations

Like any permutation, a Boolean permutation has an inverse. The inverse is also a Boolean permutation. Given a Boolean permutation $P = [f_1, f_2, \cdots, f_n]$, the inverse of P is a solution of the following equation:

$$\begin{cases} z_1 = f_1(x_1, x_2, \dots, x_n) \\ z_2 = f_2(x_1, x_2, \dots, x_n) \\ \quad \cdots\cdots \\ z_n = f_n(x_1, x_2, \dots, x_n) \end{cases} \tag{7.8}$$

i.e., an expression of each x_i in terms of z_j. Suppose we have a solution of Eq. 7.8 in the form

$$\begin{cases} x_1 = f_1^{-1}(z_1, \dots, z_n), \\ x_2 = f_2^{-1}(z_1, \dots, z_n), \\ \quad \cdots \\ x_n = f_n^{-1}(z_1, \dots, z_n), \end{cases} \tag{7.9}$$

then $P^{-1} = [f_1^{-1}, f_2^{-1}, \dots, f_n^{-1}]$ is the inverse Boolean permutation of P.

Lemma 7.3. *Let $P = [f_1, f_2, \cdots, f_n]$ and $Q = [g_1, g_2, \cdots, g_n]$ be two Boolean permutations. Then they are inverses of each other if and only if for every $i \in \{1, 2, \dots, n\}$, we have $g_i(f_1, f_2, \cdots, f_n) = x_i$ and $f_i(g_1, g_2, \cdots, g_n) = x_i$.*

Lemma 7.3 can be used to check whether two Boolean permutations are inverses of each other, especially when the number of variables of the Boolean permutations is fairly large so that it is computationally infeasible to check all the input-output pairs.

It is known that when one of the functions in Eq. 7.8 is nonlinear, to solve equation Eq. 7.8 is a hard problem, i.e., there is no efficient algorithm to solve it. However, inverses of certain special classes of Boolean permutations can easily be found. The following are the inverses of Boolean permutations from Theorems 7.3 to 7.6, respectively. Since their proofs are trivial, we only list the conclusions without any proof.

Lemma 7.4. *Let $P = [f_1, f_2, \cdots, f_n]$, σ_n and $Q = \sigma_n(P)$ be as defined in Theorem 7.3 and $P^{-1} = [f_1^{-1}(z), f_2^{-1}(z), \dots, f_n^{-1}(z)]$ be the inverse of P. Let $z' = (z_{\sigma_n^{-1}(1)}, z_{\sigma_n^{-1}(2)}, \dots, z_{\sigma_n^{-1}(n)})$. Then $Q^{-1} = [f_1^{-1}(z'), f_2^{-1}(z'), \dots, f_n^{-1}(z')]$.*

Lemma 7.5. *Let $P = [f_1, f_2, \cdots, f_n]$ and $Q = PD \oplus C$ be defined as in Theorem 7.4, where D is a nonsingular matrix. Let $z' = ((z_1, \dots, z_n) \oplus C)D^{-1}$. Then $Q^{-1} = [f_1^{-1}(z'), f_2^{-1}(z'), \dots, f_n^{-1}(z')]$, where $P^{-1} = [f_1^{-1}(z), f_2^{-1}(z), \dots, f_n^{-1}(z)]$.*

Lemma 7.6. *Let $P = [f_1, f_2, \cdots, f_n]$ and $Q = P(xD \oplus C)$ be defined as in Theorem 7.5, where D is a nonsingular matrix. Then $Q^{-1} = P^{-1}D^{-1} \oplus CD^{-1}$.*

Lemma 7.7. *Let P, Q, and $R = P(Q)$ be defined as in Theorem 7.6. Then $R^{-1} = Q^{-1}(P^{-1})$.*

Now we consider the inverse of the composed Boolean permutation obtained in Corollary 7.1, given the inverses $P^{-1} = [f_1^{-1}, f_2^{-1}, \dots, f_n^{-1}]$ and $R_i^{-1} = [g_{i,1}^{-1}, g_{i,2}^{-1}, \dots, g_{i,n_i}^{-1}]$, $i = 1, 2, \dots, n$, of the known Boolean permutations. Using

$$\begin{cases} z_1 = g_{1,1}(f_1, \ldots, f_{n_1})(x) \\ \cdots\cdots \\ z_{n_1} = g_{1,n_1}(f_1, \ldots, f_{n_1})(x) \\ z_{n_1+1} = g_{2,1}(f_{n_1+1}, \ldots, f_{n_1+n_2})(x) \\ \cdots\cdots \\ z_{n_1+n_2} = g_{2,n_2}(f_{n_1+1}, \ldots, f_{n_1+n_2})(x) \\ \cdots\cdots \\ z_{n_1+n_2+\cdots+n_{k-1}+1} = g_{k,1}(f_{n_1+n_2+\cdots+n_{k-1}+1}, \ldots, f_n)(x) \\ \cdots\cdots \\ z_n = g_{k,n_k}(f_{n_1+n_2+\cdots+n_{k-1}+1}, \ldots, f_n)(x) \end{cases} \tag{7.10}$$

and the corresponding inverse of R_i, we have

$$\begin{cases} f_1(x) = g_{1,1}^{-1}(z_1, \ldots, z_{n_1}) = y_1 \\ \cdots\cdots \\ f_{n_1}(x) = g_{1,n_1}^{-1}(z_1, \ldots, z_{n_1}) = y_{n_1} \\ f_{n_1+1}(x) = g_{2,1}^{-1}(z_{n_1+1}, \ldots, z_{n_1+n_2}) = y_{n_1+1} \\ \cdots\cdots \\ f_{n_1+n_2}(x) = g_{2,n_2}^{-1}(z_{n_1+1}, \ldots, z_{n_1+n_2}) = y_{n_1+n_2} \\ \cdots\cdots \\ f_{n_1+n_2+\ldots+n_{k-1}+1}(x) = g_{k,1}^{-1}(z_{n_1+n_2+\cdots+n_{k-1}+1}, \ldots, z_n) = y_{n_1+n_2+\ldots+n_{k-1}+1} \\ \cdots\cdots \\ f_n(x) = g_{k,n_k}^{-1}(z_{n_1+n_2+\cdots+n_{k-1}+1}, \ldots, z_n) = y_n. \end{cases} \tag{7.11}$$

By applying the inverse of P on Eq. 7.11, we have

$$\begin{cases} x_1 = f_1^{-1}(y_1, \ldots, y_n) = \varphi_1(z_1, \ldots, z_n) \\ \cdots\cdots \\ x_n = f_n^{-1}(y_1, \ldots, y_n) = \varphi_n(z_1, \ldots, z_n) \end{cases} \tag{7.12}$$

which gives an (n, n)-Boolean function with input $z = (z_1, \ldots, z_n)$ and output $x = (x_1, \ldots, x_n)$. So $Q^{-1} = [\varphi_1, \ldots, \varphi_n]$ is the inverse Boolean permutation of Q.

From the above description, for an arbitrary Boolean permutation, if it can be transformed into a concatenation of several smaller Boolean permutations by linear transforms on variables and/or component functions (refer to Theorems 7.4 and 7.5), then the complexity of finding its inverse is equivalent to the total complexity of finding the inverses of all the smaller ones.

In general, to find the inverse of a Boolean permutation is equivalent to solving a system of equations of Boolean functions over the binary field. It is well known that when at least one of the functions is nonlinear, there is no efficient (e.g., polynomial time complexity) algorithm to fulfill this task. For an arbitrary Boolean permutation, the probability that it is composed of all affine Boolean functions is $2^{n^2+n}/(2^n)!$ which becomes negligible with the increase of n. So we assume that a general

Boolean permutation is a one-way function, i.e., given any input, it is easy (with a polynomial time complexity) to generate the output, while there is not a polynomial time algorithm to find the corresponding input given any output.

However, it is possible to design a special class of Boolean permutations of which the inverses can easily be computed, as long as how the Boolean permutations are constructed is known.

Lemma 7.8. *Let* $P(x) = [f_1(x), \ldots, f_n(x)]$ *be a Boolean permutation in* n *variables with* $P^{-1}(z) = [f_1^{-1}(z), \ldots, f_n^{-1}(z)]$. *Let* $g(x) \in \mathcal{F}_n$ *be an arbitrary function and set* $f(\hat{x}) = g(x) \oplus x_{n+1}$, *where* $\hat{x} = (x_1, \ldots, x_{n+1})$. *Then* $Q(\hat{x}) = [f_1(x) \oplus f(\hat{x}), \ldots, f_n(x) \oplus f(\hat{x}), f(\hat{x})]$ *is a new Boolean permutation in* $n+1$ *variables. Moreover, let* $z' = (z_1 \oplus z_{n+1}, \ldots, z_n \oplus z_{n+1})$. *Then* $Q^{-1}(\hat{z}) = [g_1^{-1}(\hat{z}), \ldots, g_{n+1}^{-1}(\hat{z})]$, *where* $\hat{z} = (z_1, \ldots, z_{n+1})$, $(g_1^{-1}(\hat{z}), \ldots, g_n^{-1}(\hat{z})) = P^{-1}(z')$, *and* $g_{n+1}^{-1}(\hat{z}) = z_{n+1} \oplus g(g_1^{-1}(\hat{z}), \ldots, g_n^{-1}(\hat{z}))$.

From Lemma 7.8, the following is a straightforward algorithm for constructing new Boolean permutations based on old ones.

Algorithm 7.1 (Simple construction of Boolean permutations).

---

(1) Let $P = [f_1, \ldots, f_n]$ be a Boolean permutation in n variables and $g(x) \in \mathcal{F}_n$ be an arbitrary Boolean function.
(2) Set $g_i(\hat{x}) = f_i(x) \oplus x_{n+1} \oplus g(x)$, $i = 1, \ldots, n$, and $g_{n+1}(\hat{x}) = x_{n+1} \oplus g(x)$. Then $Q = [g_1, \ldots, g_{n+1}]$ is a Boolean permutation in $n+1$ variables.

Algorithm 7.1 gives an iterative method for constructing Boolean permutations based on old ones. We can also apply linear transformations on the components or the variables of the constructed Boolean permutations to get new ones. However, there has to be an initial Boolean permutation available when we use Algorithm 7.1. One way to do this is to select a Boolean permutation in small number of variables as the initial one. This makes the construction inefficient when the target permutation to construct is in a large number of variables. Another method is to construct a linear Boolean permutation as the initial one. The following gives a construction on linear Boolean permutations in an arbitrary number of variables.

Lemma 7.9. *Let* $l_i(x) = a_{i0} \oplus a_{i1}x_1 \oplus \cdots \oplus a_{in}x_n \in \mathcal{A}_n$, $i = 1, \ldots, n$. *Let* $A = [a_{ij}]$, $i, j = 1, \ldots, n$, *be the matrix of coefficients. Then* $[l_1, \ldots, l_n]$ *forms a linear Boolean permutation if and only if* A *is nonsingular.*

Note that from Lemma 7.9 we get that $[l_1, \ldots, l_n] = [a_{10}, \ldots, a_{n0}] \oplus [x_1, \ldots, x_n]A$. Hence we have the following:

Corollary 7.2. *The number of linear Boolean permutations in* n *variables is* 2^n *times the number of nonsingular matrices of order* $n \times n$.

It is known that the number of $n \times n$ nonsingular matrices is larger than $0.288 \times 2^{n^2}$. The probability that a random selection of n affine functions from $\mathcal{L}_n$ forms a Boolean permutation is $0.288 \times 2^{n^2} \times 2^n/(2^{n+1})^n = 0.288$ which is the same as the probability that a randomly chosen $n \times n$ binary matrix is nonsingular. This implies that random selection of linear Boolean permutations is acceptable.

By Lemma 7.8 we know that it is easy to find the inverse of the constructed Boolean permutation using Algorithm 7.1 provided that the inverse of the given Boolean permutation is known. So Algorithm 7.1 cannot produce trapdoor Boolean permutations even if it is repeated for several times. Another method we will use to construct trapdoor Boolean permutations is as follows:

Lemma 7.10. *Let* $P = [f_1, f_2, \cdots, f_n]$ *and* $Q = [g_1, g_2, \cdots, g_n]$ *be two Boolean permutations and* $R = P(Q)$ *be the composed Boolean function of* P *and* Q *as in Theorem 7.6. Then the inverse of this composed Boolean permutation is* $R^{-1} = Q^{-1}(P^{-1})$.

Proof. The conclusion R to be a Boolean permutation is from Theorem 7.6, and the expression of the inverse of R can easily be verified to be true. □

Now we claim that using Algorithm 7.1, we can construct a large number of Boolean permutations with known inverses. By applying Lemma 7.10, we can construct new Boolean permutations. As in general the new constructed Boolean permutations no longer have the properties as those constructed by Algorithm 7.1, we claim that there has been no efficient algorithm to find their inverses without the knowledge of the intermediate Boolean permutations. In this sense the composed Boolean permutation is a trapdoor permutation as its inverse can be found using the information of the intermediate ones.

7.5 Intractability Assumption and One-Way Trapdoor Boolean Permutations

In public key cryptography where both encryption and decryption algorithms are required, the basic idea for designing the algorithms involves the use of one-way trapdoor functions. A function $y = f_\lambda(x)$ is called a *one-way trapdoor function* with trapdoor parameter λ if it satisfies the following properties:

- **Computable:** Given any input x, it is computationally easy (e.g., in polynomial time complexity) to get the output y.
- **One-way:** Given any output y, without the knowledge of the trapdoor parameter λ or other extra information, it is computationally infeasible to trace back to the input x.
- **Trapdoor:** With the knowledge of λ, it is computationally easy to find the corresponding x given any output y.

A one-way trapdoor function is also known as a *one-way function* if the trapdoor parameter is unknown and hence the function is hard to invert. A function is called a *two-way function* if it is computationally easy to find its inverse. From these requirements, we see that a trapdoor function must be an injection (not necessarily a bijection) from the input domain to the output domain. The trapdoor parameter λ could be data, or an algorithm, or any other kind of knowledge. For instance, in the RSA public key cryptosystem [20], the encryption algorithm is a trapdoor function, where the factorization of the modulus is the trapdoor parameter. In McEliece's public key cryptosystem [11], the fast decoding algorithm is the trapdoor parameter.

When the input domain and the output domain of a trapdoor function are identical, the function is a one-to-one function and is thus a permutation on the domain. A Boolean permutation is one specific expression of such permutations.

It is well known that there have been no efficient algorithms with polynomial time complexity for solving systems of nonlinear equations in the general case. Based on this fact of intractability, we can clarify our assumption below which will be used in constructing trapdoor Boolean permutations: *There has been no efficient algorithms to find the inverse of a randomly given Boolean permutation in n variables in a polynomial time complexity in n.*

It is easy to verify that for certain subclasses of Boolean permutations, we can find their inverses easily. These subclasses include:

- Linear Boolean permutations
- Boolean permutations constructed simply using Algorithm 7.1, where P is a two-way permutation
- A linear transformation on P, where P is a two-way permutation
- Extension of several smaller Boolean permutations where no one is a one-way permutation

We can find infinite subclasses of Boolean permutations where for each specific subclass of such permutations, there is a fast algorithm to find the inverses of the permutations in the subclass. However, it does not reduce the complexity for solving a general system of nonlinear equations, because the problem to identify which subclass the permutation belongs to is by itself a hard one. At this stage, we classify a Boolean permutation as a one-way permutation if it does not belong to the above described special classes, and no efficient algorithm (in polynomial time complexity) to find its inverse is known.

It should be noted that the composition of two Boolean permutations in the above subclasses will yield a new Boolean permutation (likely to be) outside the above described subclasses if both of the initial permutations are nonlinear. This new composed permutation therefore can be treated as a one-way trapdoor permutation, and its inverse can be computed using the knowledge of the inverses of the initial Boolean permutations. It is suggested that other operations should also be used in formulating trapdoor Boolean permutations.

7.6 Construction of Boolean Permutations

Block ciphers play a very important role in contemporary cryptography. The case is more obvious in today's electronic commerce, where almost all encryption tools use block ciphers. There are two types of block ciphers, symmetric key block cipher (which is also called traditional block cipher) and asymmetric key block cipher (which is also known as public key cipher). In both of the ciphers, encryption is a one-to-one mapping from plaintext space (all possible plaintexts) to ciphertext space (all possible ciphertexts), and the corresponding decryption is the inverse mapping. Those mappings are controlled by a secret key in block ciphers or determined by the choice of public key and private key in asymmetric key ciphers.

Block ciphers are very often designed to have the same length for both plaintexts and ciphertexts, which means that the plaintext space and the ciphertext space are the same, and the one-to-one mapping becomes a invertible transformation on the plaintext space. We will denote such a space as M, and encryption and decryption are essentially a permutation and the inverse permutation on M. However, when M is very large, e.g., when it contains 2^{128} elements or more, to find such a permutation and its inverse is not an easy task. It should also have the properties that when a permutation is given, it is hard to find the inverse, and the inverse can be found given further secret information. This requires that the permutation should be presented in a concise algebraic form (e.g., RSA [20]), and without further information, from the algebraic form, it is computationally infeasible to find its inverse.

Therefore, there is a strong relationship between permutations and block ciphers. New methods of constructing block ciphers (particularly public key ciphers) are in some sense about new presentation of permutations over a large set. One of those presentations is to use Boolean functions, and this type of permutations is called Boolean permutations. Boolean permutations have been used in the design of public key cryptosystems [24].

However, it is still a hard problem as how to efficiently construct Boolean permutations. Boolean permutations composed of linear or affine Boolean functions are easy to construct, and they have little use in practical cryptographic design, because when such a permutation is given, its inverse can easily be computed given its algebraic presentation. In order to introduce nonlinear Boolean functions to the Boolean permutations, which will increase the computational complexity in finding its inverse without any further information, systematic construction methods are necessary. A probabilistic method was proposed in [18], and it is proved in [22] that the method has low successful rate. An algebraic construction method is given in [22] which can construct nonlinear Boolean permutations. However, the constructed permutations have the property that all of the coordinate Boolean functions, when XORed with a particular nonlinear Boolean function, will yield a linear or affine Boolean function. Another improved construction is proposed in [25] which is supposed to make use of the Boolean permutations constructed using the method of [22], i.e., given two Boolean permutations in $n-1$ variables, a new Boolean permutation in n variables can be constructed. This section will describe some constructions of Boolean permutations, where the inverses of the constructed Boolean permutations can be computed. These constructions are in addition to the Algorithm 7.1 and Lemma 7.10.

7.6.1 Some Primary Constructions

Algorithm 7.1 can construct new Boolean permutations based on old ones, where the newly constructed Boolean permutations have one more variable. This method can be used for recursive construction. However, the algorithm works only when there is a Boolean permutation available as the very initial one. How do we construct Boolean permutations from the scratch? Here we list the construction algorithms in [22] and [25] and will use them to give a new construction.

Algorithm 7.2 (Construction of Boolean permutations [22]).

(1) Select an arbitrary Boolean function $g(x_1, \ldots, x_{n-1})$ in $n-1$ variables. Let $f(x) = g(x_1, \ldots, x_{n-1}) \oplus x_n$.
(2) Let D be an $(n-1) \times (n-1)$ nonsingular matrix, $c = (c_1, c_2, \ldots, c_n) \in \{0,1\}^{n-1}$. Set $[l_1, l_2, \ldots, l_{n-1}] = (x_1, x_2, \ldots, x_n)D \oplus c$.
(3) Let $f_i(x) = f(x) \oplus l_i(x)$, $i = 1, 2, \ldots, n-1$; $f_n(x) = f(x)$.
(4) Output $P = [f_1, f_2, \ldots, f_n]$.

Theorem 7.9. *The (n,n)-Boolean function $P = [f_1, \ldots, f_n]$ generated by Algorithm 7.2 is indeed a Boolean permutation.*

Proof. Let $\alpha = (a_1, a_2, \ldots, a_n)$ be an arbitrary nonzero vector. If the Hamming weight of α is even, then the linear combination $a_1 f_1 \oplus \cdots \oplus a_n f_n$ becomes a linear combination of $l_1, \ldots, l_{n-1}$ and hence is an affine function in $n-1$ variables and is balanced. If the Hamming weight of α is odd, then the linear combination $a_1 f_1 \oplus \cdots \oplus a_n f_n$ can be represented as $x_n \oplus g'(x_1, \ldots, x_{n-1})$ and is also balanced. By Theorem 7.2, we know that $P = [f_1, \ldots, f_n]$ is a Boolean permutation. □

The inverses of Boolean permutations obtained by Algorithm 7.2 can be calculated as follows. From Algorithm 7.2, we have

$$\begin{cases} z_1 = f_1(x) = g(x_1, \ldots, x_{n-1}) \oplus x_n \oplus l_1(x_1, \ldots, x_{n-1}), \\ z_2 = f_2(x) = g(x_1, \ldots, x_{n-1}) \oplus x_n \oplus l_2(x_1, \ldots, x_{n-1}), \\ \cdots\cdots \\ z_{n-1} = f_{n-1}(x) = g(x_1, \ldots, x_{n-1}) \oplus x_n \oplus l_{n-1}(x_1, \ldots, x_{n-1}), \\ z_n = f_n(x) = g(x_1, \ldots, x_{n-1}) \oplus x_n. \end{cases} \tag{7.13}$$

Let $l_i(x_1, \ldots, x_{n-1}) = a_{i,1}x_1 \oplus \cdots \oplus a_{i,n-1}x_{n-1} \oplus a_i$. Since $l_1, \ldots, l_{n-1}$ and 1 are linearly independent, by Lemma 7.9 the coefficient matrix

$$A = \begin{bmatrix} a_{1,1} & a_{1,2} & \cdots & a_{1,n-1} \\ a_{2,1} & a_{2,2} & \cdots & a_{2,n-1} \\ \cdots & \cdots & \cdots & \cdots \\ a_{n-1,1} & a_{n-1,2} & \cdots & a_{n-1,n-1} \end{bmatrix}$$

must be nonsingular. By solving Eq. 7.10, we have

$$\begin{bmatrix} x_1 \\ x_2 \\ \cdots \\ x_{n-1} \end{bmatrix} = A^{-1} \begin{bmatrix} z_1 \oplus z_n \oplus a_1 \\ z_2 \oplus z_n \oplus a_2 \\ \cdots \\ z_{n-1} \oplus z_n \oplus a_{n-1} \end{bmatrix}.$$

Substituting each x_i $(i = 1, \ldots, n-1)$ in $x_n = z_n \oplus g(x_1, \cdots, x_{n-1})$ we get a representation of x_n in terms of z_j; hence, the inverse Boolean permutation is obtained which has a form as in Eq. 7.9.

Another algorithm introduced in [25] is as follows:

Algorithm 7.3 (Construction of Boolean permutations [25]).

(1) Let $P_1 = [g_1, \ldots, g_{n-1}]$, $P_2 = [h_1, \ldots, h_{n-1}]$ be two Boolean permutations in $n-1$ variables, where $g_i \oplus h_i \neq 0$, $i = 1, 2, \ldots, n-1$.
(2) Set $f_i(x) = g_i \oplus x_n(g_i \oplus h_i)$, $i = 1, 2, \ldots, n-1$; $f_n(x) = 1 \oplus x_n$.
(3) Output $P = [f_1, f_2, \ldots, f_n]$

Both Algorithms 7.2 and 7.3 can produce nonlinear Boolean permutations in n variables. The difference of the above two algorithms is that Algorithm 7.2 produces Boolean permutations based on Boolean functions that are easy to select, while Algorithm 7.3 produces Boolean permutations based on two given Boolean permutations in $n-1$ variables. In this sense, the flexibility of Algorithm 7.3 is limited.

In the following, we give a new method to construct Boolean permutations based on a given one. Although the newly generated Boolean permutations have the same number of variables as the given Boolean permutation, the method will be able to produce a large number of different Boolean permutations, instead of generating one as what the Algorithm 7.3 does. This means that the new method can be combined with Algorithm 7.3 to generate many Boolean permutations in n variables when two Boolean permutations in $n-1$ variables are given.

Theorem 7.10. *Let $[g_1(x), g_2(x), \ldots, g_n(x)]$ be a Boolean permutation and $f_i(x)$ be independent of $g_i(x)$, where $i = 1, 2, \ldots, k$, $k \leq n$, and for an arbitrary vector $(c_1, c_2, \ldots, c_k) \in \{0, 1\}^k$, $\bigoplus_{i=1}^k c_i f_i(x)$ is also independent of $\bigoplus_{i=1}^k c_i g_i(x)$. Then*

$$[f_1 \oplus g_1, \ldots, f_k \oplus g_k, g_{k+1}, \ldots, g_n]$$

is a new Boolean permutation in n variables.

Proof. By Theorem 7.2, it only needs to prove that any nonzero linear combination of Boolean functions $f_1 \oplus g_1, \ldots, f_k \oplus g_k, g_{k+1}, \ldots, g_n$ yields a balanced Boolean function. For any nonzero vector $(c_1, c_2, \ldots, c_n) \in \{0, 1\}^n$, we have the following linear combination

$$\bigoplus_{i=1}^{k} c_i\,(f_i(x) \oplus g_i(x)) \oplus \bigoplus_{j=k+1}^{n} c_j g_j(x).$$

If $c_1 = c_2 = \cdots = c_k = 0$, the above becomes $\bigoplus_{j=k+1}^{n} c_j g_j(x)$. Since $[g_1(x), g_2(x), \ldots, g_n(x)]$ is a Boolean permutation, by Theorem 7.2 it is known that $\bigoplus_{j=k+1}^{n} c_j g_j(x)$ is balanced. If $c_1, c_2, \ldots, c_k$ has at least one nonzero element, then we have

$$\bigoplus_{i=1}^{k} c_i f_i(x) \oplus \bigoplus_{j=k+1}^{n} c_j g_j(x) \oplus \bigoplus_{i=1}^{k} c_i g_i(x).$$

By the initial assumption and the properties of Boolean permutations (mainly Theorem 7.2), it is easy to show that in the above expression, the first and the second terms are all independent of the third part. By Theorem 2.24, we know that the exclusive-or of the first two parts is independent of the third one. Again by Theorem 7.2, it is known that the third part is a balanced Boolean function, and by Theorem 2.25, it is known that the exclusive-or of all the three parts is a balanced Boolean function, which proves the theorem to be true. □

Theorem 7.10 does not give an explicit construction. However, we can consider some special cases which will imply construction methods. Apparently, in order to use the method in Theorem 7.10, one needs to find Boolean functions which are independent of a given one. Using truth table, in theory it is trivial to construct such functions. However, when a Boolean function has many variables, say, 100, it is practically impossible to work on the truth table. A more practical way would be to use polynomial representation of Boolean functions. However, for a general Boolean function, even if some more information is given, say, it is balanced, it is still very hard to find another Boolean function so that they are independent of each other, unless the given Boolean function has a very special algebraic structure. It is noticed that in Theorem 7.10, a preconstructed Boolean permutation is given. We assume that this Boolean permutation is constructed using one of the algorithms introduced above, which has some algebraic structures that we know.

In practical construction, the condition of Theorem 7.10 can be made stronger, e.g., to assume that any nonzero linear combination of $f_1(x), \ldots, f_k(x)$ is independent of any nonzero linear combination of $g_1(x), \ldots, g_k(x)$. These two linear combinations do not necessarily have the same coefficients. Although this stronger assumption will yield fewer Boolean permutations, it is however practically easier to implement.

Corollary 7.3. *Let $f_1(x), f_2(x), \ldots, f_n(x)$ be k Boolean functions in n variables, and two different Boolean permutations can be made when they are combined with $g_{k+1}(x), \ldots, g_n(x)$, and $h_{k+1}(x), \ldots, h_n(x)$, respectively. Then $[f_1 \oplus h_{k+1}, f_2 \oplus h_{k+2}, \ldots, g_{k+1}, \ldots, g_n]$ is also a Boolean permutation.*

Corollary 7.3 makes use of two Boolean permutations with overlap coordinates to construct new Boolean permutations. To be precise, when $k > \frac{n}{2}$, the constructed new Boolean permutation will have the form of $f_1 \oplus h_{k+1}, \ldots, f_{n-k} \oplus h_n, f_{n-k+1}, \ldots f_k, g_{k+1}, \ldots, g_n$, and when $k \leq \frac{n}{2}$, the new constructed Boolean permutation will have the form of $f_1 \oplus h_{k+1}, \ldots, f_k \oplus h_{2k}, g_{k+1}, \ldots, g_n$.

Now we consider how to construct new Boolean permutations based on the ones constructed using Algorithm 7.2 or Algorithm 7.3. When we consider Boolean permutations constructed using Algorithm 7.2, we need to modify the algorithm as follows: for an arbitrary value k, choose two nonsingular matrices D_1 and D_2, such that their first k rows are the same and their last $n - k$ rows are different by at least one row. Then Algorithm 7.2 produces two Boolean permutations $[f_1, \ldots, f_k, g_{k+1}, \ldots, g_n]$ and $[f_1, \ldots, f_k, h_{k+1}, \ldots, h_n]$. By Corollary 7.3, these two Boolean permutations can be used to produce a new Boolean permutation.

However, there is a possibility that the newly constructed Boolean permutation may be linearly equivalent to one of the original Boolean permutations, i.e., the new Boolean permutation can be obtained from a linear transformation of Lemma 7.8 or Lemma 7.9, which is not very desirable. It is noted that Lemma 7.9 does not apply here, as the Boolean permutations constructed using Algorithm 7.2 have all their coordinate Boolean functions being nonlinear, while the method from Corollary 7.3 will construct Boolean permutations with linear coordinates. This means that if the constructed Boolean permutation is linearly equivalent to a previous one, it must be in the sense of the transformation of Lemma 7.8. If this is the case, then $f_i \oplus h_{k+i}$ can be represented as a linear combination of $[f_1, \ldots, f_k, g_{k+1}, \ldots, g_n]$, i.e.,

$$f_i \oplus h_{i+k} = c_1 f_1 \oplus \cdots \oplus c_k f_k \oplus c_{k+1} g_{k+1} \oplus \cdots \oplus c_n g_n.$$

However, from Algorithm 7.2, it is known that a Boolean permutation constructed using Algorithm 7.2 has the property that all the coordinate Boolean functions have a close relationship with a fixed nonlinear Boolean function, i.e., when they are XORed with the fixed function, the outcome is always a linear or affine Boolean function. This implies that the left-hand side of the above equation is an affine Boolean function, and for the equality to hold, the right-hand side must be the same affine function as well. This means that the coefficient vector $(c_1, c_2, \ldots, c_n)$ must have an even Hamming weight. This is not necessary and can even be made not possible by carefully choosing the matrices D_1 and D_2. By Corollary 7.3, it is known that the constructed Boolean permutation can be made not to be linearly equivalent to any of the previous Boolean permutations.

When Boolean permutations constructed using Algorithm 7.3 are further used to construct new Boolean permutations using the method of Corollary 7.3, three Boolean permutations in $n - 1$ variables are needed, where two of them have k coordinate Boolean function being the same. Similar to the idea of the construction based on the Boolean permutations constructed using Algorithm 7.2, here it is also possible to make the final Boolean permutation not to be linearly equivalent to any of the original ones.

7.6.2 On the Flexibility of the New Construction Method for Boolean Permutations

One of the measurements of good constructions is to see how many new things can be produced. This also applies to our construction method. We will give an evaluation of the number of new Boolean permutations that our new method can produce by considering very special cases. First we check how many new Boolean permutations can be constructed based on the ones constructed using Algorithm 7.2. Assume that $[f_1, f_2, \ldots, f_n]$ is a Boolean permutation constructed using Algorithm 7.2, where the algorithm used nonlinear Boolean function $f(x)$ and linear or affine functions $l_1, l_2, \ldots, l_{n-1}$. Now replace the first column of D with the bit-wise exclusive-or of the first three columns and remain the rest unchanged; then it yields a new nonsingular matrix. If this matrix is used, a new Boolean permutation is constructed using Algorithm 7.2 with only the first coordinate f_1 being different from the previous Boolean permutation; more precisely, it is $f_1' = f \oplus l_1 \oplus l_2 \oplus l_3$. By Corollary 7.3, it is known that $[f_1, \ldots, f_{n-1}, f_n \oplus f'] = [f_1, \ldots, f_{n-1}, l_1 \oplus l_2 \oplus l_3]$ is a Boolean permutation. It is easy to verify that $[f_1, \ldots, f_{n-1}, l_1 \oplus l_2 \oplus l_3]$ cannot be produced directly from Algorithm 7.2, and further analysis shows that is it not linearly equivalent to any Boolean permutation constructed using Algorithm 7.2. This means that the new method proposed in this chapter, in the very special case as just described above, can produce as many Boolean permutations as Algorithm 7.2 can, and the newly constructed Boolean permutations are not linearly equivalent to anyone constructed using Algorithm 7.2. Note that the evaluation is based on a very special case, and thus Corollary 7.3 can actually construct much more new Boolean permutations. The precise enumeration remains an open problem.

Let $[f_1, f_2, \ldots, f_n]$ be a Boolean permutation constructed using Algorithm 7.3. In its construction, if the two basis Boolean permutations in $n-1$ variables $[g_1, \ldots, g_{n-1}]$ and $[h_1, \ldots, h_{n-1}]$ are both constructed using Algorithm 7.2, then similar to the method described above, it is easy to construct another Boolean permutation $[h_1, \ldots, h_{n-2}, h_{n-1}']$. Therefore, two Boolean permutations, $P = [f_1, \ldots, f_n]$ and $P' = [f_1, \ldots, f_{n-2}, f_{n-1}', f_n]$, can be constructed. By Corollary 7.3, it is known that $[f_1, \ldots, f_{n-1}, f_n \oplus f_{n-1}']$ is also a Boolean permutation. The question concerned about is whether it can be generated from Algorithm 7.3, directly or indirectly, with the combination of a linear transformation as in Lemmas 7.8 or 7.9. Apparently, it cannot be generated directly from Algorithm 7.3, since all of the coordinate Boolean functions of the new constructed Boolean permutation are nonlinear, which is different in form from those constructed using Algorithm 7.3. From the way how new Boolean permutations are constructed based on those constructed using Algorithm 7.2, it is known that it is easy to construct h_{n-1}' such that h_{n-1}' is not a linear combination of $f_1, \ldots, f_{n-1}$. This shows that the newly constructed Boolean permutations cannot be linearly equivalent to those constructed directly from Algorithm 7.3, which also means that based on any Boolean permutation constructed using Algorithm 7.3, at least one new Boolean permutation can be constructed. Notice that this is just a very special case and in general the method of Corollary 7.3 can be used to construct much more new Boolean permutations.

7.6.3 *Construction of Trapdoor Boolean Permutations with Limited Number of Terms*

It is seen that whether a Boolean permutation is a trapdoor function depends essentially on the difficulty of computing inverses of the Boolean permutations. Because the number of variables of Boolean permutations has to be reasonably large in practice, say, 64 or larger, and a Boolean function can have as many as 2^n terms in its algebraic normal form representation, the number of terms in a Boolean permutation is an important factor in the design of trapdoor Boolean permutations. This section describes a method for constructing trapdoor Boolean permutations with limited number of terms.

Algorithm 7.4 (Construction of nonlinear Boolean permutations).

(1) Select integers $n_1, \ldots, n_k$, at random such that $n_1 + \cdots + n_k = n$.
(2) Select $g_i \in \mathcal{F}_{n_i-1}$ at random.
(3) Let $F_i = [g_i \oplus x_1 \oplus x_{n_i}, \ldots, g_i \oplus x_{n_i-1} \oplus x_{n_i}, g_i \oplus x_{n_i}]$.
(4) Then the concatenation $P = [F_i, \ldots, F_k]$ is a Boolean permutation in n variables.

Let $NT(f)$ be the number of terms of function f. It can be seen that for the above constructed Boolean permutation, we have

$$NT(P) \le \sum_{i=1}^{k} (n_i NT(g_i) + 2n_i - 1) \tag{7.14}$$

$$NT(P^{-1}) \le 2^{n_i-1} + \sum_{i=1}^{k} (2n_i + 1). \tag{7.15}$$

By choosing each g_i such that it has a small number of variables, both $NT(P)$ and $NT(P^{-1})$ can be reasonably small. However, it is also easy to compute the inverses of the Boolean permutations constructed using Algorithm 7.4. Applying another small Boolean permutation which has no component Boolean function that has degree larger than, say, 3, we can compose a Boolean permutation that is hard to invert (refer to Theorem 7.7). The lower degree of the small Boolean permutation ensures that the resulting composed Boolean permutation will only have a small number of terms and hence can be implemented effectively.

An alternative approach to constructing applicable Boolean permutations is as follows: Use Algorithm 7.4 to construct a Boolean permutation with controllable number of terms. Apply other small Boolean permutations to it to generate a new

Boolean permutation (Theorem 7.7) which may have relatively more terms and is harder to invert, while the inverse can be obtained easily using the inverses of each of the individual Boolean permutations in the process of composition.

7.7 A Small Example of Boolean Permutations

Here is a Boolean permutation in 64 variables. It is so simple that we do not suggest it be used in practice. However, it is expected to show that given any one of the Boolean permutations, to find the other (the inverse) without additional information is difficult.

In order to simplify the notation, we use the indices to represent variables. For example, x_1 is denoted as 01 and x_3x_{33} is denoted as 0333. In this way, f_9 of permutation P below represents $x_6 \oplus x_{12}x_{16} \oplus x_{16}$.

Permutation P

f_1 = 26 2631 2635 3135 35, f_2 = 18 54, f_3 = 4549 49 4958 58, f_4 = 3847 47 56, f_5 = 26 2631 2635 31 3135, f_6 = 14 1425 1448 25, f_7 = 28 2837 29, f_8 = 19 1920 20 2050, f_9 = 06 1216 16, f_{10} = 1540 1551 40 4051 51, f_{11} = 04 0411 0423 1123 23, f_{12} = 17 60 6063, f_{13} = 40 51, f_{14} = 14 2548 48, f_{15} = 3334 3339 34 3439 39, f_{16} = 33 39, f_{17} = 05 3046 46, f_{18} = 02 0208 0243 08 0843, f_{19} = 09 0944 4461, f_{20} = 07 0753 2453, f_{21} = 11 23, f_{22} = 0208 0243 08 0843 43, f_{23} = 21 42, f_{24} = 03 52, f_{25} = 26 35, f_{26} = 07 24 2453, f_{27} = 3847 3856 56, f_{28} = 08 43, f_{29} = 21 2122 2142 2242 42, f_{30} = 2829 29 2937 37, f_{31} = 45 4558 49, f_{32} = 01, f_{33} = 38 3856 4756 56, f_{34} = 05 0530 0546 30, f_{35} = 07 0724 0753 53, f_{36} = 05 0530 3046, f_{37} = 09 4461 61, f_{38} = 09 0944 0961 44, f_{39} = 03 0336 0352 36 3652, f_{40} = 18 1854 1855 54 5455, f_{41} = 10 1041 1064 41, f_{42} = 1950 20 50, f_{43} = 2732 2757 32 3257 57, f_{44} = 2837 29 2937, f_{45} = 4558 49 4958, f_{46} = 14 1425 2548, f_{47} = 1920 1950 20, f_{48} = 06 0612 1216, f_{49} = 18 1854 1855 5455 55, f_{50} = 13 1359 1362 5962 62, f_{51} = 33 3334 3339 3439 39, f_{52} = 32 57, f_{53} = 13 62, f_{54} = 06 0612 0616 12, f_{55} = 10 4164 64, f_{56} = 27 2732 2757 32 3257, f_{57} = 2122 2142 22 2242 42, f_{58} = 03 0336 0352 3652 52, f_{59} = 17 1760 1763 63, f_{60} = 17 1763 6063, f_{61} = 0411 0423 11 1123 23, f_{62} = 15 1540 1551 4051 51, f_{63} = 13 1359 1362 59 5962, f_{64} = 10 1041 4164.

Inverse permutation P^{-1}

g_1 = 32, g_2 = 18 1828 2228 28, g_3 = 2439 2458 58, g_4 = 11 1121 21 2161, g_5 = 1734 3436 36, g_6 = 0954 48 4854, g_7 = 20 2035 2635, g_8 = 1828 22 2228, g_9 = 19 1938 3738, g_{10} = 4155 4164 64, g_{11} = 1121 21 2161 61, g_{12} = 0948 54, g_{13} = 50 5053 5363, g_{14} = 0614 0646 46, g_{15} = 1013 13 1362 62, g_{16} = 09 0948 0954 48, g_{17} = 1259 5960 60, g_{18} = 0240 0249 40, g_{19} = 08 4247, g_{20} = 0842 0847 47, g_{21} = 23 2329 2357 29, g_{22} = 23 2329 2357 57, g_{23} = 1121 2161 61, g_{24} = 20 2026 26 2635, g_{25} = 06 1446, g_{26} = 01 0125 0525, g_{27} = 4352 52 5256 56, g_{28} = 07 0730 0744 44, g_{29} = 0730 3044 44, g_{30} = 1736 34, g_{31} = 0125 05

0525 25, g_{32} = 43 4352 5256, g_{33} = 1516 16 1651 51, g_{34} = 15 1516 16 1651, g_{35} = 01 0125 0525 25, g_{36} = 24 2439 2458 39, g_{37} = 0744 30, g_{38} = 0427 33, g_{39} = 1516 1651 51, g_{40} = 10 1013 13 1362, g_{41} = 41 5564, g_{42} = 2329 2357 29, g_{43} = 1828 22 2228 28, g_{44} = 1937 38, g_{45} = 0331 31 3145 45, g_{46} = 17 1734 1736 36, g_{47} = 04 0427 0433 27, g_{48} = 0614 14 1446 46, g_{49} = 0331 0345 45, g_{50} = 0842 42 4247 47, g_{51} = 10 1013 1362, g_{52} = 24 2439 2458 58, g_{53} = 2026 35, g_{54} = 02 0240 0249 40, g_{55} = 02 0240 0249 49, g_{56} = 0433 27 2733, g_{57} = 43 4352 52 5256, g_{58} = 03 3145, g_{59} = 5053 53 5363 63, g_{60} = 12 1259 1260 60, g_{61} = 19 1937 37 3738, g_{62} = 50 5053 53 5363, g_{63} = 1260 59, g_{64} = 4155 55 5564 64.

7.7.1 Linearity and Nonlinearity of Boolean Permutations

There are different definitions of nonlinearity of Boolean permutations (more generally, nonlinearity of (n, m)-Boolean functions) in public literatures. One is defined in [18] as the summation of the nonlinearities of all the coordinate functions of a Boolean permutation. It does not correctly reflect the "nonlinearity" of a Boolean permutation in terms of best linear approximation (BLA). A Boolean permutation with relatively high nonlinearity can have a very good linear approximation using this definition. In [13], the nonlinearity of a Boolean permutation is defined as the minimum nonlinearity of all possible nonzero linear combinations of the coordinate functions of the permutation and nonzero linear combinations of the coordinate functions of the inverse permutation. This does not reflect the real "nonlinearity" of a Boolean permutation either, because a Boolean permutation with one linear coordinate would have zero nonlinearity while it could be very hard to provide a linear approximation. For this reason, we tend to give a different concept called "linearity," hoping this concept describes more precisely the nonlinearity of Boolean permutations in terms of the hardness of their best linear approximations.

Although it is hard to find the inverse of an arbitrary Boolean permutation, it is possible to find another linear Boolean permutation that can be an approximation of it. The more precise the approximation is, the closer the Boolean permutation is to a linear Boolean permutation.

Let $f(x) \in \mathcal{F}_n$. If there exists an affine function $l_0(x) \in \mathcal{L}_n$ such that

$$wt(f(x) \oplus l_0(x)) = \min_{l(x)\in\mathcal{L}_n} \{wt(f(x) \oplus l(x)\},$$

then $l_0(x)$ is called the *best affine approximation* (BAA) of $f(x)$. The BAA of a Boolean function is not necessarily unique. The most efficient method for finding a BAA of a Boolean function uses Walsh techniques (see [8] for details). To find a BAA of a Boolean function in n variables, it takes on average $n \cdot 2^n$ operations.

Let $P = [f_1, \ldots, f_n]$ be a Boolean permutation of order n. If $l_i(x)$ is a BAA of $f_i(x)$ and $L = [l_1, \ldots, l_n]$ is a (linear) Boolean permutation, we call L an *optimum linear permutation approximation* (OLPA) of P. From this definition we know that $L_1 = [x_1, x_2, x_3]$ and $L_2 = [x_1, x_1 \oplus x_2, x_2 \oplus x_3 \oplus 1]$ are two OLPAs of $P = [x_3 \oplus x_1x_2 \oplus x_2x_3, x_2 \oplus x_1x_3, x_1 \oplus x_1x_2 \oplus x_2x_3]$. Note that $L_1(x) = P(x)$ if and only if $x \in \{000, 010, 011, 110\}$ and $L_2(x) = P(x)$ if and only if $x \in \{010, 011, 111\}$. We say that P is closer to L_1 than to L_2 because $|\{x : L_1(x) = P(x)\}| > |\{x : L_2(x) = P(x)\}|$. This means that an OLPA is not necessarily the best one in terms of Boolean permutation approximation. In general, if an OLPA of a Boolean permutation P is closest to P, then it is called the *best linear approximation* (BLA) of P. Likewise the BLA of a Boolean permutation is not necessarily unique. The number of coincidences between a Boolean permutation P and any one of its BLAs is called the *linearity* of P and is denoted by L_P. $L'_P = L_P/2^n$ is called the relative linearity of P. Hence, the following conclusion holds:

- For any Boolean permutation, P, $1 \leq L_P \leq 2^n$. When $L_P = 2^n$, P is a linear Boolean permutation.
- Let P^{-1} be the inverse of P. Then $L_{P^{-1}} = L_P$.

In general, given a Boolean permutation, it is difficult to construct its BAA. On the other hand, it is also difficult to construct Boolean permutations with low linearity.

References

1. Adams, C.M., Tavares, S.: The structured design of cryptographically good S-boxes. J. Cryptol. **3**(1), 27–41 (1990)
2. Armknecht, F., Krause M.: Constructing single and multi-output boolean functions with maximal immunity. In: Proceedings of ICALP 2006. LNCS 4052, pp. 162–175. Springer, Berlin Heidelberg (2006)
3. Daemen, J., Rijmen, V.: AES Proposal: Rijndael, pp. 1–45. NIST, Ventura (1998)
4. Data Encryption Standard, FIPS PUB 46, National Technical Information Services. Springerfield (1977)
5. Forre, R.: Methods and instruments for designing S-boxes. J. Cryptol. **2**, 115–130 (1990)
6. Golic, J.D.: Vectorial Boolean functions and induced algebraic equations. IEEE Trans. Inf. Theory **IT-52**(2), 528–537 (2005)
7. Gupta, K.C., Sarkar, P.: Improved construction of nonlinear resilient S-boxes. IEEE Trans. Inf. Theory **IT-51**(1), 339–348 (2005)
8. Karpovsky M.G.: Finite Orthogonal Series in the Design of Digital Devices. Wiely, New York (1976)
9. Lidle, R., Muller, W.B.: Permutation polynomials in RSA-cryptosystem. In: Advances in Cryptology – Proceedings of Crypto'83, pp. 293–301. Plenum, New York (1984)
10. Matsui, M.: On correlation between the order of S-boxes and the strength of DES. In: Advances in Cryptology – Proceedings of Eurocrypt'94. LNCS 950, pp. 366–375. Springer, Berlin (1995)
11. McEliece, R.L.: A public-key cryptosystem based on algebraic coding theory, pp. 114–116 . Deep Space Network Progress Report 42–44, Jet Propulsion Labs, Pasadena (1978)
12. Minster, S., Adams, C.: Practical S-box design. In: Proceedings of the Third Annual Workshop on Selected Areas in Cryptography, Kingston, pp. 61–76 (1996)
13. Nyberg, K.: Perfect nonlinear S-boxes. In: Advances in Cryptology – Proceedings of Eurocrypt'91. LNCS 547, pp. 378–386. Springer, Heidelberg (1991)

14. Nyberg, K.: On the construction of highly nonlinear permutations, In: Advances in Cryptology – Proceedings of Eurocrypt'92. LNCS 658, pp. 92–98. Springer, Berlin/Heidelberg (1993)
15. Nyberg, K.: Differentially uniform mappings for cryptography. In: Advances in Cryptology – Proceedings of Eurocrypt'93. LNCS 765, pp. 55–64. Springer, Berlin/Heidelberg (1994)
16. O'Connor, L.J.: Enumerating nondegenerate permutations. In: Advances in Cryptology – Proceedings of Eurocrypt'91. LNCS 547, pp. 368–377. Springer, Berlin/Heidelberg (1991)
17. Pieprzyk, J.: How to construct pseudorandom permutations from single pseudorandom functions. In: Advances in Cryptology – Proceedings of Eurocrypt'90. LNCS 473, pp. 140–150. Springer, Berlin/Heidelberg (1991)
18. Pieprzyk, J., Finkelstein, G.: Towards effective nonlinear cryptosystem design. IEE Proc. Part E **135**(6), 325–335 (1988)
19. Pieprzyk, J., Zhang, X.M.: Permutation generators of alternating groups. In: Advances in Cryptology – Proceedings of Auscrypt'90. LNCS 453, pp. 237–244. Springer, New York (1990)
20. Rivest, R.L., Shamir, A., Adleman, L.: A method for obtaining digital signatures. Commun. ACM **21**(2), 120–126 (1978)
21. Webster, A.F., Tavares, S.E.: On the design of S-boxes. In: Advances in Cryptology – Proceedings of Crypto'85. LNCS 218, pp. 523–534. Springer, Berlin (1986)
22. Wu, C.K.: Boolean functions in cryptology. Ph.D. thesis, Xidian University, Xian (1993) (in Chinese)
23. Wu, C., Wang, X.: Efficient construction of permutations of high nonlinearity. Chin. Sci. Bull. **38**(8), 679–683 (1993)
24. Wu, C.K., Varadharajan, V.: Public key cryptosystems based on Boolean permutations and their applications. Int. J. Comput. Math. **74**(2), 167–184 (2000)
25. Xing, Y.S., Yang, Y.: Construction and enumeration of Boolean permutations in cryptosystems. J. China Inst. Commun. **3**, 74–76 (1998)
26. Zhang, X.M., Zheng, Y.: Difference distribution table of a regular substitution box. In: Proceedings of the Third Annual Workshop on Selected Areas in Cryptography, kingston, pp. 57–60 (1996)

Chapter 8
Cryptographic Applications of Boolean Functions

Cryptographic applications of Boolean functions are meant to have some cryptographic properties, those properties are built to thwart cryptanalysis of certain kinds, and multiple cryptographic properties are usually required for a Boolean function to be used in cryptographic algorithm design, expected to resist some known attacks to the cryptographic algorithms. Therefore, the primary applications of cryptographic Boolean functions are the design of cryptographic algorithms, particularly stream cipher and block cipher algorithms. This chapter will discuss some applications of Boolean functions with some cryptographic properties in the areas beyond cryptographic algorithm design, where the involved Boolean functions are primary building blocks.

8.1 Applications of Degenerate Boolean Functions to Logic Circuit Representation

One of the applications of degeneracy property of Boolean functions is to simplify logic circuits. Since Boolean functions are so close to logic circuits, the Boolean operations, XOR and modulo 2 multiplication, have corresponding XOR and AND gates. We will use a notation like the letter capital "D" to denote the modulo 2 multiplication operator and the notation $\bigoplus$ to denote the XOR operation (modulo 2 addition). By Theorem 2.9 we know that, if a Boolean function $f(x) \in \mathcal{F}_n$ is degenerate, there exists $g(y) \in \mathcal{F}_k$ and an $n \times k$ binary matrix D such that $f(x) = g(xD)$ holds for all $x \in GF^n(2)$. The proof of Theorem 2.9 actually gives a way about how to find the degenerated function $g(y)$ of a given degenerate Boolean function $f(x)$. Here we will not repeat the process of how to compute the degenerated

C.-K. Wu, D. Feng, *Boolean Functions and Their Applications in Cryptography*,
Advances in Computer Science and Technology, DOI 10.1007/978-3-662-48865-2_8

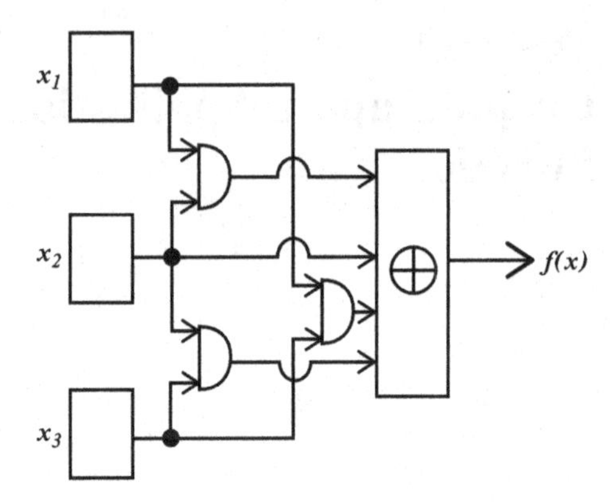

Fig. 8.1 The logic circuit representation of $f(x)$

function of a Boolean function if the given Boolean function is degenerate; instead, we claim that there is a good chance for the circuit implementation of $g(xD)$ to be simpler than that of $f(x)$. We have not yet tried on an applicable Boolean function; we only give an example here to demonstrate how it works.

Example 8.1. Boolean function $f(x) = x_1x_2 \oplus x_1x_3 \oplus x_2x_3 \oplus x_3$ represents a logic circuit (we assume the availability of *XOR* gate, although this can equivalently be implemented using *AND* and *OR* gates. The multiplication represents an *AND* gate) as shown in Fig. 8.1.

It is easy to see that the linear span of nonzero spectrum points of $f(x)$ has dimension 2, and hence, $f(x)$ can be degenerated to a function in two variables. In fact we can actually find the degenerated function $g(y_1, y_2) = y_1y_2$, since

$$g(y_1, y_2) = g((x_1, x_2, x_3)D) = f(x),$$

where $D = \begin{bmatrix} 1\,0 \\ 0\,1 \\ 1\,1 \end{bmatrix}$; hence, $y_1 = x_1 \oplus x_3$ and $y_2 = x_2 \oplus x_3$. According to the degenerated function, the logic circuit can be designed as in Fig. 8.2.

The above example shows that, when a Boolean function is degenerate, the degenerated function can be used to simplify the logic representation of the Boolean function. Although this is not always the case, given that the variables of the degenerated function are linear combinations of the original input variables, this degenerate approach of Boolean function representation has potential to simplify the hardware implementation of some Boolean functions in complex algebraic normal form representation.

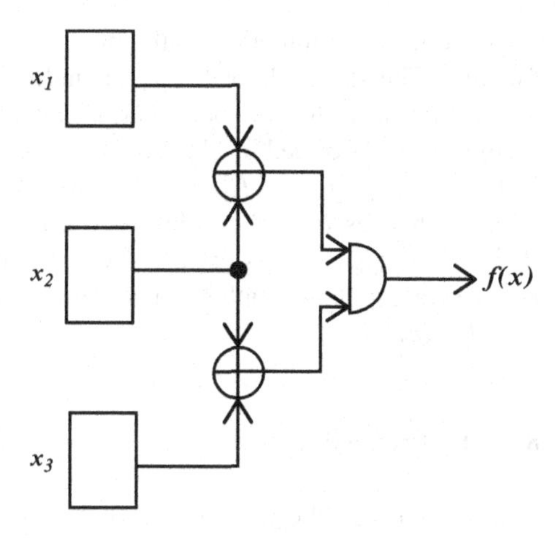

Fig. 8.2 The simplified logic circuit of $f(x)$ using its degenerated form

8.2 An Application of Boolean Permutations to Public Key Cryptosystem Design

Boolean permutations are treated as Boolean function representation of cryptographic S-boxes, and their primary applications are in the design of stream ciphers and block ciphers. Here we show how Boolean permutations can be used as primary building blocks to the design of public key cryptosystems.

Essentially any encryption algorithm without information expansion is a permutation. In the case of symmetric key systems, the permutation is hidden by a secret key. In the case of asymmetric key cryptosystems, the permutation is hidden by some special structure. For example, in the RSA cryptosystem, a special polynomial (exponentiation) is used to implement a large permutation over the integral ring Z_n, where n is the RSA modulus.

8.2.1 Public Key Cryptosystem 1 (PKC1)

A Boolean permutation can be directly used to design a public key cryptosystem if it satisfies the following properties like a one-way trapdoor function:

- Without additional knowledge, it is computationally infeasible to find the inverse of the given Boolean permutation (one way).
- With some special knowledge, it is easy to find the inverse of the Boolean permutation (trapdoor exists).
- The number of items in all the coordinate functions of the Boolean permutation is reasonably small (applicable).

A Boolean permutation with the first two properties is actually a trapdoor function. The special knowledge for finding the inverse is the trapdoor. One way to construct such Boolean permutations involves the use of composition of Boolean permutations described in Chap. 8.

Let $P = [f_1, f_2, \ldots, f_n]$ be a Boolean permutation with the above properties. User U chooses P as the public key and keeps the inverse permutation $P^{-1} = [f_1^{-1}, f_2^{-1}, \ldots, f_n^{-1}]$ as his private key. A plaintext m is a binary string of length n, and the corresponding ciphertext is then given by $c = P(m)$. Decryption is given by $m = P^{-1}(c)$.

8.2.1.1 Properties

The size of the public key is based on the number of terms in the permutation P, and the size of the private key is based on the number of terms in P^{-1}. So the number of terms of P and P^{-1} must both be reasonably small for the system to be a practical one. The best way to attack this system seems to be the determination of the BLA. In general, the cryptosystem can be made secure by choosing a Boolean permutation with low linearity.

8.2.2 Public Key Cryptosystem 2 (PKC2)

Let A be an arbitrary $k \times n$ binary matrix with $rank(A) = k$. Then there must exist an $n \times k$ binary matrix X and an $(n-k) \times n$ matrix B such that $AX = I_k$ and $BX = 0$, where I_k is the $k \times k$ identity matrix. We denote this as a triple (A, B, X). It is better to choose matrices with neither an all-zero row nor an all-zero column. One way to do this is to choose an arbitrary $n \times n$ nonsingular matrix C, then let A be composed of the first k rows of C and B be composed of the rest $n-k$ rows of C, and let X be composed of the first most left k columns of C^{-1}.

Let $P = [f_1, \ldots, f_k]$ be a Boolean permutation of order k for which the inverse P^{-1} is known only to the user. Let (A, B, X) be a triple satisfying the properties above. Let $R = [r_1, \ldots, r_{n-k}]$ be a collection of arbitrary functions from $\mathcal{F}_k$ (it should have small number of terms), and let $Q = PA \oplus RB$ be the public key. Note that as $P = QX$, the corresponding private key is given by $P^{-1}(zX)$. Hence, the public key cryptosystem is as follows:

Public:	$Q = PA \oplus RB$.
Private:	$P^{-1}(zX)$.
Message m:	binary string of length k.
Encryption:	$c = Q(m)$.
Decryption:	$m = P^{-1}(cX)$.

In this cryptosystem, the public key is a collection of n Boolean functions in k variables, while the private key is a collection of k such Boolean functions in n variables. If any k components of the public key form a Boolean permutation which is easy to invert, then the cryptosystem can be easily broken. However, determining whether any k coordinates of Q form a Boolean permutation is an *NP-complete* problem (see [5]). Even if such a Boolean permutation is occasionally found, to find its inverse seems to be as hard as to break *PKC1*. So it is infeasible to get m from c when k is fairly large. To minimize information expansion, it is proposed that n is slightly larger than k. When both the Boolean permutation P and the arbitrary function R are chosen properly, the key size can be reasonably small.

8.2.3 Public Key Cryptosystem 3 (PKC3)

Similar to the PKC2 above, in PKC3 we use a generating matrix G of an $[n, k, d]$ linear code (see [6] on error-correcting codes) with a known fast decoding algorithm (e.g., Goppa code). Let $P(x) = [f_1(x), f_2(x), \ldots, f_k(x)]$ be a Boolean permutation of order k for which the inverse P^{-1} is known only to the user U. Let $G(x) = P(x)G$ which is a collection of n Boolean functions in k variables. Note that for any $m \in F_2^k$, $G(m)$ is the code word corresponding to the message $P(m)$. Let $E(x) = [e_1(x), \ldots, e_n(x)]$ be a collection of n arbitrary Boolean functions in k variables that satisfy for any x, $wt(E(x)) \leq t = \lfloor d-1 \rfloor /2$. Then $C(x) = G(x) \oplus E(x)$ is set to be the public key and P^{-1} and the fast decoding algorithm are kept private. The public key cryptosystem is as follows:

Public:	$C(x)$.
Private:	$P^{-1}(x)$ and decoding algorithm.
Message m:	binary string of length k.
Encryption:	$c = C(m)$.
Decryption:	(1) Decoding c to get $m' = P(m)$;
	(2) $m = P^{-1}(m')$.

Similar to the case as in PKC2, if k components of the public key form a Boolean permutation, then the cryptosystem can be broken. However, this is an *NP-complete* problem and is hard when k is large. One issue with this system is how to construct the error pattern function $E(x)$. Consider the following example: let $n = 127, k = 64$, and $t = 10$ (there is a Goppa with the above parameters). Choose 4 arbitrary Boolean functions in 64 variables with a small number of terms f_1, f_2, f_3, f_4. Let $i \in \{0, 1, \ldots, 15\}$ and $(i_1 i_2 i_3 i_4)$ be the binary representation of i. Define $E_i(x) = f_1^{i_1} f_2^{i_2} f_3^{i_3} f_4^{i_4}$, where $f^1(x) = f(x)$ and $f^0 = 1 \oplus f(x)$. Then we have 16 Boolean functions in 64 variables. Repeating this construction for $t = 10$ times, we get 160 Boolean functions in 64 variables. Select from them any 127 functions. Then the Hamming weight of the error pattern function is always less than or equal to $t = 10$.

At a first glance this public key cryptosystem is similar to McEliece's [7], which is based on an error-correcting code, and may suffer the same security risks as pointed in [2]. However, there are some differences:

- In McEliece cryptosystem, matrix $G^* = SGP$ rather than G is used, where G is a generating matrix of an $[n, k]$ Goppa code, S is a $k \times k$ nonsingular matrix, and P is an $n \times n$ permutation matrix. In this cryptosystem, it is safe to use G. Nevertheless, it is better to use G^*.
- In McEliece's cryptosystem, the error pattern is a random vector, while in this cryptosystem, the error pattern is a fixed (k, n) Boolean function.
- It is inappropriate to directly use McEliece's cryptosystem to obtain signatures, while with this cryptosystem it is possible (see Sect. 8.3 below).
- In this cryptosystem, with different choices of Boolean functions/permutations, the key size varies significantly, while the security level is kept unchanged. This property can be used to have keys whose size can be very small.

8.3 Application of Boolean Permutations to Digital Signatures

Public key cryptosystems are often used in the design of digital signature schemes. For example, the well-known RSA scheme is used to create digital signatures in numerous applications. A digital signature system must satisfy the following conditions: (1) generation and verification of signatures must be computationally efficient, (2) only the owner can create his or her valid signatures, and (3) anyone should be able to verify the validity of the digital signature. Let us now consider how our Boolean permutation-based public key cryptosystems proposed in Sect. 8.2 can be used to obtain digital signatures.

It can be seen that *PKC1* can be used to obtain signatures in a straightforward manner. Here a signature is the same as decrypting a message, while verifying a signature is the same as encrypting a message.

With *PKC2*, signatures can be achieved by letting $P^{-1}(zX)$ to be the public key and letting $Q(x)$ to be the private key. Then the private key can be used to create signatures, while the public key can be used to verify them.

Now let us consider how *PKC3* can be used to obtain signatures. Without loss of generality, we will assume that the first k columns of G form a nonsingular matrix G'. Then by Theorem 7.4 we know that $PG' = [G_1, \ldots, G_k]$ is a Boolean permutation for which the inverse $[G_1^{-1}, \ldots, G_k^{-1}]$ can be obtained easily by the owner of the public key. For a message m, which is a binary string of length k, user U's signature is the pair (m', e'), where

$$m' = (G_1^{-1}(m), \ldots, G_k^{-1}(m)),$$
$$e' = (e_1(m'), \ldots, e_k(m')).$$

On receiving the signature, the verifier can validate the signature by computing

$$(C_1(m'), \ldots, C_k(m')) = (G_1(m'), \ldots, G_k(m')) \oplus (e_1(m'), \ldots, e_k(m')) = m \oplus e',$$
$$(C_1(m'), \ldots, C_k(m')) \oplus e' = (m \oplus e') \oplus e' = m.$$

It is easy to verify that this signature scheme also satisfies the required properties of normal digital signature schemes.

8.4 Application of Boolean Permutations to Shared Signatures

Suppose there is a company which has a private key for signing documents. Every member of the company shares a piece of the information relating to the private key such that a single person cannot create a valid signature; only an authorized group can generate a valid signature. This is a combination of a normal signature scheme and a secret sharing scheme. When the secret sharing scheme is a threshold scheme, it yields a threshold signature which was originally studied by Y. Desmedt [3]. Note that the main difference between secret sharing schemes and shared signatures is that in a secret sharing scheme, once the secret information is recovered, the secret is revealed forever and all of the share holders cannot use their shares later. However, in a shared signature scheme, shareholders can repeatedly use their shares for signing messages without revealing the secret key.

Assume that there is a trusted authority of a company who can generate private and public keys for the company. Let S be a collection of k Boolean functions which is the private key of the company (see digital signatures modified from PKC1 and PKC2). Let A be a $k \times n$ matrix, where $n > k$. Let α_i^T denote the i-th column of A. Then the authority distributes α_i^T and $S\alpha_i^T$ to a member U_i of the company. For a message $m \in F_2^k$, U_i's signature is α_i^T and $S(m)\alpha_i^T$. It can be seen that when k such signatures are collected such that the k α_i^T's are linearly independent, the original message m can be recovered and hence a valid signature is generated. So a collection of U_i is an authorized group if and only if their α_i's form a matrix with rank k. When k is large and we want the group to be small, every person can hold more than one column of A. It should be noted that when a message is signed, the signature together with the message itself should be sent to the receiver. When the receiver receives the signature, he/she checks if some of the α_i's can form a nonsingular matrix so that $S(m)$ can be recovered. Then by using the public key, m is recovered. By comparing the attached message with the recovered one, the validity of the signature is recognized. It is easy to verify that this shared signature has the following properties:

- Only an authorized group can generate valid signatures.
- Signing a message does not reduce the security of other signatures.

- Signatures can be verified easily.
- When new members are added to the group, their keys can be assigned by the authority without the collaboration of other members.
- When members leave the company, in order for their shares to be no longer valid, all the members' shares as well as the public key have to be changed.

8.5 An Application of Boolean Permutations to Key Escrow Scheme

Since the proposal to use key escrow based on Clipper Chips for mobile phone communications in 1994 [8], there have been many papers discussing the significance and drawbacks of key escrow schemes. Most of these proposals are based on exponentiations and discrete logarithms, and the mathematical issues are essentially similar to those in RSA [9] and Diffie-Hellman's [4]. In this paper we present a new key escrow scheme based on a different mathematical structure, namely, Boolean permutations, and analyze its security properties.

8.5.1 *Setup*

The setup process includes the following phases.

8.5.1.1 Public/Secret Keys

Assume U is a general user and his public key is a Boolean permutation $P = [f_1, f_2, \ldots, f_n]$. The inverse permutation $P^{-1} = [f_1^{-1}, f_2^{-1}, \ldots, f_n^{-1}]$ is kept secret by U as his private key.

8.5.1.2 Session Keys

Consider the situation when the user U wishes to communicate with another user, say Alice. We assume that Alice is able to get hold of the public key of the user U via some means such as using a directory service. Alice now selects a random string e of length n and sends $P(e)$ to user U. The session key is e which can be recovered by U using his private key.

8.5.1.3 Key Escrowing

In this paper, we assume that there are N different key Escrowing agencies (KEAs) and that the secret key is handed to these agencies in a secure manner so that when any K of them gets together, they are able to recover every session key transmitted using U's public key and any $K - 1$ of them is not able to recover any message encrypted using U's public key. Ideally any $K - 1$ of them is not able to get any more information than an outsider from any message encrypted using U's public key. The key escrowing procedure is as follows.

1. Extend the length of the private key (if necessary) to a multiple of K by adding zeros, i.e., $P' = [f_1, \dots, f_n, 0, \dots, 0]$.
2. Split P' into K equivalent parts $F_1, F_2, \dots, F_K$, where

$$F_i = [f_{(i-1)d+1}, f_{(i-1)d+2}, \dots, f_{id}],\ i = 1, 2, \dots, K$$

   is an (n, d)-Boolean function and d is the least integer such that $dK \geq n$ and $f_j = 0$ if $j > n$.
3. User U is to choose a $K \times N$ matrix $A = [\alpha_1^T, \dots, \alpha_N^T]$ such that any K columns of A can form a $K \times K$ nonsingular matrix.
4. Then α_i and $E_i = [F_1, \dots, F_K]\alpha_i^T$, which is an (n, d)-Boolean function, are given to key escrow agency KEA_i. These are referred to as the share of KEA_i.

8.5.2 *Escrowing Verification*

It is important for each KEA to know that their shares from the user are genuine. We assume the existence of an independent authorized *Verifier*. Escrowed keys are verified first internally by each KEA, and then each KEA passes its part to the verifier for external verification by the Verifier.

8.5.2.1 Internal Verification

Let $X_i = [x_{(i-1)d+1}, x_{(i-1)d+2}, \dots, x_{id}]$, $i = 1, 2, \dots, K$, where $x_i = 0$ if $i > n$. Using U's public key as input, KEA_i checks internally whether equality $E_i(f_1, f_2, \dots, f_n) = [X_1, \dots, X_K]\alpha_i^T$ holds. If this is not the case, then the information sent to KEA_i is fraudulent.

8.5.2.2 External Verification

Each KEA_i sends α_i to the Verifier in a secure manner (e.g., via offline). The Verifier then checks if all of the α form a matrix in which any K columns are linearly independent. If this is the case, then the Verifier informs KEAs that their shares are genuine. Otherwise, the user is asked to resubmit his secret key to each KEA before he can become a legitimate user of the system.

Note that the verification process does not reveal any information about the user's secret key.

8.5.3 Key Recovery

Upon court order, the Verifier requests at least K of KEAs to work together on a lawfully wiretapped message $c = P(e)$. Instead of presenting their original shares which were handed by the user U, each KEA_i presents $E_i(c) = [F_1(c), F_2(c), \ldots, F_K(c)]\alpha_i^T$ securely to the Verifier. Since the Verifier knows α_i, and a set of K α_i forms a nonsingular matrix, $[F_1(c), F_2(c), \ldots, F_K(c)]$ can easily be recovered which contains the session key.

8.5.4 Properties

8.5.4.1 Key Size

Each KEA needs to store the data associated with each user. In the proposed key escrow system, each KEA has to store an (n, d)-Boolean function for user U which is smaller than the secret key of U. Note that there is no need to update the shares kept by the KEAs unless the public key of user U is changed.

8.5.4.2 Forward Security

A key escrow protocol is said to be *forward secure* if the disclosure of one of the session keys does not decrease the security of other session keys. Forward security enables a user to continuously use his/her facility for further secure communications when some session keys have been compromised. Note that in the above key recovery procedure, information regarding the user's secret key (Boolean functions) is not leaked when a session key is revealed. So this key escrow protocol provides forward security.

8.5.4.3 Other Security Properties

The proposed key escrow scheme is not vulnerable to attacks by an outsider to recover the session key.e form $P(e)$. This attack is equivalent to decrypting messages encrypted by user U's public key without knowing the secret key.

Let us now consider the situation when some of the KEAs are corrupt. Let us assume that t ($t \leq K - 1$) KEAs are corrupt. When they put their shares together, they can form a $K \times t$ matrix B and $[F_1, \ldots, F_K]B$. Because B is not a nonsingular matrix, the secret key $[F_1, \ldots, F_K]$ of user U cannot be recovered. However, for a session key k, by taking $P(k)$ as an input to $[F_1, \ldots, F_K]B$, it will yield a system of equations with t independent linear equations and K unknowns. Note that each unknown of this equation is a binary vector of dimension d, and there are $2^{d(K-t)}$ solutions to this equation. Among them, one contains the session k which is the first n-bit segment. So the complexity for finding a session key with t KEAs collaborating with each other is equivalent to solving $2^{d(K-t)}$ system of linear equations with td unknowns. It is computationally infeasible when $d(K-t)$ is reasonably large.

8.5.4.4 Full Disclosure

If user U is proved to be guilty and it is required to reveal his private key, on receiving a court order, then at least K KEAs send their functions E_i to the Verifier; the Verifier can recover the private key of the user U. For example, when $i = 1, 2, \ldots, K$, the Verifier gets $(E_1, \ldots, E_K) = [F_1, \ldots, F_K][\alpha_1^T, \ldots, \alpha_K^T]$. Since the verifier knows $[\alpha_1^T, \ldots, \alpha_K^T]$ which is a nonsingular matrix, $[F_1, \ldots, F_K]$ can be recovered of which the nonzero part is the private key of the user. At this stage, no one other than the Verifier knows the private key of user U.

8.5.4.5 Partial Key Escrowing

On addressing the confidentiality of users, Shamir proposed that partial key rather than the whole key be escrowed [10] which was further supported in [1]. It is clear that our scheme can easily achieve partial key escrowing by only allowing part of the user's secret key instead of the whole key to be escrowed; all the procedures described above remain unchanged.

8.6 A Small Example of Key Escrow Scheme Based on Boolean Permutations

Here we give a small example to demonstrate how the key escrow protocol works.

8.6.1 Selecting a Boolean Permutation of Order 6

It can easily be verified that $P_1 = [x_3 \oplus x_1x_2 \oplus x_2x_3, x_2 \oplus x_1x_3, x_1 \oplus x_1x_2 \oplus x_2x_3]$ is a Boolean permutation of order 3 with $P_1^{-1} = P_1$. We can also construct another Boolean permutation of order 6 by using Algorithm 7.1. With set $g(x_1, \ldots, x_5) = x_1x_2 \oplus x_3x_4x_5$ and $l_i = x_i$ for $i = 1, \ldots, 5$, we have Boolean permutation $Q = [f_1, \ldots, f_6]$, where

$$\begin{cases} g_1 = x_1 \oplus x_1x_2 \oplus x_3x_4x_5 \oplus x_6, \\ g_2 = x_2 \oplus x_1x_2 \oplus x_3x_4x_5 \oplus x_6, \\ g_3 = x_1x_2 \oplus x_3 \oplus x_3x_4x_5 \oplus x_6, \\ g_4 = x_1x_2 \oplus x_4 \oplus x_3x_4x_5 \oplus x_6, \\ g_5 = x_1x_2 \oplus x_5 \oplus x_3x_4x_5 \oplus x_6, \\ g_6 = x_1x_2 \oplus x_3x_4x_5 \oplus x_6. \end{cases}$$

The inverse of Q can easily be computed as $Q^{-1} = [f_1^{-1}, \ldots, f_6^{-1}]$, where $f_i^{-1} = z_i \oplus x_6$ for $i = 1, \ldots, 5$, and $f_6^{-1} = z_6 \oplus z_1z_2 \oplus z_1z_6 \oplus z_2z_6 \oplus z_3z_6 \oplus z_4z_6 \oplus z_5z_6 \oplus z_3z_4z_5 \oplus z_3z_4z_6 \oplus z_3z_5z_6 \oplus z_4z_5z_6$. By Lemma 7.6, the composition of P and Q yields a new Boolean permutation $R = [P(f_1,f_2,f_3),\ P(f_4,f_5,f_6)] = [r_1, \ldots, r_6]$, where

$$\begin{cases} r_1 = x_1x_2 \oplus x_3 \oplus x_2x_3 \oplus x_1x_2x_3 \oplus x_1x_3x_4x_5 \oplus x_6 \oplus x_1x_6 \oplus x_3x_6 \\ r_2 = x_2 \oplus x_1x_2 \oplus x_1x_3 \oplus x_1x_2x_3 \oplus x_3x_4x_5 \oplus x_1x_3x_4x_5 \oplus x_1x_6 \oplus x_3x_6 \\ r_3 = x_1 \oplus x_1x_2 \oplus x_2x_3 \oplus x_1x_2x_3 \oplus x_1x_3x_4x_5 \oplus x_6 \oplus x_1x_6 \oplus x_3x_6 \\ r_4 = x_1x_2 \oplus x_1x_2x_4 \oplus x_4x_5 \oplus x_6 \oplus x_4x_6 \\ r_5 = x_1x_2x_4 \oplus x_5 \oplus x_3x_4x_5 \oplus x_4x_6 \\ r_6 = x_1x_2 \oplus x_4 \oplus x_1x_2x_4 \oplus x_4x_5 \oplus x_6 \oplus x_4x_6 \end{cases}$$

The inverse of R is also easy to compute given the inverses of P and Q; it is $R^{-1} = [r_1^{-1}, \ldots, r_6^{-1}]$, where

$$\begin{cases} r_1^{-1} = x_1x_2 \oplus x_3 \oplus x_2x_3 \oplus x_4 \oplus x_4x_5 \oplus x_5x_6, \\ r_2^{-1} = x_2 \oplus x_1x_3 \oplus x_4 \oplus x_4x_5 \oplus x_5x_6, \\ r_3^{-1} = x_1 \oplus x_1x_2 \oplus x_2x_3 \oplus x_4 \oplus x_4x_5 \oplus x_5x_6, \\ r_4^{-1} = x_4 \oplus x_6, \\ r_5^{-1} = x_4 \oplus x_5 \oplus x_4x_5 \oplus x_4x_6 \oplus x_5x_6, \\ r_6^{-1} = x_1x_2 \oplus x_1x_3 \oplus x_4 \oplus x_1x_4 \oplus x_2x_4 \oplus x_3x_4 \oplus x_1x_3x_4 \oplus x_4x_5 \\ \qquad \oplus x_2x_4x_5 \oplus x_1x_2x_4x_5 \oplus x_3x_4x_5 \oplus x_1x_3x_4x_5 \oplus x_2x_3x_4x_5 \oplus x_4x_6 \\ \qquad \oplus x_1x_4x_6 \oplus x_1x_2x_4x_6 \oplus x_2x_3x_4x_6 \oplus x_2x_5x_6 \oplus x_1x_2x_5x_6 \oplus x_3x_5x_6 \\ \qquad \oplus x_1x_3x_5x_6 \oplus x_2x_3x_5x_6 \oplus x_4x_5x_6. \end{cases}$$

Note that the composed permutation R no longer has the format similar to those functions generated by Algorithm 7.1. So there is not an efficient way to compute its inverse. We would like to point out that, for this particular example, there might exist an efficient algorithm to get the inverse of R. It is however very hard to generalize the method to arbitrary composed permutations. In general composed Boolean permutations are hard to inverse and hence can be used as trapdoor functions. Here is a method for general composition:

- Generate a Boolean permutation Q of order n having a small number of terms by Algorithm 7.1.
- Select small Boolean permutations P_i of order n_i $(i = 1, \ldots, k)$ at random such that $\sum_{i=i}^{k} n_k = n$. Concatenate them to form a Boolean permutation P of order n.
- Generate a new Boolean permutation by composition $P(Q)$ or $Q(P)$.

8.6.2 *Preparation*

Assume that the above permutations are generated by user U. U uses R^{-1} as his public key and keep R as his secret key. A session key is a random binary string of length 6 which should be encrypted by R^{-1} and sent to U. Let $e = 100110$ be an arbitrary string which is a session key. Then $c = R^{-1}(e) = 001111$ is sent to user U. U can then recover the key e using his secret key. For key escrowing, we assume that there are three KEAs and any two of them would be able to escrow the session keys. So user U chooses matrix $A = \begin{bmatrix} 0\,1\,1 \\ 1\,0\,1 \end{bmatrix}$. The secret key is split into two parts as $R = [F_1, F_2]$, where

$$\begin{aligned} F_1 = [&x_1x_2 \oplus x_3 \oplus x_2x_3 \oplus x_1x_2x_3 \oplus x_1x_3x_4x_5 \oplus x_6 \oplus x_1x_6 \oplus x_3x_6, \\ &x_2 \oplus x_1x_2 \oplus x_1x_3 \oplus x_1x_2x_3 \oplus x_3x_4x_5 \oplus x_1x_3x_4x_5 \oplus x_1x_6 \oplus x_3x_6, \\ &x_1 \oplus x_1x_2 \oplus x_2x_3 \oplus x_1x_2x_3 \oplus x_1x_3x_4x_5 \oplus x_6 \oplus x_1x_6 \oplus x_3x_6], \\ F_2 = [&x_1x_2 \oplus x_1x_2x_4 \oplus x_4x_5 \oplus x_6 \oplus x_4x_6, \\ &x_1x_2x_4 \oplus x_5 \oplus x_3x_4x_5 \oplus x_4x_6, \\ &x_1x_2 \oplus x_4 \oplus x_1x_2x_4 \oplus x_4x_5 \oplus x_6 \oplus x_4x_6]. \end{aligned}$$

The three shares of KEAs are $[E_1, E_2, E_3] = [F_1, F_2]A$, where we have $E_1 = F_1$, $E_2 = F_2$ and $E_3 = F_1 \oplus F_2$. E_i and the i-th column of matrix A are handed to KEA$_i$.

8.6.3 *Verification*

We will just demonstrate how the verification is done by KEA$_3$. With the coordinates of the public key R^{-1} as inputs, the equality

$$E_i(r_1^{-1}, \ldots, r_6^{-1}) = [x_1, x_2, x_3] \oplus [x_4, x_5, x_6] = [x_1 \oplus x_4, x_2 \oplus x_5, x_3 \oplus x_6]$$

should hold. Otherwise, the share is fraudulent. External verification is nothing but simply a check of the properties of matrix A.

8.6.4 *Key Recovery*

Normally only the session keys need to be recovered. For example, for the above message c wiretapped from a public channel, KEA$_1$ can get $E_1(c) = [1, 0, 0]$, KEA$_2$ can get $E_2(c) = [1, 1, 0]$, and KEA$_3$ can get $E_3(c) = [0, 1, 0]$. The session key e can be reformed by either (E_1, E_2) or $(E_1, E_1 \oplus E_3)$ or $(E_2 \oplus E_3, E_3)$. Also note that when E_1, E_2, E_3 are put together and matrix A is known, the secret key of user U can be fully disclosed.

8.7 Remarks

It does not need to address how wide applications that Boolean functions may have; there are many books about Boolean functions and their applications in different areas. Cryptographic Boolean functions are designed preliminarily for the use in cryptographic algorithm design. This chapter presents some other applications of Boolean functions, particularly the applications of Boolean permutations in the design of public cryptography, shared signature, and key escrow schemes. This chapter is designed to show the possibility of alternate applications of Boolean functions in the area of cryptography, and the security analyses are not very deep, since these demonstrations are not meant for practical applications. There can be many other applications of cryptographic Boolean functions apart from what have been covered by this chapter.

References

1. Bellare, M., Goldwasser, S.: Verifiable partial key escrow. In: Proceedings of the Fourth ACM Conference on Computer and Communications Security, Zurich, pp. 78–91. ACM (1997)
2. Chabaud, F.: On the security of some cryptosystems based on error-correcting codes. In: Advances in Cryptology – Proceedings of Eurocrypt'94. LNCS 950, pp. 131–139. Springer, Berlin/Heidelberg (1995)
3. Desmedt, Y.: Threshold cryptosystems. In: Advances in Cryptology – Proceedings of Auscrypt'92. LNCS 718, pp. 3–14. Springer, Berlin/Heidelberg (1993)
4. Diffie, W., Hellman, M.: New directions in cryptology. IEEE Trans. Inf. Theory **IT-22**(6), 644–654 (1976)
5. Garey, M.R., Johnson, D.S.: Computers and Intractability: A Guide to the Theory of NP-Completeness. W.H. Freeman and Company/Bell Telephone Laboratories Incorporated, New York (1978)
6. MacWilliams, F.J., Sloane, N.J.A.: The Theory of Error-Correcting Codes. North-Holland, Amsterdam (1977)
7. McEliece, R.L.: "A public-key cryptosystem based on algebraic coding theory", Deep Space Network Progress Report, Jet Propulsion Labs, Pasadena 42–44, pp. 114–116 (1978)
8. National Institute for Standards and Technology, "Escrowed Encryption Standard (EES)," Federal Information Processing Standards Publication (FIPS PUB) 185, 9 Feb 1994
9. Rivest, R.L., Shamir, A., Adleman, L.: A method for obtaining digital signatures. Commun. ACM **21**(2), 120–126 (1978)
10. Shamir, A.: Partial key escrow: a new approach to software key escrow, presented at Key Escrow Conference, Washington, D.C., 15 Sept 1995

Printed in the United States
by Bookmasters

Printed in the United States
By Bookmasters